FRACTAL CALCULUS
AND ITS APPLICATIONS
F$^\alpha$-Calculus

FRACTAL CALCULUS AND ITS APPLICATIONS

F^{α}-Calculus

Alireza Khalili Golmankhaneh

Department of Physics, Urmia Branch, Islamic Azad University, Urmia, Iran

World Scientific

EW JERSEY · LONDON · SINGAPORE · BEIJING · SHANGHAI · HONG KONG · TAIPEI · CHENNAI · TOKYO

Published by

World Scientific Publishing Co. Pte. Ltd.

5 Toh Tuck Link, Singapore 596224

USA office: 27 Warren Street, Suite 401-402, Hackensack, NJ 07601

UK office: 57 Shelton Street, Covent Garden, London WC2H 9HE

Library of Congress Control Number: 2022037228

British Library Cataloguing-in-Publication Data
A catalogue record for this book is available from the British Library.

FRACTAL CALCULUS AND ITS APPLICATIONS
Fᵅ-Calculus

ISBN 978-981-126-110-7 (hardcover)
ISBN 978-981-126-111-4 (ebook for institutions)
ISBN 978-981-126-112-1 (ebook for individuals)

For any available supplementary material, please visit
https://www.worldscientific.com/worldscibooks/10.1142/12988#t=suppl

Typeset by Stallion Press
Email: enquiries@stallionpress.com

To my parents, to whom I owe everything.

Preface

Geometry is the natural basis of a physical theory, and physical laws are a combination of geometry and physics. Throughout the history, geometry has been studied in different branches such as Euclidian geometry, Minkowski's geometry, Riemannian geometry, and recently, fractal geometry, which are used in classical mechanics, special relativity theory, and general relativity theory.

Mathematical analysis was started on sets and applied to classical mechanics. It was then generalized to vector spaces and used by physicists in quantum physics and electromagnetism. Then, with the development of differential calculus and analysis on manifolds, general relativity was formulated.

In past centuries, discussions in physics included regular objects such as straight lines, squares, spheres, cones, etc.

More recently, Benoit B. Mandelbrot (1924-2010), a French-American mathematician, introduced fractal geometry, calling it the "geometry of nature". Fractal sets or curves are irregular shapes whose Hausdorff dimension exceeds their topological dimension. Fractals are specific shapes that share a common feature, such as irregularities on a large range of scales, and whose properties, like density, length, area, and volume, are not meaningful. The surface of human lungs and snowflakes, the boundaries of clouds, and the folds of mammalian brains are in the category of fractals. The heat and wave transfer in disordered systems such as polymers, fractured and porous rocks, amorphous semiconductors, etc., was modeled by fractals and random walks on them. Clusters in nature are the points at which the density of points does not have the meaning of a quantifier. In processes with fractal structures, we note that there are no perfect Cantor sets, von Koch curves, or Sierpinski gaskets in nature. Fractal, however, can be reasonable approximations of natural shapes and are simpler to analyze since they are systematic mathematical constructions.

Motivated by these applications, mathematicians and researchers have developed the theory of "analysis on fractals". Due to the unique properties of fractals, fractal analysis is usually different from the standard and conventional methods (for example:harmonic analysis, measure theory, stochastic processes, etc). In developing the analysis on fractals, various methods have been employed, each with its advantages and disadvantages.

But never was the approach based on the Riemann method employed until Gangal et al. formulated the theory called F^α-calculus.

This excellent approach is a generalization of ordinary differential calculus, which converts standard measures to fractal measures. In other words, in this method, the Hausdorff measure is combined with the Riemann approach in order to include functions with fractal support and non-differentiable fractal curves in the ordinary calculus.This book describes fractal calculus and its applications in physics that have been formulated in recent years.

In Chapter 1, we state the motivation of fractal calculus and explain the reason for formulating fractal calculus. We compare this framework with existing methods by expressing its advantages compared to other methods. In Chapter 2, basic definitions and theorems in real analysis and measure theory are defined that help to understand fractal calculus. In Chapter 3, we study fractal Cantor-like sets such as the Cantor ternary set, the Smith-Volterra-Cantor sets, fat Cantor sets, middle-β Cantor sets, generalized Smith-Volterra-Cantor sets, and their measures and dimensions. In Chapter 4, we introduce basic definitions of fractal calculus, such as fractal derivatives and integrals. Moreover, conjugacy between fractal calculus and ordinary calculus is declared. In Chapter 5, we investigate the local fractal differential equations and the Fourier, Laplace, and Sumudu transform to solve them. In Chapter 6, the stability conditions using Lyapunov and Hyers-Ulam methods for the fractal differential equations are given. In Chapter 7, we express the generalization of fractal calculus, i.e., mean square fractal calculus, F^α-calculus on Cantor cubes and Cantor tartan, non-local fractal integrals and derivatives, Henstock-Kurzweil integrals on fractals, stochastic variables and processes on fractals, stochastic differential equations on fractals and numerical methods for solving fractal differential equations. In Chapter 8, by suggesting Fokker-Plank's equation, Langevin's equation, Maxwell's equations, quantum mechanics, Lagrange and Hamilton equations, Brownian motion, and fractional Brownian motion involving fractal derivatives, we apply them to model phenomena with fractal structure. In Appendix A, we present algorithms for the calculation of staircase function and γ-dimension that have an essential role in fractal calculus.

It is a great pleasure to thank all co-authors of my research on fractal calculus who made this book possible and readers of the book for allowing me to correct the typographical and other errors that occurred in this book.

First, I thank my colleagues Prof. C. Tunç, Prof. R.A. El-Nabulsi, Prof. Kerri Welch, Prof. Michel L. Lapidus, Prof. Marek Czachor, Prof. Alexander Balankin Cano, Prof. L. Swee Cheng and Prof. A.D. Gangal for encouragement in writing this book. I would like to thank my best friends Mr. Mohamad Afshar Urmia, Mr. Ali Moslemi Yengejeh and Siamak Pirmorad for their help and motivations. Especially I thank my uncle Prof. M. Asadi-Golmankhaneh for his useful discussions. And I also want to remember my brother (Mohehhebali) who helped me a lot during my master's degree.

Last but not least, I wish to thank my wife (S. Amir Nazmi Afshar), daughter, (Sarina) and son (Amirreza) for their patience and support during the writing of the book.

Finally, deepest thanks go to desk editor: Mr. Liu Nijia for his help during the publication of this book.

Alireza Khalili Golmankhaneh

Contents

Chapter 1

Introduction to Analysis of Fractals

Many phenomena in physics and engineering have fractal structures; then analysis on them has a vital role in the application. In this chapter, we present some frameworks of analysis on fractals [Parvate (2009); Parvate and Gangal (2011); Kigami (2001); Nottale and Schneider (1984); Mandelbrot (1982); Lapidus *et al.* (2017); Freiberg and Zähle (2002); Falconer (2004); Czachor (2019); Bunde and Havlin (2013); Barnsley (2014); Barlow and Perkins (1988); Balankin (2015); Kunze *et al.* (2019); Ri *et al.* (2021)].

1.1 Motivation of fractal analysis

Why do we study calculus on fractals?
Mathematical physics is developed in the following steps:

(1) Calculus on sets.
(2) Calculus on vector spaces.
(3) Calculus on manifolds.
(4) Calculus on fractal.

As mentioned above, analysis on fractal/calculus on fractal is a new branch of analysis. In the following, we have summarized the different approaches which formulated analysis on fractals.

1.2 Fractional calculus

Fractional calculus is the subject of derivatives and integrals with arbitrary orders, which are defined by [Podlubny (1998); Sandev (2020)]:

$$_a I_x^\alpha f(x) := \frac{1}{\Gamma(\alpha)} \int_a^x \frac{f(t)}{(x-t)^{1-\alpha}} dt, \qquad x > a, \tag{1.1}$$

where $f(x) \in C[a,b] \in R$, and

$$_x I_b^\alpha f(x) := \frac{1}{\Gamma(\alpha)} \int_x^b \frac{f(t)}{(x-t)^{1-\alpha}} dt, \quad x < b, \tag{1.2}$$

are called the left-sided and right-sided Riemann-Liouville fractional integral of order $\alpha > 0$, respectively. Next, let us define fractional derivatives as follows:

$$_a D_x^\alpha f(x) := \frac{1}{\Gamma(n-\alpha)} \left(\frac{d}{dx}\right)^n \int_a^x \frac{f(t)}{(x-t)^{-n+\alpha+1}} dt, \tag{1.3}$$

where $n - 1 \le \alpha < n$, and

$$_x D_b^\alpha f(x) := \frac{1}{\Gamma(n-\alpha)} \left(-\frac{d}{dx}\right)^n \int_x^b \frac{f(t)}{(t-x)^{-n+\alpha+1}} dt, \tag{1.4}$$

are called the left-sided and right-sided Riemann-Liouville fractional derivative of order α respectively whenever the RHS exists. For $f(x) \in C^n[a,b]$ and $n - 1 \le \alpha < n$, then the left-sided and right-sided Caputo fractional derivatives of order α (whenever the RHS exists) are respectively [Uchaikin (2013)]:

$$_a^C D_x^\alpha f(x) = \frac{1}{\Gamma(n-\alpha)} \int_a^x (x-t)^{n-\alpha-1} \left(\frac{d}{dt}\right)^n f(t) dt, \quad a < x < b, \tag{1.5}$$

and

$$_x^C D_b^\alpha f(x) = \frac{1}{\Gamma(n-\alpha)} \int_x^b (t-x)^{n-\alpha-1} \left(-\frac{d}{dt}\right)^n f(t) dt, \quad a < x < b. \tag{1.6}$$

Using fractional derivatives, fractional diffusion equations which models anomalous diffusion, is suggested as follows [Klafter *et al.* (2012)]:

$$\frac{\partial}{\partial t} f(x,t) = K_\beta \, _0 D_t^{1-\beta} \frac{\partial^2}{\partial x^2} f(x,t), \quad f(x,0) = \delta(x), \quad 0 < \beta < 1, \tag{1.7}$$

where $f(x,t)$ the probably density and $[K_\beta] = cm^2/sec^\beta$ is a constant. One can obtain the mean square displacement of the random walk by using Eq.(1.7) in following form [Klafter *et al.* (2012)]:

$$< x(t)^2 >= \int x(t)^2 f(x,t) dt = \frac{2 K_\beta t^\beta}{\Gamma(1+\beta)}. \tag{1.8}$$

The fractional generalization of the Fokker-Planck-Kolmogorov (FPK) equation is suggested as follows [Zaslavsky (1994); Yanovsky *et al.* (2000)]:

$$\frac{\partial f(x,t)}{\partial t} + \gamma \frac{\partial f(x,t)}{\partial x} = -A \,_{-\infty}D_t^{\alpha/2}\,_tD_\infty^{\alpha/2}f(x,t)$$
$$-B\tan\frac{\pi\alpha}{2}\frac{\partial}{\partial x}\,_{-\infty}D_t^{(\alpha-1)/2}\,_tD_\infty^{(\alpha-1)/2}f(x,t),$$

(1.9)

where γ, A and B are constant. The Riemann-Liouville fractional derivatives are nonlocal , therefore, Eq.(1.9) not suitable for the study of local scaling behavior. Because of this disadvantage local fractional derivative is defined by [Kolwankar and Gangal (1998)]:

$$\mathbb{D}^\alpha f(y) = \lim_{x\to y}\,_yD_x^\alpha\left(f(x) - f(y)\right), \qquad 0 < \alpha < 1. \qquad (1.10)$$

Using of a local fractional derivative local fractional Fokker-Planck equation is derived as [Kolwankar and Gangal (1998)]:

$$\mathbb{D}^\alpha f(x,t) = \frac{\Gamma(\alpha+1)}{4}\chi_C(t)\frac{\partial^2}{\partial x^2}f(x,t), \qquad (1.11)$$

where $\chi_C(t)$ is characteristic function of Cantor set [Kolwankar and Gangal (1998)]. Local fractional derivatives lead to a new measure on fractals such as Cantor set while fractional derivatives are defined base on the real-line measure viz length [Kolwankar and Gangal (1999)]. This contradiction leads us to conclude that the local fractional derivative can only be defined on sets with a fraction dimension [Tarasov (2018); Valério *et al.* (2022)].
In this approach, the following questions remain unanswered, are still unanswered despite many types of r in this field [Podlubny (1998); Ben-Avraham and Havlin (2000); Bouchaud and Georges (1990)].

(1) There is no physical or geometrical meaning to the order of fractional derivatives.
(2) Fractional diffusion equation violates simplest form of central limit theorem
(3) The order of fractional derivatives and the relation to underlying fractal space is not clear and is adhoc.
(4) Non-local operators are not suitable to formulate causal behavior.

Remark 1.1. Recently, many researchers have tried to use fractional derivatives on real line for the fractals [Valério *et al.* (2022)].

1.3   Fractional space

A seminal paper expressed axiomatic bases for spaces with noninteger dimensions. Four of the hypotheses are topological, and the fifth specifies an integration measure [Stillinger (1977)].

Let $\mathcal{S}_D$ contains points $x_1, x_2, \ldots$ with dimension D also it has topological structure specified by the following axioms [Stillinger (1977)]:

(1) $\mathcal{S}_D$ is a metric space.
(2) $\mathcal{S}_D$ is dense in itself.
(3) $\mathcal{S}_D$ is metrically unbounded.
(4) $\forall\ x_2, x_3 \in \mathcal{S}_D$, and $\forall\ \epsilon > 0$, $\exists\ x_1 \in \mathcal{S}_D$ such that:
   - $r(x_1, x_2) + r(x_1, x_3) = r(x_2, x_3)$.
   - $r(x_1, x_2) - r(x_1, x_3) < \epsilon\, r(x_2, x_3)$.
(5) For any positive integer n,

$$\int dx_0 \exp\left(-\sum_{j=1}^{n} \alpha_j r_{0j}^2\right)$$

$$= \left(\frac{\pi}{\tau}\right)^{D/2} \exp\left(-\frac{1}{\tau} \sum_{j<k=1}^{n} \alpha_j \alpha_k r_{jk}^2\right), \quad (1.12)$$

where $r_{0j} = r(x_0, x_j)$ is the standard metric, $\tau = \sum_{j=1}^{n} \alpha_j$, and α_j are constants.

According to the above formulation, we provide some applications.

1.3.1   *The Laplace equation in fractional space*

The Laplace equation was derived in polar coordinates in [Stillinger (1977)]:

$$\nabla^2 g = \left[\frac{\partial^2}{\partial r^2} + \frac{D-1}{r}\frac{\partial}{\partial r} + \frac{1}{r^2 \sin^{D-2}\theta}\frac{\partial}{\partial \theta}\sin^{D-2}\theta\frac{\partial}{\partial \theta}\right] g. \quad (1.13)$$

1.3.2   *Schrödinger's equation in fractional space*

The time-dependent Schrödinger's equation in fractional space $\mathcal{S}_D$ is of the form [Stillinger (1977)]:

$$\left(-\frac{1}{2}\nabla^2 + U\right)\Psi = i\frac{\partial\Psi}{\partial t}. \quad (1.14)$$

Here, we choose different potential energy and substitute into Eq.(1.14) for obtaining their corresponding energy eigenvalues [Stillinger (1977)].

(1) The simple harmonic oscillator the potential energy is $U = K/2r^2$ where K is a constant. Then, the corresponding energy eigenvalues are

$$E_n = K^{1/2}\left(\frac{1}{2}D + \Lambda + 2n\right), \quad \Lambda = 0, 1, \ldots, n - 1. \tag{1.15}$$

(2) The hydrogen atom with the potential energy $U = -Z/r$, where Z is a constant. Thus, the corresponding spectrum of bound state energies are

$$E_n = -\frac{Z^2}{2(n + \frac{1}{2}D - \frac{3}{2})^2}. \tag{1.16}$$

1.3.3 *Diffusion equation in fractional space*

The halved diffusion equation was suggested in [Metzler *et al.* (1994)] by

$$\frac{\partial^{1/d_w}}{\partial t^{1/d_w}} f(r, t) = -G\frac{1}{r^k}\frac{\partial}{\partial r}(r^k f(r, t)). \tag{1.17}$$

where $G > 0$ is a constant and $k = d_f/d_w - 1/2$ here d_f is the anomalous diffusion exponent and d_f the fractal dimension of the underling object. The solution of Eq.(1.17) is

$$f(r, t) = Br^{-d_f} H_{1,1}^{1,0}\left[\frac{r}{Gt^{1/d_w}}\bigg|_{(d_f-k,1)}^{(1,1/d_w)}\right], \tag{1.18}$$

where $H_{1,1}^{1,0}$ is an H-function and B is a constant. The purpose of this formulation is to obtain the following equation:

$$< r^2(t) > \sim t^{2/d_w}, \tag{1.19}$$

which is the mean-square displacement of a random walker. Eq.(1.19) for the Euclidean spaces with dimension d is of the form:

$$< r^2(t) > \sim t. \tag{1.20}$$

The drawbacks of this model are the following:

(1) The dimension D in Eq.(1.13) is not defined mathematically.
(2) One can not apply Eq.(1.13) to all fractals i.e. Cantor-like sets.

1.4 Probabilistic approach

The Sierpinski gasket is defined by:

$$K = f_1(K) \cup f_2(K) \cup f_3(K), \tag{1.21}$$

where $f_i(z) = (z - p_i)/2 + p_i$, $i = 1, 2, 3$, and $\{p_1, p_2, p_3\}$ is a set of vertices of an equilateral [Kigami (2001)]. Also, let $V_0 = \{p_1, p_2, p_3\}$ then one can define a sequence of finite sets $\{V_m\}_{m\geq0}$ inductively by $V_{m+1} = f_1(V_m) \cup f_2(V_m) \cup f_3(V_m)$. Here, G_m is defined a graph whose set of vertices is V_m. For $p \in V_m$, suppose $V_{m,p}$ be the collection of the direct neighbors of p in V_m. The simple random walk on G_m indicates by G_m, and it means that if a particle is at p at time t, it will jumps to one of direct neighbors with the probability $\neq (V_{m,p})^{-1}$ at time $t + 1$, where $\neq (A)$ indicates the number of elements in a set A. The diffusion process on or Brownian motion the Sierpinski gasket is defined by:

$$X^m_{5^{-m}t} \to X_t \quad \text{as} \quad m \to \infty. \tag{1.22}$$

In this approach, the Laplacian is defined as an infinitesimal generator of Brownian motion on fractal sets. The solutions of equations were obtained by utilizing the self-similarities of fractals [Barlow and Perkins (1988)].

Advantage of this approach:
This approach can applied to infinitely ramified self-similar sets, i.e. the Sierpinski carpet.

1.5   Harmonic analysis approach

In this method, a sequence of discrete Laplacian is considered instead of the sequence of random walks. Then by choosing a proper scaling, the discrete Laplacian converges to the Laplacian on the Sierpinski gasket. In mathematical language, one can write:

$$l(V_m) = \{f : V_m \to R\}. \tag{1.23}$$

Then let us define

$$L_m : l(V_m) \to l(V_m)$$
$$(L_m u)(p) = \sum_{q \in V_{m,p}} (u(p) - u(p)), \quad \forall\, u \in l(V_m),\, \forall\, p \in V_m. \tag{1.24}$$

where L_m is called the discreet Laplacian on G_m. Then, the Laplacian on Sierpinski gasket is denoted by Δ, and defined by:

$$5^m(L_m u)(p) \to (\Delta u)(p), \quad \text{as} \quad m \to \infty. \tag{1.25}$$

Advantage of this approach:
It provides concrete and direct description of harmonic functions, Green's functions, Dirichlet forms and Laplacians Self-similar sets.[1]

---

[1] Self-similar sets are the simplest fractals

1.6　Measure theory approach

Let $[a, b] \subset R$ be a closed interval and μ be a measure on $[a, b]$, and $L_2 :=$ $L_2(K, \mu)$, where $K := \sup \mu$, be the separable Hilbert space with scalar product $< f, g >= \int_a^b f g d\mu$. Then we define

$$D_1^\mu := \{f \in L_2 : \exists\, f' \in L_2,\ f(x) = f(a) + \int_a^x f'(y) d\mu;\quad x \in K\} \quad (1.26)$$

where f' is μ derivative of f [Freiberg and Zähle (2002)].

Disadvantage of this approach:
The function $f'(y)$ does not always exist and is not unique for a given $f(x)$. Even though the measure theoretical approach is elegant but Riemann-like approach has its own place in the case of fractals which is described in this book.

1.7　F^α-calculus/fractal calculus approach

A type of Riemann-like calculus formulated for functions supported on fractal totally disconnected sets such as fractal Cantor sets , fractal Cantor cubes [Parvate (2009); Parvate and Gangal (2009, 2011); Golmankhaneh and Welch (2021)], fractal Koch curve [Golmankhaneh and Welch (2021)], and Cantor tartan spaces [Golmankhaneh and Welch (2021)]. F^α-Calculus is a simple, constructive, and algorithmic approach to performing analysis on fractals. It is a generalization of ordinary calculus which applies in cases where standard calculus is not applicable [Golmankhaneh and Welch (2021)].

Advantages of this approach:

(1) The fractal derivative is local, which is very important in physics: firstly so as not to violate causality
(2) The order of the fractal derivative is non-integer and has geometrical meaning , being equal to the dimension of the support of the function.
(3) The order of the fractal derivative also has physical meaning by having a relationship with the spectral dimension

With this introduction and comparison of different methods, we want to express the fractal calculus in this book.

Chapter 2

Basic Tools

This chapter gives the basics definitions of real analysis and measure theory.

2.1 Real analysis

In this chapter, we give basic definitions of analysis.

Definition 2.1. A set is a collection of objects called members of the set, e.g. the real numbers (R), the natural number (N), the integers, the rational numbers , the irrational numbers, etc [Spiegel (1969); Bruckner *et al.* (1997)].

Definition 2.2. If each element of a set A also belongs to a set B, we call A a subset of B ($A \subset B$), or mathematically we write [Royden and Fitzpatrick (1988)]:

$$(\forall a \in A \Rightarrow a \in B) \Leftrightarrow A \subset B \qquad (2.1)$$

Definition 2.3. A set A equal to set a B if we have [Royden and Fitzpatrick (1988)]

$$(A \subset B \wedge B \subset A) \Leftrightarrow (A = B). \qquad (2.2)$$

Definition 2.4. A real number a is called an upper bound of a set $A \subset R$ if we have

$$\forall x \in A \Rightarrow x \leq a. \qquad (2.3)$$

Definition 2.5. A real number b is called a lower bound of a set $A \subset R$ if we have [Royden and Fitzpatrick (1988)]

$$\forall x \in A \Rightarrow b \leq x. \qquad (2.4)$$

Definition 2.6. If there exists an upper bound p of set real numbers ($A \subset R$) such that

$$p \leq a, \tag{2.5}$$

for all upper bounds of a set A, then p is called least upper bound or supremum of A, abbreviated by $\sup A$.

If a real number set $A \neq \emptyset$ has an upper bound, then it has a supremum. This property is called completeness of real numbers. [Royden and Fitzpatrick (1988)]

Definition 2.7. If there exists a lower bound q of set real numbers ($B \subset R$) such that

$$b \leq q, \tag{2.6}$$

for all lower bounds b of a set B, then q is called greatest lower bound or infimum of B, abbreviated by $\inf B$.

If a real number set $B \neq \emptyset$ has a lower bound, then it has a infimum [Royden and Fitzpatrick (1988)].

Definition 2.8. A set A has both upper and lower bound that are called to be bounded.

Definition 2.9. The union of sets A and B is defined by

$$A \cup B = \{x | x \in A \vee x \in B\}. \tag{2.7}$$

Definition 2.10. The intersection of sets A and B is defined by

$$A \cap B = \{x | x \in A \wedge x \in B\}. \tag{2.8}$$

Definition 2.11. The difference of sets A and B is defined by

$$A - B = \{x | x \in A \wedge x \notin B\}. \tag{2.9}$$

Definition 2.12. The complement of a set A relative to B is defined by

$$A \subset B \Rightarrow B - A = \tilde{A}_B. \tag{2.10}$$

If $B = U$ where U is the universal set , then we have

$$A \subset U \Rightarrow U - A = \tilde{A}. \tag{2.11}$$

Definition 2.13. The Cartesian product of A and B is defined by

$$A \times B = \{(x, y) | x \in A \wedge y \in B\}. \tag{2.12}$$

Definition 2.14. A set of pairs $f \subset A \times B$ is called a function if we have

$$x \in A, \ y \in B, \ (x,y) \in f \wedge (x,z) \in f \Rightarrow y = z. \tag{2.13}$$

The domain of f is defined by

$$D(f) = \{a \in A | (a,y) \in f, y \in B\}. \tag{2.14}$$

The range of f is defined by

$$R(f) = \{b \in B | (x,b) \in f, x \in A\}. \tag{2.15}$$

Definition 2.15. A function f is called one to one if we have

$$f(x_1) = f(x_2) \Rightarrow x_1 = x_2. \tag{2.16}$$

Definition 2.16. A function f is called onto if we have

$$f : A \to B \Rightarrow R(f) = B \tag{2.17}$$

Definition 2.17. The inverse function f^{-1} for a one to one function f is defined by

$$(x,y) \in f^{-1} \Rightarrow (x,y) \in f. \tag{2.18}$$

Definition 2.18. If a function f be one to one and onto, then we have one to one correspondence between A and B, and they are equivalent or denoted by $(A \sim B)$.

Definition 2.19. Let set A be equivalent $B = \{1,2,3,\ldots,n\}$ then A is called finite ; otherwise it is called infinite .

Definition 2.20. An infinite set A is equivalent to the natural numbers is called denumerable ; otherwise it is called non-denumerable .

Definition 2.21. If a set A is either empty, finite or denumerable, then it is called countable , otherwise it is called non-countable .

Definition 2.22. A set $\{1,2,3,\ldots,n\}$ as well as any set equivalent to it has cardinal number n

Definition 2.23. The Cartesian product $R^n = R \times R \cdots \times R$ is called n dimensional Euclidean space, and a point in this space is denoted by an ordered n-tuplet $(x_1, x_2, \ldots, x_n)$ of real numbers.

Definition 2.24. Let x and y be two points in the Euclidean space , then their the Euclidean distance is defined by

$$d(x,y) = \sqrt{(x_1 - y_1)^2 + (x_2 - y_2)^2 + \cdots + (x_n - y_n)^2} \tag{2.19}$$

Definition 2.25. The set of points is given by

$$\{x | d(x, y) < r\}, \quad r > 0 \tag{2.20}$$

is called an open sphere/ball with radius r and center at y.

Definition 2.26. The set of points is given by

$$\{x | d(x, y) \leq r\}, \quad r > 0 \tag{2.21}$$

is called a closed sphere/ball with radius r and center at y.

Definition 2.27. A metric space is the Euclidean space with the following conditions:

(1) $f(x, y) \geq 0$
(2) $d(x, y) = d(y, x)$
(3) $d(x, y) = 0 \Leftrightarrow x = y$
(4) $d(x, y) \leq d(x, y) + d(y, z)$

Definition 2.28. A $\delta > 0$ neighborhood of a point a is a set of all points such that $|x - a| < \delta$.

Definition 2.29. A deleted $\delta > 0$ neighborhood of a point a is a set of all points such that $0 < |x - a| < \delta$.

Definition 2.30. A point a is an interior point of set M if there exists a δ neighborhood of a all of whose points belong to M.

Definition 2.31. A set M is called an open set if each of its points is an interior point .

Definition 2.32. A point a is called an exterior point of a set M set if there exists a δ neighborhood of a all of whose points belong to $\tilde{M}$.

Definition 2.33. A point a is called a boundary point of a set M set if every δ neighborhood of a contains at leat one point belonging to M and at leat one point belonging to $\tilde{M}$.

Definition 2.34. A point a is called a limit point of a set M if every deleted δ neighborhood of a contains points of M.

Definition 2.35. The set of all limit points of M is called the derived set and is denoted by M'.

Definition 2.36. Closure of a set M is denoted by $\bar{M}$ and defined by

$$\bar{M} = M \cup M' \tag{2.22}$$

Definition 2.37. A set M is called closed if $\bar{M} = M$.

Theorem 2.1 (Weierstrass-Bolzano [Spiegel (1969); Royden and Fitzpatrick (1988)]). *Every bounded infinite set in R has at least one limit point.*

Definition 2.38. A collection of sets C is called a open covering of M if $M \subset \{\cup_i O_i : O \in C\}$ where O_i is an open set.

Theorem 2.2 (Heine-Borel[Spiegel (1969); Royden and Fitzpatrick (1988)]). *Every open covering of a closed and bounded set M contains a finite open subcovering.*

Definition 2.39. A set M is called compact if every open covering of M has a finite subcovering.

Definition 2.40. A set M is called dense in R if $\bar{M} = R$.

2.2 Measure theory

Measure theory is a generalization of the concept of length to arbitrary sets which is called their measures such as area and volume. Here, we give a summery of the measure theory [Spiegel (1969); Royden and Fitzpatrick (1988)].

Definition 2.41. The power set of a set M is denoted by $P(M)$ and it is the sets of all subsets of M.

Definition 2.42. A set $\mathcal{A} \subseteq P(M)$ is called a σ-algebra if we have

(1) $\emptyset, M \in \mathcal{A}$,
(2) $A \in \mathcal{A} \Rightarrow \tilde{A}_M \in \mathcal{A}$,
(3) $A_i \in \mathcal{A} \Rightarrow \bigcup_{i=1}^{\infty} A_i \in \mathcal{A}$.

Definition 2.43. A set A is called an $\mathcal{A}$-measurable set if it is σ-algebra.

Definition 2.44. For $N \subseteq P(M)$, there is a smallest σ-algebra that contains N, namely,

$$\sigma(N) = \bigcap_{\mathcal{A}_i \supseteq N} \mathcal{A}_i. \tag{2.23}$$

Example 2.1. Let us consider

$$M = \{a, b, c, d\}, \quad N = \{\{a\}, \{b\}\}, \tag{2.24}$$

then σ-algebra that contains N is

$$\sigma(N) = \{\emptyset, M, \{a\}, \{b\}, \{a, b\}, \{b, c, d\}, \{a, c, d\}, \{c, d\}\} \tag{2.25}$$

Definition 2.45. Let M be a metric space the Borel σ-algebra $B(M)$ on M is generated by open sets subset of M.

Definition 2.46. Let M be a metric space and $\mathcal{A}$ be a σ-algebra on M, that is, $(M, \mathcal{A})$ is a measurable set .

Definition 2.47. A map $\mu : \mathcal{A} \to [0, \infty] = [0, \infty) \cup \{\infty\}$ on a measurable space $(M, \mathcal{A})$ is called a measure if it satisfies

(1) $\mu(\emptyset) = 0$,
(2) $\mu\left(\bigcup_{i=1}^{\infty} A_i\right) = \sum_{i=1}^{\infty} \mu(A_i)$ for all $A_i \in \mathcal{A}$, with $A_i \cap A_j = \emptyset$, $i \neq j$.

$(M, \mathcal{A}, \mu)$ is called a measure space [Hartman and Mikusinski (2014); Spiegel (1969); Royden and Fitzpatrick (1988)].

Example 2.2. Counting measure is defined as

$$\mu(A) = \begin{cases} \#A, & \text{if } A \text{ has finitely many elements;} \\ \infty, & \text{otherwise.} \end{cases} \tag{2.26}$$

where $\#$ indicates number of set.

Remark 2.1. Calculation of $[0, \infty]$ is as follows:

(1) $x + \infty = \infty$, for all, $x \in [0, \infty]$,
(2) $x * \infty = \infty$, for all, $x \in (0, \infty]$,
(3) $0 * \infty = 0$, it is in most case in measure theory.

Example 2.3. For a point $p \in M$ and $A \subset M$, the Dirac measure is defined by

$$\delta_p(A) = \begin{cases} 1, & p \in M, \\ 0, & \text{otherwise.} \end{cases} \tag{2.27}$$

Definition 2.48. A map $\varphi : P(M) \to [0, \infty]$ is called an outer measure if [Hartman and Mikusinski (2014); Spiegel (1969); Royden and Fitzpatrick (1988)]

(1) $\varphi(\emptyset) = 0$,

(2) $A \subseteq B \Rightarrow \varphi(A) \le \varphi(B)$, (monotoicity),
(3) $A_1, A_2, A_3, \ldots \in P(M)$,
$\varphi(\bigcup_{n=1}^{\infty} A_n) \le \sum_{n=1}^{\infty} \varphi(A_n)$, ($\sigma$ − subadditivity).

Definition 2.49. Let φ be an outer measure, $A \in P(M)$ is called φ-measurable if for $Q \in P(m)$ we have:

$$\varphi(Q) = \varphi(Q \cap A) + \varphi(Q \cap \tilde{A}_M) \qquad (2.28)$$

where Q is called test sets since they are used to test measurability [Hartman and Mikusinski (2014); Spiegel (1969); Royden and Fitzpatrick (1988)].

Example 2.4. Let $\varphi : P(R) \to [0, \infty]$ which is defined by

$$\varphi(A) = \begin{cases} 0, & A = \emptyset; \\ 1, & A \ne \emptyset. \end{cases} \qquad (2.29)$$

is an outer measure but not a measure.

Example 2.5. Let $\varphi : P(N) \to [0, \infty]$ which is called counting measure and defined by

$$\varphi(A) = \begin{cases} |A|, & A \text{ is finite}; \\ \infty, & A \text{ is not finite}. \end{cases} \qquad (2.30)$$

where N is the natural numbers, $|*|$ is number of element set, and $\varphi(A)$ is an outer measure but also a measure.

Definition 2.50. On a measurable set $(M, \mathcal{A})$, $L : \mathcal{A} \to R$ is called the Lebesgue measure if:

(1) $L([a, b]) = b - a$,
(2) $L(x + A) = L(A)$, $A = [a, b] \in \mathcal{A}$, $x \in R$.

Definition 2.51. The Lebesgue outer/exterior measure of a set A is defined by

$$L_e(A) = \inf L(O) \text{ for all open sets } A \subset O. \qquad (2.31)$$

where $L(*)$ is the lengths of all open sets which contain A. If $A = I$ where I is interval, then we have $L_e(I) = L(I)$.

Definition 2.52. The Lebesgue interior measure of a set A is defined by

$$L_i(A) = \sup L(C) \text{ for all closed sets } C \subset A. \qquad (2.32)$$

Definition 2.53. A set A is measurable with respect to the Lebesgue outer measure if we have $L_e(A) = L(A)$ [Hartman and Mikusinski (2014); Spiegel (1969); Royden and Fitzpatrick (1988)].

Definition 2.54. A property which is true for a set of measure zero is said to hold almost everywhere.

Definition 2.55. The countable unions or intersection of open and closed sets is called the class of Borel sets .

Theorem 2.3. *Any Borel set is measurable.*

Definition 2.56. Let $(M_1, \mathcal{A}_1)$ and $(M_2, \mathcal{A}_2)$ be measurable spaces, then $f : M_1 \to M_2$ is called measurable function if we have

$$f^{-1}(A_1) \in \mathcal{A}_2, \quad \text{for all} \quad A_2 \in \mathcal{A}_2. \tag{2.33}$$

Example 2.6. Suppose $(M, \mathcal{A})$ and $(R, B(R))$ be measurable sets, the characteristic function $\chi_A : M \to R$ is defined by

$$\chi_A(\omega) = \begin{cases} 1, & \omega \in A, \\ 0, & \omega \notin A. \end{cases} \tag{2.34}$$

For all measurable $A \in \mathcal{A}$, $\chi_A(\omega)$ is a measurable function.

Definition 2.57. The Lebesgue integral of a measurable and bounded function $K_1 < f(x) < K_2$, K_1, $K_2 \in R$ is as follows:

(1) Divided the range K_1 to K_2 into subintervals as follows:

$$K_1 = y_0, y_1, \ldots, y_n = K_2 \tag{2.35}$$

(2) Find $A_k \subset [a, b]$ such that

$$A_k = \{x : y_{k-1} \le f(x) < y_k\}, \quad k = 1, 2, \ldots, n. \tag{2.36}$$

(3) Define upper and lower sum as follows:

$$S = \sum_{k=1}^{n} y_k L(A_k) \tag{2.37}$$

$$s = \sum_{k=1}^{n} y_{k-1} L(A_k) \tag{2.38}$$

(4) Taking infimum and supremum of S and s for all possible partitions of $[K_1, K_2]$, respectively:

$$\mathcal{I} = \inf \sum_{k=1}^{n} y_k L(A_k), \tag{2.39}$$

$$\mathcal{J} = \sup \sum_{k=1}^{n} y_{k-1} L(A_k). \tag{2.40}$$

If $\mathcal{I} = \mathcal{J}$, we say that $f(x)$ is the Lebesgue integrable on $[a, b]$ and denote the common value by:

$$\int_a^b f(x)dL < \infty. \tag{2.41}$$

In Figure 2.1 we present geometric interpretation of the Lebesgue integral.

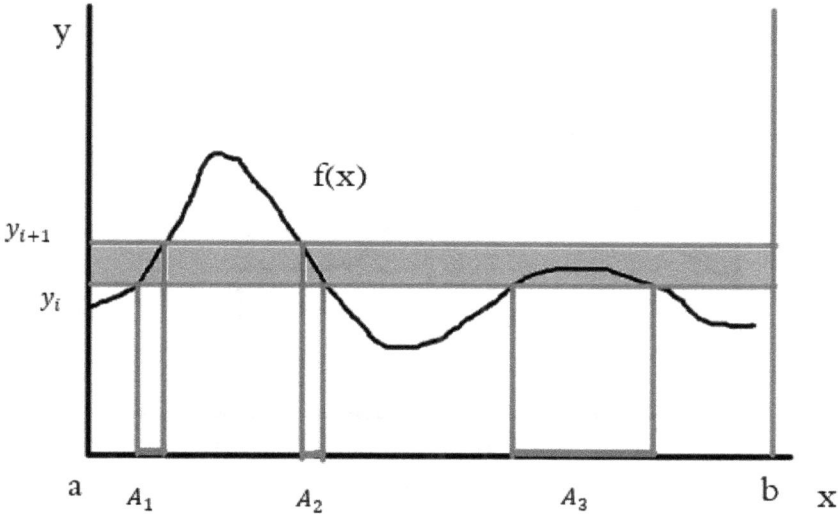

Fig. 2.1: Geometric interpretation of the Lebesgue integral

Definition 2.58. The Riemann integral of a bounded function $f(x)$ is defined on the interval $[a, b]$ as follows (See Figure 2.2):

(1) Consider a partition of the interval $[a, b]$

$$a = x_0 < x_1 < x_1 \cdots < x_n = b. \tag{2.42}$$

(2) Defined the Riemann sums of f with respect to partition as

$$S = \sum_{k=1}^n f(x_i)\Delta x \tag{2.43}$$

$$\tag{2.44}$$

where $\Delta x = \frac{b-a}{n}$, and $x_i = a + i\Delta x$.

The Riemann integral on $[a,b]$ is defined by:

$$\int_a^b f(x)dx = \lim_{n\to\infty} \sum_{k=1}^n f(x_i)\Delta x < \infty. \qquad (2.45)$$

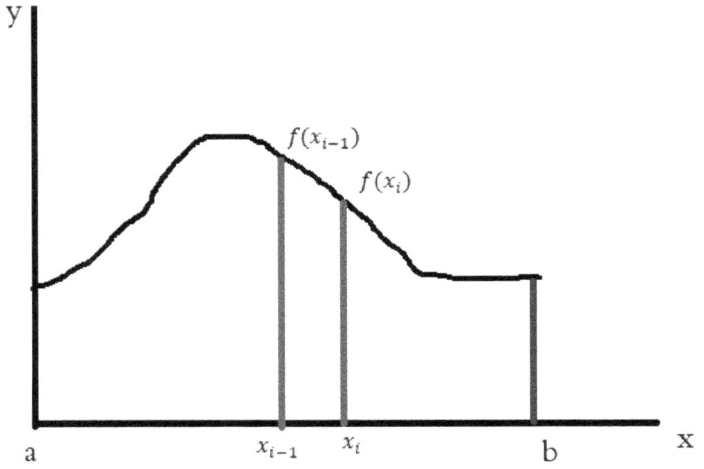

Fig. 2.2: Geometric interpretation of the Riemann integral

Definition 2.59. The Darboux integral of a bounded function $f(x)$ on $[a,b]$ is similar to the Riemann integral but based on the following sums

$$\mathcal{P} = \sum_{i=1}^n \inf_{t\in[x_{i-1},x_i]} f(t)(x_i - x_{i-1}), \qquad (2.46)$$

$$\mathcal{Q} = \sum_{i=1}^n \sup_{t\in[x_{i-1},x_i]} f(t)(x_i - x_{i-1}). \qquad (2.47)$$

If $\mathcal{P} = \mathcal{Q}$, then we call the common value the Darboux integral. We also say that $f(x)$ is the Darboux integrable of f.

Definition 2.60. The Riemann-Stieltjes integral of a bounded function $f(x)$ on the interval $[a,b]$ with respect to another bonded function $g(x)$ is defined based on the following sums:

$$\mathcal{U} = \sum_{i=1}^{n} \sup_{t\in[x_{i-1},x_i]} f(t)(g(x_i) - g(x_{i-1})), \tag{2.48}$$

$$\mathcal{T} = \sum_{i=1}^{n} \inf_{t\in[x_{i-1},x_i]} f(t)(g(x_i) - g(x_{i-1})). \tag{2.49}$$

If $\mathcal{U} = \mathcal{T}$, then we call the common value the Riemann-Stieltjes integral of f [Fonda *et al.* (2018); Swartz (2001)]. We also say that $f(x)$ is the Riemann-Stieltjes integrable with respect to $g(x)$.

Remark 2.2. Let $g(x)$ be continuous and differentiable over $[a, b]$. Then we can write

$$\int_a^b f(x)dg(x) = \int_a^b f(x)\frac{dg}{dx}dx = \int_a^b f(x)g'(x)dx. \tag{2.50}$$

Definition 2.61. The Henstock-Kurzweil integral or gauge integral of a function $f(x)$ on $[a, b]$ is defined as follows [Fonda *et al.* (2018); Swartz (2001)]:

(1) Define tagged partition P of $[a, b]$, that is,

$$a = x_0 < x_1 < \cdots < x_n = b, \quad t_i \in [x_{i-1}, x_i]. \tag{2.51}$$

(2) Define a gauge $\delta : [a, b] \to (0, \infty)$ such that

$$\forall i [x_{i-1}, x_i] \subset [t_i - \delta(t_i), t_i + \delta(t_i)], \tag{2.52}$$

that is called the tagged partition P is δ-fine.
(3) Define the generalized Riemann sums as

$$\mathcal{W} = \sum_{i=1}^{n} \sup_{t_i\in[x_{i-1},x_i]} f(t_i)\Delta x_i, \quad \Delta x_i = x_i - x_{i-1}, \tag{2.53}$$

$$\mathcal{M} = \sum_{i=1}^{n} \inf_{t_i\in[x_{i-1},x_i]} f(t_i)\Delta x_i. \tag{2.54}$$

If $\mathcal{W} = \mathcal{M}$, then we call the common value the Henstock-Kurzweil integral of f. We also say that $f(x)$ is the Henstock-Kurzweil integrable [Fonda *et al.* (2018)] .

Definition 2.62. A cover C of a set M is defined by

$$C = \{U_\alpha : \alpha \in A\} \tag{2.55}$$

where $M \subseteq \bigcup_{\alpha\in A} U_\alpha$ and A is index set.

If each member of C is an open set, then it is called an open cover.

Definition 2.63. A topological space is an ordered pair (T, X) where T is a set and X is a collection of subsets of T, satisfying the following axioms:

(1) The empty set $\emptyset$ and T itself belong to X.
(2) The finite or infinite union of members of X belongs to X.
(3) The intersection of any finite number of members of X belongs to X

The elements of X are called open sets and the collection X is called a topology on T.

Definition 2.64. Let (M, ρ) be a metric space so we can define

$$H_\delta^d(S) = \inf\left\{ \sum_{i=1}^\infty (diam\ U_i)^d : \bigcup_{i=1}^\infty \supseteq S,\ diam\ U_i < \delta \right\} \qquad (2.56)$$

where the infimum is over all countable covers of S by sets U_i, and

$$diam\ U = \sup\{\rho(x, y) : x, y \in U\},\ diam\ \emptyset = 0. \qquad (2.57)$$

Then we can define

$$H^d(S) = \lim_{\delta \to 0} H_\delta^d(S) = \sup_{\delta > 0} H_\delta^d(S), \qquad (2.58)$$

which is called the d-dimensional Hausdorff measure of Borel set S. This leads to definitions of the Hausdorff dimension as follows:

$$\dim_H(S) = \inf\{d \geq 0 : H^d(S) = 0\}$$
$$= \sup(\{d \geq 0 : H^d(S) = \infty\} \cup \{0\}), \qquad (2.59)$$

where $\inf \emptyset = \infty$.

Remark 2.3. For the case of $\rho > 0$ the Hausdorff measure [Edgar (1998); Falconer (2004)] of R^ρ we have

$$L_\rho(S) = 2^{-\rho} \frac{\pi^{\rho/2}}{\Gamma(\frac{\rho}{2} + 1)} H^\rho(S) \qquad (2.60)$$

where $L_\rho(S)$ is the ρ-dimensional Lebesgue measure of S.

Chapter 3

Fractal Cantor Like Sets

In this chapter, we define the Cantor like sets and their properties. The Cantor set was constructed by George Cantor in 1883, which is a simple example of fractal sets. It is a perfect nowhere-dense set in the real line. In the section, we present the generalized Cantor sets and their properties [DiMartino and Urbina (2014b,a)].

3.1   Cantor ternary set

The Cantor ternary set C is obtained from closed interval $C_0 = [0, 1]$ by removing of middle-third open intervals from it and repeating the process as the following form [DiMartino and Urbina (2014b,a)]

$$C_1 = \left[0, \frac{1}{3}\right] \cup \left[\frac{2}{3}, 1\right]$$
$$C_2 = \left[0, \frac{1}{9}\right] \cup \left[\frac{2}{9}, \frac{1}{3}\right] \cup \left[\frac{2}{3}, \frac{7}{9}\right] \cup \left[\frac{8}{9}, 1\right]$$
$$\vdots \tag{3.1}$$
$$C_n = \frac{C_{n-1}}{3} \cup \left(\frac{2}{3} + \frac{C_{n-1}}{3}\right), \quad n \geq 1.$$
$$\vdots$$

The Cantor ternary set contains all points in the interval that are not deleted at any step in this infinite process, namely,

$$C = \bigcap_{n=0}^{\infty} C_n. \tag{3.2}$$

The Cantor ternary set include numbers in $[0, 1]$ that can be represented in base 3 as follows

$$C = \left\{ x \in [0, 1] : x = \sum_{k=1}^{\infty} \frac{\varepsilon_k(x)}{3^k}, \quad \varepsilon_k(x) = 0, 2 \text{ for all } k = 1, 2, \ldots \right\} \quad (3.3)$$

The Cantor ternary C is a nowhere-dense set, namely, there are no intervals included in C. The Cantor ternary C has the Lebesgue's measure zero [DiMartino and Urbina (2014b,a)], since

$$L(C) = L([0, 1]) - L(\tilde{C}) = 1 - \sum_{n=1}^{\infty} \frac{2^{n-1}}{3^n} \quad (3.4)$$

$$= 1 - \frac{1}{3} \sum_{n=0}^{\infty} \left(\frac{2}{3} \right)^n = 1 - \frac{1/3}{1 - 2/3} = 0. \quad (3.5)$$

Additionally, the Cantor set is also presented using an Iterated Function System (IFS). Consider the following map:

$$\omega_0(x) = \frac{x}{3}, \quad \omega1(x) = \frac{x}{3} + \frac{2}{3}, \quad (3.6)$$

then

$$C = \omega_0([0, 1]) \cup \omega_1([0, 1]) \quad (3.7)$$

3.2   The Devil's staircase function

The Devil's staircase or Cantor function is defined on $[0, 1]$ as follows [DiMartino and Urbina (2014b,a)]:

$$G(x) = \sum_{k=1}^{\infty} \frac{a_k(x)}{2^k} \quad (3.8)$$

where

$$a_k(x) = \begin{cases} \frac{\varepsilon_k(x)}{2}, & \text{for } k < N(x) \\ 0, & \text{for } k = N(x) \\ 1, & \text{for } k > N(x) \end{cases} \quad (3.9)$$

and

$$N(x) = \min\{k : \varepsilon_k(x) = 1\}. \quad (3.10)$$

This function is a monotone non-decreasing continuous function on $[0, 1]$ and increasing only in the Cantor set, so $G(x)$ is called a singular function [DiMartino and Urbina (2014b,a)].

Now, we want to define generalized ternary Cantor sets which are bounded, perfect, and nowhere-dense sets.

3.3 The Smith-Volterra-Cantor sets

The Smith-Volterra-Cantor sets $(SVC(n))$ are similarly constructed Cantor ternary sets by removing open intervals from closed intervals, such that, in the kth step of the iteration, an open interval of length $1/n^k, n \geq 3$ from the center of each of the remaining closed intervals. The $SVC(n)$ is the intersection of the countably infinite collection of sets that remain after each iteration[DiMartino and Urbina (2014b,a)].

The $SVC(n)$ sets are also perfect sets and their Lebesgue measures are as follows[DiMartino and Urbina (2014b,a)]:

$$L([0,1]) - L(S\tilde{V}C(n)) = 1 - \frac{1}{n} - \frac{2}{n^2} - \frac{4}{n^3} - \cdots$$

$$= 1 - \frac{1}{n}\frac{1}{1 - 2/n} = \frac{n-3}{n-2}. \qquad (3.11)$$

The $SVC(n)$ sets have positive Lebesgue measure, thus they are called fat Cantor sets.

3.4 The fat Cantor sets

The ν-fat Cantor sets are constructed by taking out $0 < \nu \leq 1$ of open intervals from closed $[0,1]$ in the kth, interval of length $\nu/3^k$ from it[DiMartino and Urbina (2014b,a)], namely,

$$C_{\nu,1} = \left[0, \frac{3-\nu}{6}\right] \cup \left[\frac{3+\nu}{6}, 1\right],$$

$$C_{\nu,2} = \left[0, \frac{9-5\nu}{36}\right] \cup \left[\frac{9-\nu}{36}, \frac{3-\nu}{6}\right]$$

$$\cup \left[\frac{3+\nu}{6}, \frac{27+\nu}{36}\right] \cup \left[\frac{27+5\nu}{36}, 1\right],$$

$$\vdots$$

$$C_\nu = \bigcap_{k=1}^{\infty} C_{\nu,k} \qquad (3.12)$$

The ν-fat Cantor set C_ν is a perfect, nowhere-dense set with Lebesgue positive measure [DiMartino and Urbina (2014b,a)], that is:

$$
\begin{aligned}
L([0,1]) - L(C_\nu) &= 1 - \frac{\nu}{3} - 2\frac{\nu}{9} - 4\frac{\nu}{27} + \cdots \\
&= 1 - \frac{\nu}{3}\left[1 + \frac{2}{3} - \frac{4}{9} + \cdots\right] \\
&= 1 - \frac{\nu}{3}\sum_{k=0}^{\infty}\left(\frac{2}{3}\right)^k \\
&= 1 - \frac{\nu}{3}\frac{1}{1-\frac{2}{3}} = 1 - \frac{\nu}{3}\frac{1}{\frac{1}{3}} = 1 - \nu
\end{aligned}
\tag{3.13}
$$

If we reduce size of $C_{\nu,1}$ by $(3-\nu)/6$ we will have two $C_{\nu,1}$ then by using definition of similarity dimension we can compute the simplify dimension of C_ν as follows[DiMartino and Urbina (2014b,a)]:

$$
\left(\frac{6}{3-\nu}\right)^d = 2, \quad d = \frac{\ln 2}{\ln[\frac{6}{(3-\nu)}]} = \frac{\ln 2}{\ln 6 - \ln(3-\nu)}.
\tag{3.14}
$$

3.5   The middle-β Cantor sets

Middle-β Cantor sets or thin Cantor sets are constructed by removing open subinterval of length β of $[0,1]$ from the middle of the interval as follows [DiMartino and Urbina (2014b,a)]:

$$
\begin{aligned}
C_1^\beta &= \left[0, \frac{1}{2}(1-\beta)\right] \cup \left[\frac{1}{2}(1+\beta), 1\right] \\
C_2^\beta &= \left[0, \frac{1}{4}(1-\beta)^2\right] \cup \left[\frac{1}{4}(1-\beta)(1+\beta), \frac{1}{2}(1-\beta)\right] \\
&\cup \left[\frac{1}{2}(1+\beta), \frac{1}{2}((1+\beta) + \frac{1}{2}(1-\beta)^2)\right] \\
&\cup \left[\frac{1}{2}(1+\beta)\left(1 + \frac{1}{2}(1-\beta)\right), 1\right] \\
&\vdots
\end{aligned}
\tag{3.15}
$$

Continuing iterating by removing an open subinterval of length β-proportional of the disjoint intervals from the middle of each of the intervals [DiMartino and Urbina (2014b,a)]. Then we can define Middle-β Cantor set as

$$
C^\beta = \bigcap_{k=1}^{\infty} C_k^\beta.
\tag{3.16}
$$

Middle-β Cantor set has Lebesgue measure zero, since

$$L([0,1]) - L(C^\beta) = 1 - \beta - 2\left(\frac{1}{2}(1-\beta)\beta\right) - 4\left(\frac{1}{4}(1-\beta)^2\beta\right) - \cdots$$

$$= 1 - \beta\frac{1}{1-(1-\beta)} = 1 - 1 = 0. \tag{3.17}$$

Middle-$\{\beta_i\}$ Cantor sets are sets which are constructed similar to Middle-β Cantor sets with the difference that in each iteration we remove $0 < \{\beta_i\} < 1$ from the middle of the interval as the following

(1) Let

$$C_0^{\{\beta_i\}} = [0,1],$$

(2) then first step gives

$$C_1^{\{\beta_i\}} = \left[0, \frac{1}{2}(1-\beta_1)\right] \cup \left[\frac{1}{2}(1+\beta_1), 1\right],$$

(3) and second step leads

$$C_2^{\{\beta_i\}} = \left[0, \frac{1}{4}(1-\beta_1)(1-\beta_2)\right] \cup \left[\frac{1}{4}(1-\beta_1)(1+\beta_2), \frac{1}{2}(1-\beta_1)\right]$$

$$\cup \left[\frac{1}{2}(1+\beta_1), \frac{1}{2}((1+\beta_1) + \frac{1}{2}(1-\beta_1)(1-\beta_2))\right]$$

$$\cup \left[\frac{1}{2}(1+\beta_1)\left(1 + \frac{1}{2}(1-\beta_2)\right), 1\right].$$

$$\vdots$$

Finally, we can write

$$C^{\{\beta_i\}} = \Pi_{j=1}^\infty C_j^{\{\beta_j\}} \tag{3.18}$$

which is called Middle-$\{\beta_i\}$ Cantor sets [DiMartino and Urbina (2014b,a)] and their Lebegue measures are

$$L(C_0^{\{\beta_i\}}) = 1$$
$$L(C_1^{\{\beta_i\}}) = (1-\beta_1)$$
$$L(C_2^{\{\beta_i\}}) = (1-\beta_1)(1-\beta_2)$$

$$\vdots$$

$$L(C_n^{\{\beta_i\}}) = (1-\beta_1)(1-\beta_2)(1-\beta_3)\cdots(1-\beta_n), \tag{3.19}$$

$$\tag{3.20}$$

therefore we can write

$$L(C^{\{\beta_i\}}) = \Pi_{j=1}^{\infty}(1 - \beta_j). \tag{3.21}$$

Note that, by choosing different values of $\{\beta_i\}$, we can obtain the fat or thin Cantor sets.

3.6   Generalized Smith-Volterra-Cantor sets

Generalized $SVC(\alpha, \beta)$, $0 < \alpha < 1$, $0 < \beta < 1$, $0 < \beta \leq 1 - 2\alpha$ are sets that construct as follows [DiMartino and Urbina (2014b,a)]:
Let us consider an interval $[0, 1]$, in the first iteration, we pick up an open interval of length β centered at $1/2$ from $[0, 1]$, arriving at

$$C_1^{(\alpha,\beta)} = \left[0, \frac{1}{2}(1 - \beta)\right] \cup \left[\frac{1}{2}(1 + \beta), 1\right] \tag{3.22}$$

In the second iteration, we delete an open interval of length $\alpha\beta$ of each of the two disjoint intervals from middle of each of the intervals in $C_1^{(\alpha,\beta)}$ [DiMartino and Urbina (2014b,a)], namely,

$$\begin{aligned}
C_2^{(\alpha,\beta)} = &\left[0, \frac{1}{4}(1 - \beta) - \frac{1}{2}\alpha\beta\right] \cup \left[\frac{1}{4}(1 - \beta) + \frac{1}{2}\alpha\beta, \frac{1}{2}(1 - \beta)\right] \\
&\cup \left[\frac{1}{2}(1 + \beta) + \frac{1}{2}\left((1 + \beta) + \frac{1}{2}(1 - \beta) - \alpha\beta\right)\right] \\
&\cup \left[\frac{1}{2}(1 + \beta) + \frac{1}{2}\left((1 + \beta) + \frac{1}{2}(1 - \beta) + \alpha\beta\right), 1\right]
\end{aligned}$$

By iterating this procedure in the kth-step we remove an open interval of length $\alpha^{k-1}\beta$ from middle of each of the disjoint intervals. In the other hands we remove 2^{k-1} subintervals of length $\alpha^{k-1}\beta$ each time [DiMartino and Urbina (2014b,a)]. Then the generalized $SVC(\alpha, \beta)$ is defined by [DiMartino and Urbina (2014b,a)]:

$$C^{(\alpha,\beta)} = \bigcap_{k=1}^{\infty} C_k^{(\alpha,\beta)} \tag{3.23}$$

The Lebesgue measure of $SVC(\alpha, \beta)$ is

$$L(C^{(\alpha,\beta)}) = 1 - \beta - 2\alpha\beta - 4\alpha^2\beta - \cdots = 1 - \frac{\beta}{1 - 2\alpha}. \tag{3.24}$$

From here we have additions conditions as follows:

$$1 - 2\alpha > 0, \quad 0 < \beta \leq 1 - 2\alpha. \tag{3.25}$$

We can see that $C^{(\alpha,\beta)}$ are fat fractal sets unless $\beta = 1 - 2\alpha$.

Note that we can conclude:

(1) If $\alpha = \beta = 1/3$ then $C^{(\alpha,\beta)}$ becomes C Cantor ternary set.
(2) If $\alpha = \beta = 1/3$ then $C^{(\alpha,\beta)}$ turns into C_λ Cantor sets.
(3) If $\alpha = (1 - \beta)/2$, $\beta = \beta$ then $C^{(\alpha,\beta)}$ taks form C^β.
(4) If $\alpha = \beta = 1/n$ then $C^{(\alpha,\beta)}$ leads to $SVC(n)$.

The Hausdorff dimension of different kind the Cantor sets are [DiMartino and Urbina (2014b,a)] :

$$dim_H(C^\beta) = \frac{\log 2}{\log 2 - \log(1 - \beta)} \tag{3.26}$$

$$dim_H(C^{\{\beta_i\}}) = \frac{\log 2}{\log 2 - \log(\lim_{n\to\infty}(1 - \beta_n)} \tag{3.27}$$

$$dim_H(C^{(\alpha,\beta)}) = 1. \tag{3.28}$$

$$dim_H(C_\lambda) = \frac{\ln 2}{\ln 6 - \ln(3 - \lambda)}. \tag{3.29}$$

Chapter 4

Fractal Calculus

In this chapter, we present fractal calculus on thin Cantor sets. This chapter is based on the seminal paper in which the fractal calculus was formulated by A. Parvate and A.D. Gangal [Parvate (2009); Parvate and Gangal (2009, 2011)]. We must note that we have changed the definition of the mass function by replacing the gamma function so that the value of the step function does not exceed one. As a result, in some cases, the corresponding formulas and theorems have been rewritten under this definition.

We present the basic definitions in fractal calculus as follows:

4.1 The mass function of fractal sets

Definition 4.1. The flag function for a thin fractal sets F and a closed interval $I = [a, b], a < b \in R$ is defined by

$$\theta(F, I) = \begin{cases} 1, & \text{if } F \cap I \neq \emptyset; \\ 0, & \text{otherwise.} \end{cases} \tag{4.1}$$

Definition 4.2. Let $P_{[a,b]}$ or (P) be a subdivision of the interval I, which is a finite set of points $\{a = x_0, x_1, \ldots, x_n\}$, $x_i < x_{i+1}$. Here, $[x_i, x_{i+1}]$ is called a component interval of the P. If Q is any subdivision of $[a, b]$ and $P \subset Q$, then Q is called a refinement of P.

Note that if $a = b$, then $P = \{a\}$ is only subdivision of $[a, b]$.

Definition 4.3. For a set F and a subdivision P one can define [Parvate (2009); Parvate and Gangal (2011); Golmankhaneh *et al.* (2018)]:

$$\sigma^\alpha[F, P] = \sum_{i=0}^{n-1} \Gamma(\alpha + 1)(x_{i+1} - x_i)^\alpha \theta(F, [x_i, x_{i+1}]). \tag{4.2}$$

Note that $\sigma^\alpha[F, P] \geq 0$, and if $a = b$, then we have $\sigma^\alpha[F, P] = 0$. The sum in Eq.(4.2) contains a contribution from a component interval if and only if that component include at least one point of F.

Definition 4.4. The coarse-grained mass of $F \cap [a, b]$ for given $\delta > 0$ and $a \leq b$ is defined by:

$$\gamma_\delta^\alpha(F, a, b) = \inf_{\{P_{[a,b]} : |P| \leq \delta\}} \sigma^\alpha[F, P], \tag{4.3}$$

where

$$|P| = \max_{0 \leq i \leq n-1} (x_{i+1} - x_i), \tag{4.4}$$

and the infimum in Eq.(4.3) is taken over all subdivisions P of $[a, b]$ satisfying $|P| \leq \delta$.

Lemma 4.1. *If $0 < \delta_1 < \delta_2$, then $\gamma_{\delta_1}^\alpha(F, a, b) \geq \gamma_{\delta_2}^\alpha(F, a, b)$.*

Proof. The proof is straightforward. □

Lemma 4.2. *If $\delta > 0$, and $a < b < c$, then $\gamma_\delta^\alpha(F, b, c) \leq \gamma_\delta^\alpha(F, a, c)$ and $\gamma_\delta^\alpha(F, a, b) \leq \gamma_\delta^\alpha(F, a, c)$.*

Proof. According to the definition Eq.(4.3), and if there exists a subdivision $P_{[a,c]} = \{x_0 = a, x_1, \ldots, x_n = c\}$ such that $|P| \leq \delta$ we have:

$$\sigma^\alpha[F, P] < \gamma_\delta^\alpha(F, a, c) + \epsilon \tag{4.5}$$

Suppose $Q_{[b,c]} = \{x \in P : b \leq x < c\} \cup \{c\}$ viz $Q_{[b,c]} = \{y_0, y_1, \ldots, y_m\}$ where $y_i = x_i$ if $b \leq x < c$ and $y_m = c$. It follows that $|Q_{[b,c]}| \leq |P_{[a,c]}| \leq \delta$ and $\theta(F, [y_{m-1}, y_m]) \leq \theta(F, [x_{m-1}, x_m])$ since $[y_{m-1}, y_m] \subset [x_{m-1}, x_m]$. As a result, we have

$$\sigma^\alpha[F, Q_{[b,c]}] \leq \sigma^\alpha[F, P_{[a,c]}] < \gamma_\delta^\alpha(F, a, c) + \epsilon \tag{4.6}$$

As we know $\gamma_\delta^\alpha(F, b, c) \leq \sigma^\alpha[F, Q_{[b,c]}]$, thus we have

$$\gamma_\delta^\alpha(F, b, c) \leq \gamma_\delta^\alpha(F, a, c), \tag{4.7}$$

which completes the proof of the first part. The proof of the second part is similar to the proof of the first case. □

Theorem 4.1. *$\gamma_\delta^\alpha(F, a, b)$ is continuous in b and c.*

Proof. For proving continuous in a we suppose b, δ, α are constant. For a given $\epsilon > 0$, suppose

$$\Delta' = (\epsilon \frac{1}{\Gamma(\alpha+1)})^\alpha, \quad and, \Delta = \min(\Delta', \delta). \tag{4.8}$$

For $\epsilon_1 > 0$, there exists a subdivision P, as follows:

$$\sigma^\alpha[F, P] < \gamma_\delta^\alpha(F, b, c) + \epsilon_1. \tag{4.9}$$

Now $Q = P \cup \{a - \Delta\}$ is a subdivision of $[a - \Delta, b]$. Thus we have:

$$\begin{aligned}
\gamma_\delta^\alpha(F, a - \Delta, b) &\leq \sigma^\alpha[F, Q] \\
&= \sigma^\alpha[F, P] + \theta(F, [a - \Delta, b])\Gamma(\alpha+1)\Delta^\alpha \\
&\leq \sigma^\alpha[F, P] + \epsilon \\
&< \gamma_\delta^\alpha(F, a, b) + \epsilon_1 + \epsilon.
\end{aligned} \tag{4.10}$$

As ϵ_1 is arbitrary, then we get:

$$\gamma_\delta^\alpha(F, a - \Delta, b) < \gamma_\delta^\alpha(F, a, b) + \epsilon. \tag{4.11}$$

Since $\gamma_\delta^\alpha(F, a, b)$ is a nondecreasing function of a, hence we can write

$$\gamma_\delta^\alpha(F, a - t, b) < \gamma_\delta^\alpha(F, a, b) + \epsilon \tag{4.12}$$

for $0 < t < \Delta$. Sum upping, for given $\epsilon > 0$, there exists a $\Delta > 0$ we obtain:

$$a - c < \Delta \Rightarrow \gamma_\delta^\alpha(F, c, b) - \gamma_\delta^\alpha(F, a, b) < \epsilon \tag{4.13}$$

which mentions that $\gamma_\delta^\alpha(F, a, b)$ is continuous in a from left. The continuity from right can be obtained by replacement of a by $a + \Delta$ in the above proof. $\qquad\square$

Definition 4.5. The mass function is defined by:

$$\gamma^\alpha(F, a, b) = \lim_{\delta \to 0} \gamma_\delta^\alpha(F, a, b). \tag{4.14}$$

Note that $\gamma^\alpha(F, a, b)$ increases by decreasing δ, therefore $\gamma^\alpha(F, a, b)$ always exists and $0 \leq \gamma^\alpha(F, a, b) \leq +\infty$.

Lemma 4.3. *If $F \cap [a, b] = \emptyset$, then $\gamma^\alpha(F, a, b) = 0$.*

Proof. If $F \cap [a, b] = \emptyset$, then $\gamma_\delta^\alpha(F, a, b) = 0$ for any δ, consequently $\gamma^\alpha(F, a, b) = 0$. $\qquad\square$

Theorem 4.2. *If $a < b < c$ and $\gamma^\alpha(F, a, c) < \infty$, then*

$$\gamma^\alpha(F, a, c) = \gamma^\alpha(F, a, b) + \gamma^\alpha(F, b, c). \tag{4.15}$$

Proof. For given $\delta > 0$, let P_1 and P_2 be any subdivisions of $[a, b]$ and $[b, c]$, respectively, such that $|P_1| < \delta$ $|P_2| < \delta$. Then, $P_1 \cup P_2$ is a subdivision of $[a, c]$, namely,

$$\sigma^\alpha[F, P_1 \cup P_2] = \sigma^\alpha[F, P_1] + \sigma^\alpha[F, P_1]. \tag{4.16}$$

By taking infimum over all subdivisions P_1 and P_2, we obtain:

$$\gamma_\delta^\alpha(F, a, c) \leq \inf_{|P_1| \leq \delta, |P_2| \leq \delta} \sigma^\alpha[F, P_1 \cup P_2]$$
$$= \gamma_\delta^\alpha(F, a, b) + \gamma_\delta^\alpha(F, b, c). \tag{4.17}$$

Let us consider a subdivision $P_{[a,c]}$, with $|P| \leq \delta$, then we can set up a subdivision $P' = P \cup \{b\}$. It is obvious that $P' = P_1 \cup P_2 = P$ if $b \in P$, and $\sigma^\alpha[F, P] = \sigma^\alpha[F, P']$. Otherwise, suppose $[x_k, x_{k+1}]$ be the interval which contains b. Therefore we can write:

$$\sigma^\alpha[F, P \cup \{b\}] - \sigma^\alpha[F, P] \leq 3\Gamma(\alpha + 1)\delta^\alpha. \tag{4.18}$$

This connote that

$$3\Gamma(\alpha + 1)\delta^\alpha + \sigma^\alpha[F, P] \geq \sigma^\alpha[F, P \cup \{b\}] \tag{4.19}$$
$$= \sigma^\alpha[F, P_1] + \sigma^\alpha[F, P_2] \tag{4.20}$$
$$\geq \gamma_\delta^\alpha(F, a, b) + \gamma_\delta^\alpha(F, b, c), \tag{4.21}$$

for all P. By taking infimum over all subdivision such that $|P| \leq \delta$, we obtain

$$3\Gamma(\alpha + 1)\delta^\alpha + \gamma_\delta^\alpha(F, a, c) \geq \gamma_\delta^\alpha(F, a, b) + \gamma_\delta^\alpha(F, b, c). \tag{4.22}$$

By taking the limit from both side of Eq.(4.22), and considering Eq.(4.17), we complete the proof. $\square$

Corollary 4.1. $\gamma^\alpha(F, a, b)$ *is increasing in b and decreasing in a.*

Proof. As regards each term of Eq.(4.15) is nonnegative for $a \leq b \leq c$, then it says that $\gamma^\alpha(F, a, b)$ is the monotonic respect to a, b, that complete the proof. $\square$

Theorem 4.3. *Let $a < b$ and $\gamma^\alpha(F, a, b) \neq 0$ be finite. If $0 < y < \gamma^\alpha(F, a, b)$, then there exists c such as*

$$\gamma^\alpha(F, a, b) = y, \quad a < c < b. \tag{4.23}$$

Proof. Let $z = \gamma^\alpha(F, a, b) - y$. Given $\delta > 0$, consider the set of all x of $[a, b]$ such that $\gamma_\delta^\alpha(F, x, b) \leq z$ which is $[s_\delta, b]$, where $a \leq s_\delta < b$. Likewise, the set of all points x of $[a, b]$ such that $\gamma_\delta^\alpha(F, a, x) \leq y$ that is $[a, t_\delta]$, where $a < s_\delta \leq b$. If $x \in (a, b)$ applying Theorem 4.2 we can write

$$\gamma^\alpha(F, a, b) = \gamma^\alpha(F, a, x) + \gamma^\alpha(F, x, c) \geq \gamma_\delta^\alpha(F, a, x) + \gamma_\delta^\alpha(F, x, b). \quad (4.24)$$

As $y, z < \gamma^\alpha(F, a, b)$ there exists δ_0 such that $\delta < \delta_0$. Since $\gamma^\alpha(F, a, b) > y$ and $\gamma^\alpha(F, a, u)$ is continuous and increasing in u, so there exists $x \in (a, b)$ such that $\gamma^\alpha(F, a, x) = y$. This point that $x \in [a, t_\delta]$. Also, in view of Eq.(4.24) we have

$$z = \gamma^\alpha(F, a, b) - y = \gamma^\alpha(F, a, b) - \gamma_\delta^\alpha(F, a, x) \geq \gamma_\delta^\alpha(F, x, b) \quad (4.25)$$

which implies that $x \in [s_\delta, b]$. This can happen only when $s_\delta \leq t_\delta$. Now, we set

$$0 < \delta < \delta_0 \Rightarrow \gamma_\delta^\alpha(F, a, b) > y, \quad \text{and,} \quad \gamma_\delta^\alpha(F, a, b) > z. \quad (4.26)$$

Therefore for every δ there exists an interval $[s_\delta, t_\delta]$ such that

$$x \in [s_\delta, t_\delta] \Rightarrow \gamma_\delta^\alpha(F, x, b) \leq z, \quad \text{and,} \quad \gamma_\delta^\alpha(F, a, x) \leq y. \quad (4.27)$$

Let us define:

$$s = \sup_{0 < \delta < \delta_0} s_\delta, \quad \text{and,} \quad t = \inf_{0 < \delta < \delta_0} t_\delta. \quad (4.28)$$

We note as $\delta \to 0$ then s_δ increases and t_δ decreases. Since $s_\delta \leq t_\delta$ for every δ so that $s \leq t$, and

$$[s, t] = \bigcap_{0 < \delta < \delta_0} [s_\delta, t_\delta] \quad (4.29)$$

Consequently, we can conclude

$$x \in [s, t] \Rightarrow \gamma^\alpha(F, x, b) \leq z, \quad \text{and,} \quad \gamma^\alpha(F, a, x) \leq y \quad (4.30)$$

But as we know

$$\gamma^\alpha(F, a, x) + \gamma^\alpha(F, x, b) = \gamma^\alpha(F, a, b) \quad (4.31)$$

In view of Eqs.(4.31), and (4.30) , we arrive at that there exists a set $[s, t] \subset [a, b]$ such that

$$x \in [s, t] \Rightarrow \quad \gamma^\alpha(F, a, x) = y \quad (4.32)$$

which completes the proof. $\qquad \square$

Corollary 4.2. *If $\gamma^\alpha(F, a, b)$ is finite, $\gamma^\alpha(F, a, x)$ is continuous for $x \in (a, b)$.*

Proof. Since $\gamma^\alpha(F, a, b)$ is a monotonic function respect to a and b, then it is continuous for $x \in (a, b)$. $\qquad \square$

4.1.1 *The mass function under translation and scaling*

Let us define translation and scaling as follows [Parvate (2009); Parvate and Gangal (2011)]:

$$F + \lambda = \{x + \lambda : x \in F\}$$
$$\lambda F = \{\lambda x : x \in F\}, \quad \lambda \in R. \tag{4.33}$$

Then, the translation property of the mass function is given by:

$$\gamma^\alpha(F + \lambda, a + \lambda, b + \lambda) = \gamma^\alpha(F, a, b). \tag{4.34}$$

The scaling property of the mass function can be written as:

$$\gamma^\alpha(\lambda F, \lambda a, \lambda b) = \lambda^\alpha \gamma^\alpha(F, a, b). \tag{4.35}$$

Example 4.1. Let F be the middle-$\frac{1}{3}$ Cantor set, with $a = 0, b = 1$, and $\alpha = \ln(2)/\ln(3)$ then for this case we have $\lambda = \frac{1}{3^n}, n \in N (=$ positive integer).

4.2 The integral staircase function

In this section, we want to define the main and important part of fractal calculus [Parvate (2009); Parvate and Gangal (2011); Wibowo *et al.* (2021)].

Definition 4.6. Let a_0 be an arbitrary and fixed real number. The integral staircase function $S_F^\alpha(x)$ of order α for a set F is defined by:

$$S_F^\alpha(x) = \begin{cases} \gamma^\alpha(F, a_0, x), & \text{if } x \geq a_0 \\ -\gamma^\alpha(F, x, a_0), & \text{otherwise.} \end{cases} \tag{4.36}$$

4.2.0.1 *Properties of the integral staircase function*

If $\gamma^\alpha(F, a, b)$ is finite then for all $x, y \in (a, b)$ such that $x < y$ the following statements hold [Parvate (2009); Parvate and Gangal (2011)]:

(1) $S_F^\alpha(x)$ is increasing in x.
(2) If $F \cap (x, y) = \emptyset$, then $S_F^\alpha(x)$ is constant in $[x, y]$.
(3) $S_F^\alpha(y) - S_F^\alpha(x) = \gamma^\alpha(F, x, y)$.
(4) $S_F^\alpha(x)$ is continuous on (a, b).

4.3 The γ-dimension of fractal sets

The γ-dimension can be defined by recalling the Hausdorff dimension which defined by Eq.(2.59) [Parvate (2009); Parvate and Gangal (2011)]. Let $0 < \alpha < \beta \leq 1$ then we can write

$$\sigma^\beta[F, P] \leq |P|^{\beta - \alpha} \sigma^\alpha[F, P] \frac{\Gamma(\beta + 1)}{\Gamma(\alpha + 1)}, \tag{4.37}$$

it follows that

$$\gamma_\delta^\beta(F, a, b) \leq \delta^{\beta - \alpha} \gamma_\delta^\alpha(F, a, b) \frac{\Gamma(\beta + 1)}{\Gamma(\alpha + 1)}, \tag{4.38}$$

where if $\delta \to 0$ we have:

$$\gamma^\beta(F, a, b) = \lim_{\delta \to 0} \gamma_\delta^\beta(F, a, b) = \begin{cases} 0, & \text{if } \alpha < \beta \\ \infty, & \text{if } \alpha > \beta \end{cases} \tag{4.39}$$

where $\gamma^\beta(F, a, b)$ also can have finite number which is motivation of the γ-dimension.

Definition 4.7. The γ-dimension of $F \cap [a, b]$ is defined by:

$$dim_\gamma(F \cap [a, b]) = \inf\{\alpha : \gamma^\beta(F, a, b) = 0\} \tag{4.40}$$

$$= \sup\{\alpha : \gamma^\beta(F, a, b) = \infty\}. \tag{4.41}$$

4.4 The α-perfect sets

In this section, we express that the correspondence between sets F and their staircase function $S_F^\alpha(x)$ is many to one [Parvate (2009); Parvate and Gangal (2011)].

Definition 4.8. Let $F \subset R$ and $G \subset R$ be such that $\dim_\gamma F = \dim_\gamma G = \alpha$, $0 < \alpha \leq 1$. Then F and G are said to be staircasewise congruent if

$$S_F^\alpha(x) = S_G^\alpha(x). \tag{4.42}$$

All sets which have same staircase function with set F, are in the equivalence class and is denoted by $\mathcal{E}_F^\alpha$.

Definition 4.9. A point x is a point of change of a function f, if f is not constant over any open interval (c, d) containing x. The set of all points of change of f is called the set of change of f and is denoted by Sch$f$

Theorem 4.4. *Let $F \subset R$ be such that $S_F^\alpha(x)$ is finite for all $x \in R$ for $\alpha = dim_\gamma F$ and $H = Sch(S_F^\alpha)$. Then H belongs to $\mathcal{E}_F^\alpha$, namely, $S_F^\alpha = S_H^\alpha$.*

Proof. It is sufficient to prove that $\gamma^\alpha(F,a,b) = \gamma^\alpha(H,a,b)$ for any $a,b \in R$. First, let $F \cap [u,v] = \emptyset$, then $\gamma^\alpha(F,u,v)$. As a result S_F^α is constant on (u,v), so that $(u,v) \cap H = \emptyset$. Then for any $\epsilon' > 0$ we have

$$\theta(H, [u+\epsilon', v-\epsilon']) = 0. \tag{4.43}$$

Subsequent, for a given $\epsilon > 0$ there exists a subdivision $P_{[a,b]} = \{y_0, y_1, \ldots, y_n\}$ such that $|P| \leq \delta$ and

$$\sigma^\alpha[F,P] \leq \gamma^\alpha(F,a,b) + \frac{\epsilon}{2}. \tag{4.44}$$

In the case of $\theta(F,I) = 1$ for all components I of P, then we have:

$$\sigma^\alpha[H,P] \leq \sigma^\alpha[F,P]. \tag{4.45}$$

Or else, let K be the set of all points as follows:

$$c' = c + \left(\frac{\epsilon}{\Gamma(\alpha+1)2n}\right)^{\frac{1}{\alpha}}, \quad d' = d + \left(\frac{\epsilon}{\Gamma(\alpha+1)2n}\right)^{\frac{1}{\alpha}}, \tag{4.46}$$

where c and d are endpoints of those components I of P such that $\theta(F,I) = 0$. Then $Q_{[a,b]} = P \cup K$ is a refined subdivision and $|Q| \leq |P|$. If $I = [c,d]$ is a component of P such that $\theta(F,I) = 0$. Hence, it involves three components of Q, that is $[c,c']$, $[c',d']$ and $[d',d]$. The terms in $\sigma^\alpha[H,Q]$ corresponding to three components contribute $0, \epsilon/2n$ for $[c',d']$ and $[c,c'],[d',d]$, respectively.

$$\sigma^\alpha[H,Q] \leq \sigma^\alpha[F,P] + \frac{\epsilon}{2} \tag{4.47}$$

In view of Eqs.(4.44), (4.45), and (4.47), there is a subdivision Q, as:

$$\sigma^\alpha[H,Q] \leq \gamma\delta^\alpha(F,a,b) + \epsilon, \quad |Q| \leq \delta. \tag{4.48}$$

Since ϵ is arbitrary we can write:

$$\gamma_\delta^\alpha(H,a,b) \leq \gamma_\delta^\alpha(F,a,b). \tag{4.49}$$

Now we want to check the validity of the following inequality

$$\gamma_\delta^\alpha(H,a,b) < \gamma_\delta^\alpha(F,a,b). \tag{4.50}$$

Using Eq.(4.50) we conclude that there is a subdivision $P_1 = \{x_0, \ldots, x_n\}$ such that $|P_1| \leq \delta|$ and

$$\sigma^\alpha[H,P_1] < \gamma_\delta^\alpha(F,a,b). \tag{4.51}$$

Using Eq.(4.17) we get:

$$\gamma\delta^\alpha(F,a,b) \leq \sum_{i=0}^{n-1} \gamma_\delta^\alpha(F,x_i,x_{i+1}), \tag{4.52}$$

thus

$$\sigma^{\alpha}[H, P_1] < \sum_{i=0}^{n-1} \gamma_{\delta}^{\alpha}(F, x_i, x_{i+1}). \tag{4.53}$$

If Eq.(4.53) holds, then we arrive at:

$$\Gamma(\alpha + 1)(x_{k+1} - x_k)^{\alpha}\theta(H, [x_k, x_{k+1}]) < \gamma_{\delta}^{\alpha}(F, x_k, x_{k+1})$$
$$\leq \gamma^{\alpha}(F, x_k, x_{k+1}). \tag{4.54}$$

where $0 \leq k \leq n-1$. Since $\gamma^{\alpha}(F, x_k, x_{k+1}) > 0$ and S_F^{α} is not constant in $[x_k, x_{k+1}]$. So that $\text{Sch}(S_F^{\alpha} \cap [x_k, x_{k+1}] \neq \emptyset$, viz $H \cap [x_k, x_{k+1}] \neq \emptyset$. Hence $\theta(H, [x_k, x_{k+1}]) = 1$, then Eq.(4.54) turns into

$$\Gamma(\alpha + 1)(x_{k+1} - x_k)^{\alpha} < \gamma_{\delta}^{\alpha}(F, x_k, x_{k+1}). \tag{4.55}$$

As $Q = \{x_k, x_{k+1}\}$ is a subdivision of $[x_k, x_{k+1}]$ such that $|Q| \leq \delta$

$$\Gamma(\alpha + 1)(x_{k+1} - x_k)^{\alpha} = \sigma^{\alpha}[H, Q]$$
$$< \gamma_{\delta}^{\alpha}(F, x_k, x_{k+1}) \tag{4.56}$$

which is a contradiction by the definition of $\gamma_{\delta}^{\alpha}(F, x_k, x_{k+1})$ that mentioning that our assumption Eq.(4.50) is wrong. Then Eq.(4.49) is an equality for any $\delta > 0$, namely, $\gamma^{\alpha}(H, a, b) = \gamma^{\alpha}(F, a, b)$. $\qquad\square$

Lemma 4.4. *Let $F \subset R$ be such that $S_F^{\alpha}(x)$ is finite for all $x \in R$ for $\alpha = dim_{\gamma}F$. Then the set $H = Sch(S_F^{\alpha})$ is perfect i.e. H is closed and every point of H is its limit point.*

Proof. If y is a limit point of H, then any open interval (c, d) containing y contains a point z of $H = \text{Sch}(S_F^{\alpha})$. Then S_F^{α} is not constant on (c, d). Thus $y \in H$ implying that F is closed. If $x \in H$ is not a limit point of H, then there exists an open interval (c, d) containing x that is $F \cap (c, x) = \emptyset$ and $F \cap (x, d) = \emptyset$. And using Section 4.2.0.1 we have

$$S_F^{\alpha}(d) - S_F^{\alpha}(c) = (S_F^{\alpha}(d) - S_F^{\alpha}(x)) - (S_F^{\alpha}(d) - S_F^{\alpha}(x)) = 0. \tag{4.57}$$

Thus x is not in $H = \text{Sch}(S_F^{\alpha})$ which is a contradiction. $\qquad\square$

Definition 4.10. If $S_F^{\alpha}(x)$ is finite for $x \in R$ and $\alpha = dim_{\gamma}F$. Then $\text{Sch}(S_F^{\alpha})$ is said α-perfect set and α-perfect representative of $\mathcal{E}_F^{\alpha}$.

Remark 4.1. We note that every $\mathcal{E}_F^{\alpha}$ contains a unique α-perfect set.

Remark 4.2. We note that $\text{Sch}(S_F^{\alpha})$ is called canonical representative of $\mathcal{E}_F^{\alpha}$.

Theorem 4.5. *An α-perfect set F is the intersection of all the closed sets G in $\mathcal{E}_F^\alpha$. That is, it is the minimal closed set in $\mathcal{E}_F^\alpha$.*

Proof. Let $\mathcal{G}$ be the class of all closed sets in $\mathcal{E}_F^\alpha$. Since F is perfect, it is closed. Then we have

$$\bigcap_{G \in \mathcal{G}} G \subset F. \tag{4.58}$$

Let $G_0 \in \mathcal{G}$ and $x \notin G_0$. So that there is an open interval (c, d) containing x but no point of G_0 as G_0 is closed. This says that $S_{G_0}^\alpha(c) = S_{G_0}^\alpha(d)$ by Section 4.2.0.1, and $S_F^\alpha(c) = S_F^\alpha(d)$ as $G_0 \in \mathcal{E}_F^\alpha$. Since F is α-perfect, we have $F \cap (c, d) = \emptyset$ implying that $x \notin F$ [Parvate (2009); Parvate and Gangal (2011)]. Hence, $x \in F \Rightarrow x \in G_0$ for all $G_0 \in \mathcal{G}$, therefore we have:

$$F \subset \bigcap_{G \in \mathcal{G}} G, \tag{4.59}$$

by considering Eqs.(4.58) and (4.59) the proof is completed. $\qquad\square$

Lemma 4.5. *Let $F \subset R$ be α-perfect and $x \in F$. If $y < x < z$, then either $S_F^\alpha(y) < S_F^\alpha(x)$ or $S_F^\alpha(x) < S_F^\alpha(z)$ or both.*

Proof. The proof is left to the reader. $\qquad\square$

Remark 4.3. We note that Lemma 4.5 confirms that if $x \in F$, then the values of $S_F^\alpha(y)$ is different from $S_F^\alpha(x)$ at all points y at leat one side of x.

Example 4.2. The ternary Cantor set (C) is α-perfect set, for $\alpha = \frac{\log(2)}{\log(3)}$, and $C = \text{Sch}(S_F^\alpha)$.

Definition 4.11. Let $F \subset R$, $f : R \to R$ and $x \in F$. A number l is called the limit of f through the points of F, or simply $F - limit$ of f, as $y \to x$, if for given any $\epsilon > 0$, there exist δ such that:

$$y \in F, \text{ and } |y - x| < \delta \Rightarrow |f(y) - \le| < \epsilon. \tag{4.60}$$

If such a number exists, then it is denoted by:

$$l = F - \lim_{y \to x} f(y) \tag{4.61}$$

Remark 4.4. We note that in the definition (4.60):

(1) It does not include y if $y \notin F$.
(2) It is not defined for the points $x \in F$.
(3) In some points of F we have only a one-sided limit.

Definition 4.12. A function $f : R \to R$ is said to be F-continuous at $x \in F$ if

$$f(x) = F - \lim_{y \to x} f(y). \tag{4.62}$$

Remark 4.5. We note that F-continuity

(1) is not defined at $x \notin F$.
(2) If f is continuous, then it is also F-continuous but inverse is not correct.

Definition 4.13. A function $f : R \to R$ is called uniformly F-continuous on $E \subset F$ if for any $\epsilon > 0$ there exists $\delta > 0$ such that:

$$x \in F, \ y \in E, \ \text{and} \ |y - x| < \delta \Rightarrow |f(y) - f(x)| < \epsilon. \tag{4.63}$$

Remark 4.6. We note that:

(1) Uniform F-continuity on E implies F-continuity on E.
(2) F-continuity on E implies uniform F-continuity on E only in certain cases.

Theorem 4.6. *If a function $f : R \to R$ is F-continuous on a compact set $E \subset F$, then it is uniformly F-continuous on E.*

4.5 F^α-integration on fractal sets

The F^α-integration is similar to Riemann approach for defining integral as it more direct and algorithmically advantageous [Parvate (2009); Parvate and Gangal (2011)].

Definition 4.14. The class of functions $f : R \to R$ which are bounded on F is denoted by $B(F)$. That is

$$f \in B(F) \Leftrightarrow -\infty < \inf_{x \in F} f(x) \leq \sup_{x \in F} f(x) < +\infty. \tag{4.64}$$

Definition 4.15. Let $f \in B(F)$ and $I = [a, b]$ be a closed interval. Then we define

$$M[f, F, I] = \begin{cases} \sup_{x \in F \cap I} f(x), & \text{if } F \cap I = \emptyset, \\ 0, & \text{otherwise.} \end{cases} \tag{4.65}$$

and likewise

$$m[f, F, I] = \begin{cases} \inf_{x \in F \cap I} f(x), & \text{if } F \cap I = \emptyset, \\ 0, & \text{otherwise.} \end{cases} \tag{4.66}$$

Definition 4.16. Let $S_F^\alpha(x)$ be finite for $x \in [a,b]$ and $P_{[a,b]} = \{x_0, x_1, \ldots, x_n\}$ be a subdivision of $[a,b]$. The upper F^α-sum and lower F^α-sum for the function f over the subdivision P are given respectively by:

$$U^\alpha[f, F, P] = \sum_{i=0}^{n-1} M[f, F, [x_i, x_{i+1}]](S_F^\alpha(x_{i+1}) - S_F^\alpha(x_i), \quad (4.67)$$

and

$$L^\alpha[f, F, P] = \sum_{i=0}^{n-1} m[f, F, [x_i, x_{i+1}]](S_F^\alpha(x_{i+1}) - S_F^\alpha(x_i). \quad (4.68)$$

Remark 4.7. We note that the advent of $F \cap I$ in the definition of M and m, also the use of $S_F^\alpha(x_{i+1}) - S_F^\alpha(x_i$ as in a Riemann-Stieltjes sum instead of $(x_{i+1} - x_i)$.

Lemma 4.6. *Let $F \subset R$ and $f \in B(F)$. If Q is a refinement of a subdivision P, then $U^\alpha[f, F, Q] \leq U^\alpha[f, F, P]$ and $L^\alpha[f, F, Q] \geq L^\alpha[f, F, P]$.*

Proof. Let $P = \{x_0, x_1, \ldots, x_n\}$, and $Q = P \cup \{x'\}$ where $x' \in (x_i, x_{i+1})$. Let $I = [x_i, x_{i+1}], I' = [x_i, x']$, and $I'' = [x', x_{i+1}]$. If there are no points of F in I, then $M[f, F, I] = M[f, F, I'] = M[f, F, I''] = 0$. Otherwise we have

$$I' \cap F \neq \emptyset, \quad I'' \cap F \neq \emptyset, \quad (4.69)$$

or

$$I' \cap F \neq \emptyset, \quad I'' \cap F = \emptyset. \quad (4.70)$$

In the case of Eq.(4.69), we have

$$M[f, F, I'] \leq M[f, F, I], \quad \text{and} \quad M[f, F, I''] \leq M[f, F, I]. \quad (4.71)$$

Hence we have

$$M[f, F, I](S_F^\alpha(x_{i+1}) - S_F^\alpha(x_i))$$
$$= M[f, F, I]\{(S_F^\alpha(x') - S_F^\alpha(x_i)) + (S_F^\alpha(x_{i+1}) - S_F^\alpha(x'))\}$$
$$\geq M[f, F, I'](S_F^\alpha(x') - S_F^\alpha(x_i)) + M[f, F, I''](S_F^\alpha(x_{i+1}) - S_F^\alpha(x')). \quad (4.72)$$

In the case of Eq.(4.70), we have:

$$S_F^\alpha(x_{i+1}) = S_F^\alpha(x'), \quad \text{and}, \quad M[f, F, I] = M[f, F, I'] \quad (4.73)$$

Therefore, we get

$$M[f, F, I](S_F^\alpha(x_{i+1}) - S_F^\alpha(x_i)) = M[f, F, I'](S_F^\alpha(x') - S_F^\alpha(x)), \quad (4.74)$$

where we use $M[f, F, I''] = 0$. In view of Eqs.(4.72), and (4.74), one can obtain

$$M[f, F, I](S_F^\alpha(x_{i+1}) - S_F^\alpha(x_i))$$
$$\geq M[f, F, I'](S_F^\alpha(x') - S_F^\alpha(x_i)) + M[f, F, I''](S_F^\alpha(x_{i+1}) - S_F^\alpha(x')). \quad (4.75)$$

It follows that

$$U^\alpha[f, F, Q] \leq U^\alpha[f, F, P], \quad (4.76)$$

which holds for any refinement of P. The same conclusion can be drawn for

$$L^\alpha[f, F, Q] \geq U^\alpha[f, F, P], \quad (4.77)$$

which finishes the proof. □

Lemma 4.7. *If P and Q are any two subdivisions of $[a, b]$, then*

$$U^\alpha[f, F, P] \geq L^\alpha[f, F, Q] \quad (4.78)$$

Proof. Since $P \cup Q$ is a refinement of both P and Q, and using Lemma 4.6 and,

$$U^\alpha[f, F, P] \geq L^\alpha[f, F, F], \quad (4.79)$$

we can write

$$U^\alpha[f, F, P] \geq U^\alpha[f, F, P \cup Q] \geq L^\alpha[f, F, P \cup Q] \geq L^\alpha[f, F, Q]. \quad (4.80)$$

□

Definition 4.17. Let F be such that $S_F^\alpha(x)$ is finite on $[a, b]$. For $f \in B(F)$, the lower F^α-integral is given by:

$$\underline{\int_a^b} f(x) d_F^\alpha x = \sup_{P_{[a,b]}} L^\alpha[f, F, P], \quad (4.81)$$

and the upper F^α-integral is defined by:

$$\overline{\int_a^b} f(x) d_F^\alpha x = \inf_{P_{[a,b]}} U^\alpha[f, F, P]. \quad (4.82)$$

Here in both Eqs.(4.81), and (4.82) the supremum and infimum are taken over all the subdivision P of $[a, b]$.

Remark 4.8. The $d_F^\alpha x$ appearing in Eqs.(4.81), and (4.82) is just the notation without separate meaning.

It is obvious that

$$\underline{\int_a^b} f(x)d_F^\alpha x \le \overline{\int_a^b} f(x)d_F^\alpha x. \qquad (4.83)$$

Definition 4.18. If $f \in B(F)$, we say that f is F^α-integrable on $[a,b]$ if

$$\underline{\int_a^b} f(x)d_F^\alpha x = \overline{\int_a^b} f(x)d_F^\alpha x. \qquad (4.84)$$

In that case the F^α-integral of f on $[a,b]$, denoted by $\int_a^b f(x)d_F^\alpha x$ is given by the common value.

Lemma 4.8. *If $f \in B(F)$, then f is F^α- integrable on $[a,b]$ if and only if, for any given ϵ, there exists a subdivision P of $[a,b]$ such that:*

$$U^\alpha[f,F,P] < L^\alpha[f,F,P] + \epsilon \qquad (4.85)$$

Theorem 4.7. *Let F be such that $F \cap [a,b]$ is compact and $S_F^\alpha(x)$ is finite on $[a,b]$. Let $f \in B(F)$, and $a < b$. If is F-continuous on $F \cap [a,b]$, then f is F^α-integrable on $[a,b]$.*

Proof. We give the proof for two cases:

(1) If $S_F^\alpha(a) = S_F^\alpha(b)$, then the F^α-integral is zero, and the result is obvious.
(2) Next we consider $S_F^\alpha(a) \neq S_F^\alpha(b)$. The function f is uniformly F-continuous on $F \cap [a,b]$ since $F \cap [a,b]$ is compact. Hence, for a given $\epsilon > 0$, there is a $\delta > 0$ such that:

$$x, y \in F \cap [a,b], \quad \text{and} \quad |y - x| < \delta \Rightarrow |f(y) - f(x)| < \frac{\epsilon}{S_F^\alpha(b) - S_F^\alpha(a)} \qquad (4.86)$$

Let P be a subdivision such that $|P| < \delta$ which leads to

$$U^\alpha[f,F,P] < L^\alpha[f,F,P] + \epsilon. \qquad (4.87)$$

Therefore, in view of Lemma 4.8 the proof is completed. $\qquad \square$

Theorem 4.8. *Let $a < b$ and f be an F^α-integrable on $[a,b]$. Let $c \in [a,b]$, then f is F^α-integrable on $[a,c]$ and $[c,b]$, and we have [Parvate (2009); Parvate and Gangal (2011)]:*

$$\int_a^b f(x)d_F^\alpha x = \int_a^c f(x)d_F^\alpha x + \int_c^b f(x)d_F^\alpha x. \qquad (4.88)$$

Proof. This can be proved in the manner analogous to Riemann integral. $\qquad \square$

4.5.1 *A few properties of F^α-integral on fractal sets*

(1) If f is F^α-integrable on $[a, b]$, and λ is any real number, then

$$\int_a^b \lambda f(x) d_F^\alpha x = \lambda \int_a^b .f(x) d_F^\alpha x. \tag{4.89}$$

(2) If f and g are F^α-integrable function on $[a, b]$, then

$$\int_a^b (f(x) + g(x)) d_F^\alpha x = \int_a^b .f(x) d_F^\alpha x + \int_a^b .g(x) d_F^\alpha x. \tag{4.90}$$

(3) Let $f : [a, b] \to [a', b']$ be F^α-integrable and let $s : [a', b'] \to R$ be a continuous function. Thus $s \circ f$ is F^α-integrable $[a, b]$ where $\circ$ indicates composition of s and f.

(4) If f and g are F^α-integrable over $[a, b]$, and $f(x) \geq g(x)$ for all $x \in [a, b]$, then

$$\int_a^b f(x) d_F^\alpha x \geq \int_a^b g(x) d_F^\alpha x. \tag{4.91}$$

Definition 4.19. If f is F^α-integrable on $[a, b]$, $a < b$, then

$$\int_b^a f(x) d_F^\alpha x = - \int_a^b f(x) d_F^\alpha x. \tag{4.92}$$

Lemma 4.9. *If $\chi_F(x)$ is the characteristic function of $F \subset R$, then*

$$\int_b^a \chi_F(x) d_F^\alpha x = S_F^\alpha(b) - S_F^\alpha(a). \tag{4.93}$$

Proof. For a closed interval $I \subset [a, b]$,

$$M[\chi_F, F, I] = m[\chi_F, F, I] = \begin{cases} 1, & \text{If } F \cap I \neq \emptyset \\ 0, & \text{otherwise.} \end{cases} \tag{4.94}$$

So that $M[\chi_F, F, I]$ is zero for a closed interval $I = [c, d]$ only when $S_F^\alpha(d) - S_F^\alpha(c) = 0$ [Parvate (2009); Parvate and Gangal (2011)]. Therefore we have

$$U^\alpha[\chi_F, F, P] = L^\alpha[\chi_F, F, P] = S_F^\alpha(b) - S_F^\alpha(a), \tag{4.95}$$

for any subdivision P of $[a, b]$. $\qquad \square$

4.6 Differences and similarities between F^α-integral and Riemann-Stieltjes integral

(1) Similar to Riemann-Stieltjes integral since $S_F^\alpha(x_{i+1}) - S_F^\alpha(x_i) = 0$ is used in definitions 4.67 and 4.68[Parvate (2009); Parvate and Gangal (2011)].
(2) Difference with Riemann-Stieltjes integral since we are taking supremum and infimum over $x \in I \cap F$ not over $x \in I$.
(3) The inverse of the Riemann-Stieltjes integral in the sense of fundamental theorem of calculus is not generally defined unless the weight function is strictly monotonic but in the case of the fractal calculus, even if the weight function $S_F^\alpha(x)$ has a zero ordinary almost everywhere, F^α-derivatives which will be defined in the following acts as the inverse of F^α-integral [Parvate (2009); Parvate and Gangal (2011)].

4.7 F^α-differentiation on fractal sets

Definition 4.20. If F is an α-perfect set [Parvate (2009); Parvate and Gangal (2011)], then the F^α-derivative(=fractal derivative) of f at x is

$$D_F^\alpha f(x) = \begin{cases} F - \lim_{y \to x} \frac{f(y)-f(x)}{S_F^\alpha(y)-S_F^\alpha(x)}, & \text{if } x \in F; \\ 0, & \text{otherwise.} \end{cases} \tag{4.96}$$

If the limit exists [Parvate (2009); Parvate and Gangal (2011)].

Remark 4.9.

(1) In Eq.(4.96), we have F-limit and the difference of the staircase function S_F^α in the denominator which are differences with definition of ordinary derivatives .
(2) If $x \in F$, then we can find such points y which are arbitrarily close to x at least one side of x thus the denominator in Eq.(4.96) is not zero.
(3) Since Eq.(4.96) involves F-limit, then we may have a one-side derivative.

Theorem 4.9. *If $D_F^\alpha f(x)$ exists for all x in (a,b), then $f(x)$ is F-continuous in (a,b).*

4.7.1 *The linearity of the fractal derivatives*

(1) Let f be a function on $[a,b]$ and $D_F^\alpha f(x)$ exists for all $x \in [a,b]$, then we have

$$D_F^\alpha \lambda f(x) = \lambda D_F^\alpha f(x). \tag{4.97}$$

(2) Let f and g be function on $[a, b]$ [Parvate (2009); Parvate and Gangal (2011)]. If $D_F^\alpha f(x)$ and $D_F^\alpha g(x)$ exists for all $x \in [a, b]$, then we have

$$D_F^\alpha(f + g)(x) = D_F^\alpha f(x) + D_F^\alpha g(x). \qquad (4.98)$$

Example 4.3. Consider $f : R \to R$, $f(x) = k \in R$ then F^α-derivative of $f(x)$ is

$$D_F^\alpha f = 0. \qquad (4.99)$$

Remark 4.10. This result is to be differed from fractional Riemann-Liouville derivatives.

Example 4.4. The fractal derivative of $S_F^\alpha(x)$ is

$$D_F^\alpha S_F^\alpha(x) = \chi_F(x). \qquad (4.100)$$

where

$$\chi_F(x) = \begin{cases} \frac{1}{\Gamma(\alpha+1)}, & x \in F \\ 0, & \text{otherwise.} \end{cases} \qquad (4.101)$$

Theorem 4.10. *If $f : R \to R$ is a continuous function such that $Sch \subset F$ where F is α-perfect set, and $D_F^\alpha f(x)$ is defined for all $x \in [a, b]$, and $f(a) = f(b) = 0$. Then there is a point $c \in F \cap [a, b]$ such that $D_F^\alpha f(x) \geq 0$ and a point $d \in F \cap [a, b]$ such that $D_F^\alpha f(x) \leq 0$.*

Proof. If $f(x) = k(constant)$, $\forall x \in [a, b]$, then $D_F^\alpha f(x) = 0$ Therefore c can be any number in $F \cap [a, b]$ and this theorem holds true. If $f(y) > 0$, $\exists\ y \in (a, b)$, since f is continuous then $\exists\ y \in (c, d) \subset (a, b)$ such that $f(z) > 0, \forall z \in (c, d)$. Suppose (c_0, d_0) is largest interval so we have $f(c_0) = f(d_0) = 0$. The point $c_0 \in Sch f \subset F$ as f is positive on the right side of c_0. Therefor we have

$$D_F^\alpha f(c_0) = F - \lim_{z \to c_0} \frac{f(z) - f(c_0)}{S_F^\alpha(z) - S_F^\alpha(c_0)} \geq 0. \qquad (4.102)$$

Likewise $d_0 \in F$ and we can write

$$D_F^\alpha f(d_0) = F - \lim_{z \to d_0} \frac{f(z) - f(d_0)}{S_F^\alpha(z) - S_F^\alpha(d_0)} \leq 0. \qquad (4.103)$$

The point c_0 and d_0 can be recognized as point c and d in the statement of the theorem. If there are no point y such that $f(y) > 0$ and neither is the function zero throughout, then one can use a points such that $f(y) < 0$ and kept on in a similar method. $\qquad \square$

Example 4.5. Let us consider a function as

$$f(x) = \begin{cases} S_F^\alpha(x), & 0 \le x \le 0.5; \\ \Gamma(\alpha+1) - S_F^\alpha(x), & 0.5 < x \le 1. \end{cases} \qquad (4.104)$$

and $f(0) = f(1) = 0$ [Parvate (2009); Parvate and Gangal (2011)]. The set of change of this function is C. The C^α-derivative is given by

$$D_F^\alpha f(x) = \begin{cases} \chi_C(x), & 0 \le x \le 0.5; \\ -\chi_C(x), & 0 < x \le 1. \end{cases} \qquad (4.105)$$

which is continuous in the interval $[0,1]$. Then we have

$$x \in C \Rightarrow D_F^\alpha f(x) = \mp \frac{1}{\Gamma(\alpha+1)} \ne 0. \qquad (4.106)$$

Remark 4.11. We note that the fragmented nature of F does not allow us to fined the analogue of Rolle's theorem/Fractal Rolle's theorem as strong as its ordinary calculus version.

Corollary 4.3. *Let $f : R \to R$ be a continuous function such that its set of change is contained in an α-perfect set $F \subset R$, $D_F^\alpha f(x)$ exists at all points $x \in [a,b]$ and $S_F^\alpha(b) \ne S_F^\alpha(a)$. Thus there exists a point $c \in F$ such that*

$$D_F^\alpha f(c) \ge \frac{f(b) - f(a)}{S_F^\alpha(b) - S_F^\alpha(a)} \qquad (4.107)$$

and a point $d \in F$ such that

$$D_F^\alpha f(c) \le \frac{f(b) - f(a)}{S_F^\alpha(b) - S_F^\alpha(a)}. \qquad (4.108)$$

Corollary 4.4. *Let $f : R \to R$ be a continuous function such that its set of change is contained in an α-perfect set $F \subset R$, $D_F^\alpha f(x)$ exists at all points $x \in [a,b]$ and $S_F^\alpha(a) \ne S_F^\alpha(b)$ [Parvate (2009); Parvate and Gangal (2011)]. Thus there exists a point $c \in F$ such that*

$$D_F^\alpha f(c) \ge \frac{f(b) - f(a)}{S_F^\alpha(b) - S_F^\alpha(a)} \qquad (4.109)$$

and a point $d \in F$ such that

$$D_F^\alpha f(d) \le \frac{f(b) - f(a)}{S_F^\alpha(b) - S_F^\alpha(a)}. \qquad (4.110)$$

Proof. Since we have

$$g(x) = (f(x) - f(a)) - \frac{f(b) - f(a)}{S_F^\alpha(b) - S_F^\alpha(a)} S_F^\alpha(x) - S_F^\alpha(a). \qquad (4.111)$$

If $g(a) = g(b) = 0$, then using corollary 4.4, there exists a point $c \in F$ such that $D_F^\alpha f(c) \geq 0$ and a point $d \in F$ such that $D_F^\alpha f(d) \leq 0$. This follows that

$$D_F^\alpha f(c) - \frac{f(b) - f(a)}{S_F^\alpha(b) - S_F^\alpha(a)} \geq 0, \qquad (4.112)$$

for some $c \in F$, and

$$D_F^\alpha f(d) - \frac{f(b) - f(a)}{S_F^\alpha(b) - S_F^\alpha(a)} \leq 0, \qquad (4.113)$$

for some $d \in F$, which lead to the required relations [Parvate (2009); Parvate and Gangal (2011)]. □

Corollary 4.5. *Let $f : R \to R$ be a continuous function such that $Sch(f) \subset F$ and $D_F^\alpha f(x) = 0$ for all $x \in [a, b]$. Then $f(x) = k$ for any $x \in [a, b]$ where k is some real constant.*

Proof. If f is not a constant, then there exist y and z, $y < z$ such that $f(y) \neq f(z)$. Thus we can write either $f(y) < f(z)$ or $f(y) > f(z)$.

(1) If $f(y) < f(z)$ then we have

$$\exists\, c \in F \cap (y, z) \Rightarrow D_F^\alpha f(c) \geq \frac{f(z) - f(y)}{S_F^\alpha(z) - S_F^\alpha(y)} > 0 \qquad (4.114)$$

(2) If $f(y) > f(z)$, then we can write

$$\exists\, d \in F \cap (y, z) \Rightarrow D_F^\alpha f(d) \leq \frac{f(z) - f(y)}{S_F^\alpha(z) - S_F^\alpha(y)} < 0. \qquad (4.115)$$

In both the cases, we show that there exist a point where the derivative is not zero which contradicts our assumption. □

Remark 4.12. Because of the fragmented nature of F the F^α-differentiability of f is not sufficient to the above results. Further, we need that f be a continuous and $Sch(f) \subset F$.

Theorem 4.11. *If the functions $f : R \to R$ and $g : R \to R$ are differentiable, then $h(x) = f(x)g(x)$ is F^α-differentiable [Parvate (2009); Parvate and Gangal (2011)], and we have*

$$D_F^\alpha(f(x)g(x)) = D_F^\alpha f(x)g(x) + f D_F^\alpha g(x). \qquad (4.116)$$

Proof. The proof is straightforward. □

4.8   Fundamental theorem of F^α-calculus of fractal sets

In the we study the relations of F^α-integration and F^α-differentiation [Parvate (2009); Parvate and Gangal (2011)].

Theorem 4.12. *Let $F \subset R$ be an α-perfect set. If $f \in B(f)$ is an F-continuous function on $F \cap [a, b]$, and*

$$g(x) = \int_a^x f(y) d_F^\alpha y \tag{4.117}$$

for all $x \in [a, b]$, then

$$D_F^\alpha g(x) = f(x)\chi_F(x). \tag{4.118}$$

Proof. If $x \notin F$, then $D_F^\alpha g(x) = 0$ by definition. For $x \in F$, if there are points in F arbitrarily close to x on both sides of x, then we have to consider both the following cases:

(1) The set $F \cap (x, z)$ is never empty for $z > x$ and

$$g(z) - g(x) = \int_x^z f(y) d_F^\alpha y. \tag{4.119}$$

(2) The set $F \cap (z, x)$ is never empty for $z < x$ and

$$g(x) - g(z) = \int_z^x f(y) d_F^\alpha y. \tag{4.120}$$

In the first case, $F \cap (x, z)$ is not empty for any $z > x$. Taking the F-limit as $z \to x$, we obtain

$$D_F^\alpha g(x) = F - \lim_{z \to x} \frac{\int_x^z f(y) d_F^\alpha y}{S_F^\alpha(z) - S_F^\alpha(y)}, \tag{4.121}$$

Since we have

$$m[f, F, [x, z]] \int_x^z \chi_F(y) d_F^\alpha y \leq \int_x^z f(y) d_F^\alpha y$$

$$\leq M[f, F[x, z]] \int_x^z \chi_F(y) d_F^\alpha y, \tag{4.122}$$

and

$$\int_x^z \chi_F(y) d_F^\alpha y = (S_F^\alpha(z) - S_F^\alpha(x)). \tag{4.123}$$

Therefore, in view of Eqs. (4.122) and (4.123) we obtain

$$m[f, F, [x, z]] \leq \frac{\int_x^z f(y) d_F^\alpha y}{S_F^\alpha(z) - S_F^\alpha(x)} \leq M[f, F[x, z]]. \tag{4.124}$$

As f is F-continuous,

$$F - \lim_{z \to x} m[f, F, [x, z]] = F - \lim_{z \to x} M[f, F[x, z]] = f(x). \qquad (4.125)$$

Taking into account Eqs.(4.121), (4.124), and (4.125) we get required result.

□

Theorem 4.13. *Let $f : R \to R$ be a continuous, F^α-differentiable function such that $Sch(f)$ is contained in an α-perfect set F and $h : R \to R$ be F-continuous, such that*

$$h(x)\chi_F(x) = D_F^\alpha f(x). \qquad (4.126)$$

Then we have

$$\int_a^b h(x)d_F^\alpha x = f(b) - f(a). \qquad (4.127)$$

Proof. Let

$$g(x) = \int_a^x h(x)d_F^\alpha x \qquad (4.128)$$

then we can write $D_F^\alpha g(x) = h(x)\chi_F(x)$ by using Theorem 4.12. Thus we have $D_F^\alpha(g(x) - f(x)) = 0, \forall\, x \in [a, b]$, and it follows that $g(x) - f(x) = k$, a constant by using Corollary 4.5. Therefore we can write as

$$f(b) - f(a) = g(b) - g(a) = \int_a^b h(x)d_F^\alpha x, \qquad (4.129)$$

which completes the proof.

□

Theorem 4.14. *Let the function $f : R \to R$, $g : R \to R$ be such that*

(1) f is continuous on $[a, b]$, and $Sch(f) \subset F$.
(2) $D_F^\alpha f(x)$ exists and is F-continuous on $[a, b]$.
(3) $g(x)$ is F-continuous on $[a, b]$ [Parvate (2009); Parvate and Gangal (2011)].

Therefore we have

$$\int_a^b f(x)g(x)d_F^\alpha x$$

$$= \left[f(x) \int_a^x g(x')d_F^\alpha x'\right]_a^b - \int_a^b D_F^\alpha f(x) \int_a^x g(x')d_F^\alpha x' d_F^\alpha x. \qquad (4.130)$$

Proof. The proof is left to the reader.

□

4.9   Function spaces in F^α-calculus on fractal sets

In this section, we preset analogues of various function spaces for F^α-calculus [Parvate (2009); Parvate and Gangal (2011)].

4.9.1   *Spaces of F^α-differentiable functions on fractal sets*

In this subsection, we note that SAF is short-form of set of all function, F is assumed to be bounded and α-perfect set, and $a = \inf F$, $b = \sup F$.

Definition 4.21. We introduce the following spaces:

(1) $B(F)$: SAF bounded on F.
(2) $C^0(F)/C(F)$ which are F-continuous.
(3) $C^k(F)$, $k \in N$ is SAF u such that

$$(D_F^\alpha)^n u \in C^0(F) \ \ \forall \ n \leq k. \tag{4.131}$$

A norm on $C^k(F)$ is defined by:

$$||u|| = \sum_{0 \leq n \leq k} \sup |(D_F^\alpha)^n u(x)|, \quad u \in C^k(F). \tag{4.132}$$

where $C^k(F)$ is complete with respect to this norm .

Definition 4.22. If a function u satisfy the following condition

$$S_F^\alpha(x) = S_F^\alpha(y) \Rightarrow u(x) = u(y) \tag{4.133}$$

it is called to be S_F^α-concordant.
The class of all S_F^α-concordant function u in $C^0(F)$ is denoted by $\tilde{C}^0(F)$.
The class of all functions u in $C^k(F)$ such that

$$0 \leq n \leq k \Rightarrow (D_F^\alpha)^n u \in \tilde{C}^0(F) \tag{4.134}$$

is denoted by $\tilde{C}^k(F)$.

We note that $\tilde{C}^k(F)$ is complete with respect to norm defined in Eq.(4.132).

Definition 4.23. For a given function $u \in \tilde{C}^0(F)$ one can uniquely define a function as follows:

$$v(S_F^\alpha(x)) = u(x) \tag{4.135}$$

where $v \in C^0([c,d])$ and $c = S_F^\alpha(a)$, $d = S_F^\alpha(b)$ [Parvate (2009); Parvate and Gangal (2011)].

Remark 4.13. We note that map given in Eq.(4.135) is a one-to-one from $\tilde{C}^0(F)$ to $C^0([c,d])$, therefore $\tilde{C}^0(F)$ is separable since $C^0([c,d])$.

Lemma 4.10. *The class of all functions which is denoted by $C^k(F)$ is separable.*

Proof. Let Y be the Cartesian product of $k+1$ copies of $\tilde{C}^0(F)$, namely,

$$Y = \Pi_{i=0}^{k} Y_i, \quad Y_i = \tilde{C}^0(F), \quad 0 \le i \le k. \tag{4.136}$$

The element of Y is written in the form $\mathbf{u} = (u_0, \dots, u_k)$. Since the Cartesian product of separable spaces is separable space then the space Y is separable which proves the lemma. $\qquad\square$

Let Y' be the subspace of Y, is defined by

$$Y' = \{\mathbf{u} = (u_0, \dots, u_k) \in Y | u_n = D_F^\alpha)^n u_0\}. \tag{4.137}$$

then Y' is separable since it a subspace of Y, also is isomorphic to $\tilde{C}^k(F)$.

4.9.2 *Spaces of F^α-integrable functions on fractal sets*

Let us consider the set F^α-integrable functions and denote by $\mathcal{L}(F)$. It is easy to see that $\mathcal{L}(F)$ is a vector space with usual operations of addition and scalar multiplication [Parvate (2009); Parvate and Gangal (2011)].

Definition 4.24. The norm in $\mathcal{L}(F)$ space is defined by

$$\mathcal{N}_p(u) = ||u||_p = \left[\int_a^b |u(x)|^p d_F^\alpha x \right]^{1/p}, \quad 1 \le p < \infty. \tag{4.138}$$

with the following property

$$||\lambda u||_p = |\lambda| ||u||_p, \quad \lambda \in R. \tag{4.139}$$

Theorem 4.15. *For $u, v \in \mathcal{L}(F)$ and $p \in (1, \infty)$,*

$$\int_a^b |f(x)g(x)| d_F^\alpha x \le \mathcal{N}_p(f)\mathcal{N}_{p'}(g) \tag{4.140}$$

which is called analogue of Hölder's inequality and where we have Young's inequality , namely,

$$\frac{1}{p} + \frac{1}{p'} = 1, \quad ab \le \frac{a^p}{p} + \frac{a^{p'}}{p'}, \tag{4.141}$$

and $a, b \ge 0$, $p \in (1, \infty)$.

Proof. If either $\mathcal{N}_p(f)$ or $\mathcal{N}_{p'}(g)$ is zero, the result is clear. If not, using Inequality (4.141) with

$$a = \frac{|f(x)|}{\mathcal{N}_p(f)}, \quad b = \frac{|g(x)|}{\mathcal{N}_{p'}(g)}, \tag{4.142}$$

we have

$$\frac{|f(x)|}{\mathcal{N}_p(f)} \cdot \frac{|g(x)|}{\mathcal{N}_{p'}(g)} \leq \frac{1}{p} \frac{|f(x)|^p}{\mathcal{N}_p(f)} + \frac{1}{p'} \frac{|g(x)|^{p'}}{\mathcal{N}_{p'}(g)}, \tag{4.143}$$

for all $x \in F$. By taking F^α-integrating from both side of Eq.(4.143) which completes the proof of the theorem. $\qquad \square$

Theorem 4.16. *For $1 \leq p < \infty$ and $f, g \in \mathcal{L}(F)$, we have*

$$\mathcal{N}_p^p(f + g) \leq \mathcal{N}_p(f) + \mathcal{N}_p(g), \tag{4.144}$$

which is called analogue of Minskowksi's inequality.

Proof. The case $p = 1$ is obvious. For $p > 1$

$$\mathcal{N}_p(f + g) \leq \int_a^b |f(x)||f(x) + g(x)|^{p-1} d_F^\alpha x$$

$$+ \int_a^b |g(x)||f(x) + g(x)|^{p-1} d_F^\alpha x. \tag{4.145}$$

From Theorem 4.15 we have

$$\int_a^b |f(x)||f(x) + g(x)|^{p-1} d_F^\alpha x$$

$$\leq \mathcal{N}_p(f)\mathcal{N}_{p'}(|f(x) + g(x)|^{p-1})$$

$$= \mathcal{N}_p(f) \left[\int_a^b |f(x) + g(x)|^{(p-1)p'} d_F^\alpha x \right]^{1/p'}$$

$$= \mathcal{N}_p(f) \left[\int_a^b |f(x) + g(x)|^p d_F^\alpha x \right]^{(p-1)/p}$$

$$= \mathcal{N}_p(f)\mathcal{N}_p^{p-1}(f + g). \tag{4.146}$$

Likewise we can obtain

$$\int_a^b |g(x)||f(x) + g(x)|^{p-1} d_F^\alpha x \leq \mathcal{N}_p(g)\mathcal{N}_p^{p-1}(f + g). \tag{4.147}$$

Therefore we get

$$\mathcal{N}_p^p(f + g) \leq \mathcal{N}_p(f)\mathcal{N}_p^{p-1}(f + g) + \mathcal{N}_p(g)\mathcal{N}_p^{p-1}(f + g) \tag{4.148}$$

which proves the theorem. $\qquad \square$

In view of Theorem 4.16, then $\mathcal{N}_p$ is a semi-norm.

Lemma 4.11. *For two functions $f, g \in \mathcal{L}(F)$, $\mathcal{N}_p(f - g) = 0$ for $p > 1$, if and only if $\mathcal{N}_1(f - g) = 0$.*

Proof. The proof is straightforward. $\qquad\qquad\qquad\qquad\qquad\square$

Definition 4.25. We say that two function $f, g \in \mathcal{L}(F)$ are $\mathcal{N}_p$-equivalent if

$$\mathcal{N}_p(f - g) = 0. \qquad (4.149)$$

Definition 4.26. The space of equivalence $L'(F)$ is a vector space with addition and scalar multiplication. The function $||.||_p = \mathcal{N}_p$ acts as a norm on $L'(F)$. Also $L'_p(F)$ indicates the $L'(F)$ with norm $\mathcal{N}_p$.

Definition 4.27. Two Cauchy function sequences $\{f_n\}, \{g_n\}$ are called $\mathcal{N}_p$-equivalent if

$$\lim_{n \to \infty} ||f_n - g_n|| = 0. \qquad (4.150)$$

This equivalence relation partitions the set of sequences in $L'_p(F)$ into equivalence classes . The set of the equivalence classes of sequences in $L'_p(F)$ is denoted by $L_p(F)$.

Remark 4.14. Thus $L_p(F)$ is complete by definition, and therefore is a Banach space [Parvate (2009); Parvate and Gangal (2011)].

4.10 Analogues of abstract Sobolev spaces on fractal sets

Definition 4.28. Let J be a finite set of nonnegative integers $\{j_1, \ldots, j_m\}$ not exceeding a fixed integer $k \geq 0$, and containing 0. Let $\{X_j, ||.||_{X_j}\} = \{X_j\}_{j \in J}$ be a family of Banach spaces X_j with norms $||.||_{X_j}$. Then one can define X Cartesian product as follows [Parvate (2009); Parvate and Gangal (2011)]:

$$X = \prod_{j \in J} X_j. \qquad (4.151)$$

The members of X are tuples written in the form $\mathbf{u} = (u_{j_1}, \ldots, u_{j_m})$.

The set X is a vector space with usual addition and scalar multiplication.

Definition 4.29. A norm is defined on X as follows:

$$||u||_X = \sum_{j \in J} ||u_j||_{X_j}. \qquad (4.152)$$

where $\mathbf{u} = (u_{j_1}, \ldots, u_{j_m}) \in X$.

Therefore X is a Banach space, and is separable if and only if each of the $X_j, j \in J$ is separable. Set $X_j = L_p(F)$ for each $j \in J$ where $p \in [1, \infty)$ is fixed. Since $\tilde{C}^\infty(F) \subset L_p(F)$, and if $u \in \tilde{C}^\infty(F)$, then $D_F^\alpha u \in \tilde{C}^\infty(F)$.

Definition 4.30. A norm on $u \in \tilde{C}^\infty(F)$is defined as follows [Parvate (2009); Parvate and Gangal (2011)]

$$||u||_J = \sum_{j \in J} ||(D_F^\alpha)^j u||_p. \tag{4.153}$$

The space $\tilde{C}^\infty(F)$ is not complete under this norm.

Definition 4.31. A mapping $I_J : \tilde{C}^\infty(F) \to X$ is defined by

$$I_J(u) = ((D_F^\alpha)^{j_1} u, \dots, (D_F^\alpha)^{j_m} u). \tag{4.154}$$

Definition 4.32. A projection operator $P_n : X \to X_n, n \in J$ is defined by

$$P_n(\mathbf{u} = (u_{j_1}, \dots, u_{j_m})) = u_n. \tag{4.155}$$

The mapping I_J is linear and isometric.

Definition 4.33. For $u \in \tilde{C}^\infty(F)$ a norm is defined by [Parvate (2009); Parvate and Gangal (2011)]:

$$||I_J(u)||_X = ||u||_J. \tag{4.156}$$

Let the image of $I_J(\tilde{C}^\infty(F))$ is denoted by $[\tilde{Y}^{k,p}(F)]$. Then I_J is an isometric isomorphism between $\tilde{C}^\infty(F)$ and $[\tilde{Y}^{k,p}(F)]$. The map P_n is a continuous linear map from X to X_n. Let us consider closure of $[\tilde{Y}^{k,p}(F)]$ in the topology of X by $[\tilde{W}^{k,p}(F)]$. Since $[\tilde{W}^{k,p}(F)]$ is closed in X, it is a Banach and separable space because X is separable .

Definition 4.34. Abstract Sobolev space $[\tilde{W}^{k,p}(F)]$ is defined as

$$\tilde{W}^{k,p}(F) = P_0([\tilde{W}^{k,p}(F)]). \tag{4.157}$$

It is a Banach space and separable since $L_p(F)$ is separable.

Definition 4.35. The abstract Sobolev F^α-derivative of $u \in \tilde{W}^{k,p}(F)$ is defined as

$$(D_F^\alpha)_W^j u = P_j(P_0^{-1}(u)). \tag{4.158}$$

Example 4.6. Consider the F^α-differential equation

$$D_F^\alpha f = g(x), \tag{4.159}$$

where

$$g(x) = \text{sgn}(x)\chi_F(x), \tag{4.160}$$

and

$$\text{sgn}(x) = \begin{cases} +1, & \text{if } x > 0; \\ -1, & \text{if } x < 0; \\ 0, & \text{if } x = 0. \end{cases} \tag{4.161}$$

The solution of Eq.(4.159) is

$$f(x) = \begin{cases} S_F^\alpha(x) - S_F^\alpha(0), & \text{if } x \geq 0; \\ S_F^\alpha(0) - S_F^\alpha(x), & \text{if } x < 0. \end{cases} \tag{4.162}$$

But f is not F^α-differentiable at $x = 0$. Thus we need to consider the abstract Sobolev derivative of f.

Example 4.7. Consider the following fractal differential equation

$$D_F^\alpha f = g(x) \tag{4.163}$$

where

$$g_\lambda(x) = \frac{2}{\pi}\chi_F(x)\tan^{-1}(\lambda S_F^\alpha(x)) \tag{4.164}$$

The solution of Eq.(4.163) is

$$f_\lambda(x) = \frac{2}{\pi}S_F^\alpha(x)\tan^{-1}(\lambda S_F^\alpha(x)) - \frac{1}{\lambda\pi}\ln(1 + (\lambda S_F^\alpha(x))^2). \tag{4.165}$$

In a Sobolev space $\tilde{W}^{1,p}$, $1 \leq p < \infty$ we have

$$\lim_{\lambda \to \infty} g_\lambda(x) = g(x) \tag{4.166}$$

and

$$\lim_{\lambda \to \infty} f_\lambda = f(x), \tag{4.167}$$

which implies that g is the abstract Sobolev derivative of f and f is a solution of Eq.(4.163) in the Sobolev sense [Parvate (2009); Parvate and Gangal (2011)].

4.11   Conjugacy between fractal calculus and ordinary calculus

The staircase function S_F^α,when restricted to the set $F \subset [a,b]$ maps the set F onto the interval $K = [S_F^\alpha(a), S_F^\alpha(b)]$ [Parvate (2009); Parvate and Gangal (2011)]. A transformation $\xi : B(K) \to B(F)$ which takes functions $g : K \to R$ to functions $f : F \to R$, namely,

$$f(x) = \xi[g(x)] = g(S_F^\alpha(x)). \tag{4.168}$$

This is a transform which creates a conjugacy between the ordinary calculus and F^α-calculus. In other words, F^α-derivatives and F^α-integrals of functions can be calculated by the ordinary integrals and derivatives of transformed functions. Then the results from ordinary calculus can be translated into results for F^α-calculus. A class of F^α-differential equations can be solved by transforming them into ordinary differential equations and transforming the solutions back. In this section we give the conjugacy between the two calculi.

As we know that $F \subset R$ is an α-perfect bounded set for some $\alpha \in (0,1]$, $a = \inf F$, $b = \sup F$, and K is the image of F under S_F^α viz. $K = [S_F^\alpha(a), S_F^\alpha(b)]$. Let $S_F^\alpha(a) \neq S_F^\alpha(b)$. We give some basic definition in the following:

(1) $B(F)$ denotes the class of functions bounded on F. Likewise $B(K)$ denotes the class of functions on K.
(2) $\mathcal{L}(F)$ denotes the class of functions which are F^α-integrable over $[a,b]$, and $\mathcal{L}(K)$ denotes the class functions in $B(K)$ which are Riemann integrable over the interval $[S_F^\alpha(a), S_F^\alpha(b)]$.
(3) A function space $\mathcal{A}$ containing all the S_F^α-concordant functions which satisfying the condition

$$S_F^\alpha(x_1) = S_F^\alpha(x_2) \Rightarrow h(x_1) = h(x_2), \quad \forall\, x_1, x_2 \in F. \tag{4.169}$$

We denote this subspace by $\tilde{\mathcal{A}}$, as a convention. For example, $\tilde{B}(F)$ denotes the subspace of $B(F)$ containing all functions in $B(F)$ satisfying condition 4.169.

Definition 4.36. The map $\phi : \tilde{B}(F) \to B(K)$ takes $h \in \tilde{B}(F)$ to $\phi[h] \in B(K)$ such that for each $x \in F$

$$\phi[h(S_F^\alpha(x))] = h(x). \tag{4.170}$$

where ϕ is inverse of ξ which defined in Eq.(4.168).

Remark 4.15. The domain of h in $\tilde{B}(F)$ is larger than F. However the values of the function h at points outside F are of no relevance.

Theorem 4.17. *The map* $\phi : \tilde{B}(F) \to B(K)$ *is one-to-one and onto.*

Proof. First, let us to prove that ϕ is one-to-one. If h_1 and h_2 are two distinct functions in $\tilde{B}(F)$. Then there exists $x \in F$ such that $h_1(x) \neq h_2(x)$. Then using Eq.(4.170) we have $\phi[h_1] = \phi[h_2]$. Second, we prove that ϕ is onto. Let g is any function in $B(K)$. Define function h by $h(x) = g(S_F^\alpha(x))$ for any $x \in F$. Then, we have $h \in \tilde{B}(F)$ and $g = \phi[h]$. $\square$

Definition 4.37. The map $\eta : B(F) \to \tilde{B}(F)$ takes $f \in B(f)$ to $g = \eta[f] \in \tilde{B}(F)$ such that

$$g(x) = \eta[f(x)] = \begin{cases} \min_{\{z \in F : S_F^\alpha(z) = S_F^\alpha(x)\}} f(z), & \text{if } x \in F \\ 0, & \text{if } x \notin F, \end{cases} \quad (4.171)$$

where we use minimum instead of infimum because Lemma 4.5 implies that there are at the most two points of F where the value of S_F^α is the same.

Remark 4.16. By Definition 4.37, we extend the domain of ϕ from $\tilde{B}(F)$ to $B(F)$. For $f \in B(F)$, $\eta[f]$ is pointwise the largest function in $\tilde{B}(F)$ which is nowhere greater than f on F. Further, if $f \in \tilde{B}(F)$, then $\eta[f] = f$, e.g. η restricted to $\tilde{B}(F)$ is the identity map.

Example 4.8. Consider the function $f(x) = x\chi_F(x)$ [Parvate (2009); Parvate and Gangal (2011)]. Let $g = \eta[f]$. Then for $x \in F$ we have

$$g(x) = \min_{\{z \in F : S_F^\alpha(z) = S_F^\alpha(x)\}} f(z)$$

$$= \min_{\{z \in F : S_F^\alpha(z) = S_F^\alpha(x)\}} z$$

$$= \min\{z \in F : S_F^\alpha(z) = S_F^\alpha(x)\}. \quad (4.172)$$

In particular case when $F = C$, the middle-$\frac{1}{3}$ Cantor set, $g(\frac{1}{3}) = g(\frac{2}{3}) = \frac{1}{3}$.

Definition 4.38. The map $\psi : B(F) \to B(K)$ is the composite map $\phi \circ \eta$

Remark 4.17. We note that ψ is an extension of ϕ from domain $\tilde{B}(F)$ to domain $B(F)$, since η is an identity over $\tilde{B}(F)$.

4.11.1 *Review of the Riemann integral on R*

In this section we review the Riemann integral to prove the conjugacy between F^α-calculus and ordinary calculus. Let $g \in B([c,d])$ and $I \subset [c,d]$ be any closed interval [Parvate (2009); Parvate and Gangal (2009, 2011)].

Definition 4.39. In the following we give some definitions as follows:

$$M'[g, I] = \sup_{x \in I} g(x) \tag{4.173}$$

and

$$m'[g, I] = \inf_{x \in I} g(x). \tag{4.174}$$

Definition 4.40. For a subdivision $P = \{y_0, \ldots, y_n\}$, the upper and the lower sums are defined respectively by

$$U'[g, P] = \sum_{i=0}^{n-1} M'[g, [y_i, y_{i+1}]](y_{i+1} - y_i) \tag{4.175}$$

and

$$L'[g, P] = \sum_{i=0}^{n-1} m'[g, [y_i, y_{i+1}]](y_{i+1} - y_i) \tag{4.176}$$

Definition 4.41. The upper and the lower Riemann integrals are defined respectively as

$$\overline{\int_c^d} g(y)dy = \inf_P U'[g, P], \tag{4.177}$$

and

$$\underline{\int_c^d} g(y)dy = \sup_P L'[g, P], \tag{4.178}$$

where the supremum and the infimum are taken over all the subdivision P of $[c, d]$. The function g is said to be Riemann integrable over $[c, d]$ if the upper and the lower integrals are equal, and in that case the Riemann integral of g over $[c, d]$ is defined to be the common value.

4.11.2 *Conjugacy between fractal integral on fractal sets and the ordinary integrals*

We present the equality between F^α-integrals of functions in $\tilde{\mathcal{L}}(F)$, the Riemann integrals of their images under ϕ [Parvate (2009); Parvate and Gangal (2009, 2011)].

Theorem 4.18. *A function $h \in \tilde{B}(F)$ is F^α-integrable over $[a, b]$ if and only if $g = \phi[h]$ is Riemann integrable over $K = [S_F^\alpha(a), S_F^\alpha(b)]$, namely, a function $h \in \tilde{B}(F)$ belongs to $\tilde{\mathcal{L}}(F)$ if and only if g belongs to $\mathcal{L}(K)$. Further, in that case*

$$\int_a^b h(x)d_F^\alpha x = \int_{S_F^\alpha(a)}^{S_F^\alpha(b)} g(u)du. \tag{4.179}$$

Proof. First, suppose h is F^α-integrable over $[a, b]$. Then for any given $\epsilon > 0$, there exists a subdivision $P = \{x_0, x_1, \ldots, x_n\}$ of $[a, b]$. Then such that

$$U^\alpha[f, F, P] - L^\alpha[f, F, P] < \epsilon \qquad (4.180)$$

Let $y_i = S_F^\alpha(x_i)$. Then $Q = \{y_i, 0 \le i \le n\}$ is a subdivision of $[S_F^\alpha(a), S_F^\alpha(b)]$. We note the possibility that $y_i = y_{i+1}$ for some values of i, in which case the number of points m in Q would be less than n. But for notation convenience, we use y_i just as defined above. Further, we denote $\mathcal{I} = \{i : 0 \le i \le n - 1; y_i \ne y_{i+1}\}$. For any $i \in \mathcal{I}$

$$
\begin{aligned}
M[h, F, [x_i, x_{i+1}]] &= \sup_{x \in F \cap [x_i, x_{i+1}]} h(x) \\
&= \sup_{x \in F \cap [x_i, x_{i+1}]} g(S_F^\alpha(x)) \\
&= \sup_{y \in [y_i, y_{i+1}]} g(y), \quad (S_F^\alpha \text{ being continuous, monotonic}) \\
&= M'[g, [y_i, y_{i+1}]] \qquad (4.181)
\end{aligned}
$$

Therefore,

$$
\begin{aligned}
U^\alpha[h, F, P] &= \sum_{i=0}^{n-1} M[h, F, [x_i, x_{i+1}]](S_F^\alpha(x_{i+1}) - S_F^\alpha(x_i)) \\
&= \sum_{i=0}^{n-1} M[h, F, [x_i, x_{i+1}]]((y_{i+1} - y_i)) \\
&= \sum_{i=0}^{n-1} M'[g, [y_i, y_{i+1}]]((y_{i+1} - y_i)), \qquad (4.182) \\
&= U'[g, Q]. \qquad (4.183)
\end{aligned}
$$

Note that the sum in the second and third step is over $i \in \mathcal{I}$ since the remaining terms do not contribute to the sum. In the same manner we can see that

$$L^\alpha[h, F, P] = L'[g, Q] \qquad (4.184)$$

Then from Eqs.(4.180), (4.182), and (4.184) we arrive at

$$U'[g, Q] - L'[g, Q] < \epsilon \qquad (4.185)$$

which implies that g is Riemann integrable over $[S_F^\alpha(a), S_F^\alpha(b)]$ and recalling the definition of F^α-integral and Riemann integral,

$$\int_{S_F^\alpha(a)}^{S_F^\alpha(b)} g(x)dx = \int_a^b h(x)d_F^\alpha x. \qquad (4.186)$$

Conversely, if g is Riemann integrable, then for a given $\epsilon > 0$ there exists a subdivision $Q' = \{z_0, z_1, \ldots, z_m\}$ of $[S_F^\alpha(a), S_F^\alpha(b)]$ such that $U'[g, Q'] - L'[g, Q'] < \epsilon$ [Parvate (2009); Parvate and Gangal (2009, 2011)]. Further, because of the continuity and monotonicity of S_F^α, there exist distinct $w_i, 0 \le i \le m$, such that $z_i = S_F^\alpha(w_i)$. Then the converse follows in a similar manner and we complete the proof. $\qquad\square$

Example 4.9. Consider the integral

$$\int_0^b h(x) d_F^\alpha x, \quad h(x) = (S_F^\alpha(x))^n, \quad S_F^\alpha(0) = 0. \tag{4.187}$$

It is obvious $h \in \tilde{B}(F)$, and further $h \in \tilde{\mathcal{L}}(F)$ using Theorem 4.7. Now if $g = \phi[h]$, then $g(u) = u^n$ by Eq.(4.170), and Theorem 4.179 we can write

$$\int_0^b h(x) d_F^\beta x = \int_0^{S_F^\alpha(b)} g(u) du = \frac{1}{n+1} (S_F^\alpha(b)))^{n+1}. \tag{4.188}$$

Theorem 4.19. *If $v_1, v_2 \in \tilde{\mathcal{L}}(F)$ are $\mathcal{N}_p$-equivalent, then $\phi[v_1]$ and $\phi[v_2]$ are $\mathcal{N}_p$-equivalent in $\mathcal{L}(K)$.*

Proof. By using the following

(1) $\phi[v_1 - v_2] = \phi[v_1] - \phi[v_2]$
(2) $\phi[|v|] = |\phi[v]|$
(3) $\phi[|v|^p] = (\phi[|v|])^p$

and Theorem 4.179 we conclude the results and complete the proof. $\qquad\square$

Remark 4.18. Let $f \in B(F)$ and $h = \eta[f]$. Then for a closed interval $I = [c, d]$, it is obvious from the definitions of M, m and η that

$$m[f, F, I] \le M[f, F, I] \tag{4.189}$$
$$m[h, F, I] \le M[h, F, I] \tag{4.190}$$

and also

$$M[h, F, I] \le M[f, F, I] \tag{4.191}$$
$$m[h, F, I] \le m[f, F, I]. \tag{4.192}$$

Lemma 4.12. *Let $f \in B(F)$ and $h = \eta[f]$. Then for a closed interval $I = [c, d]$ such that $S_F^\alpha(c) \ne S_F^\alpha(d)$ we have*

$$m[f, F, I] \le M[h, F, I]. \tag{4.193}$$

Proof. For $x \in F \cap I$,

$$f(x) \geq \inf_{z \in F \cap I} f(z) = m[f, F, I] \tag{4.194}$$

Let $x \in F$ such that $S_F^\alpha(c) < S_F^\alpha(x) < S_F^\alpha(d)$. Then by Lemma 4.5 there is at the most one other points in two cases [Parvate (2009); Parvate and Gangal (2009, 2011)]:

(1) $y \in F, y \neq x$ and $S_F^\alpha(y) = S_F^\alpha(x)$. Then $y \in I$ since S_F^α is monotonic. Further from Eqs. (4.171) and (4.194) we have

$$h(x) = h(y) = \min(f(x), f(y)) \geq m[f, F, I]. \tag{4.195}$$

(2) $y \in F, S_F^\alpha(y) = S_F^\alpha(x) \Rightarrow y = x$. Then from Eq.(4.194)

$$h(x) = f(x) \geq m[f, F, I]. \tag{4.196}$$

Therefore in the both cases we have

$$h(x) \geq m[f, F, I], \quad x \in F \cap I, \tag{4.197}$$

hence

$$M[h, F, I] = \sup_{x \in F \cap I} h(x) \geq m[f, F, I] \tag{4.198}$$

which completes the proof. $\qquad\qquad\qquad\qquad\qquad\qquad\qquad\qquad\square$

Remark 4.19. Now we summarize the above results as

$$m[h, F, I] \leq m[f, F, I] \leq M[h, F, I] \leq M[f, F, I] \tag{4.199}$$

where $I = [c, d]$ such that $S_F^\alpha(c) \neq S_F^\alpha(d)$.

Lemma 4.13. *Let $f \in B(F)$ and $h = \eta[f]$. Further, let $p < q < r < s \in [a, b]$ be such that*

$$S_F^\alpha(p) < S_F^\alpha(q) < S_F^\alpha(r) < S_F^\alpha(s). \tag{4.200}$$

Denote $I = [p, s]$ and $I' = [q, r]$ then

$$\omega[f, F, I] \geq \omega[h, F, I'] \tag{4.201}$$

where

$$\omega[f, F, I] = M[f, F, I] - m[f, F, I]. \tag{4.202}$$

which is called F-oscillation of $f \in B(F)$ over I [Parvate (2009); Parvate and Gangal (2009, 2011)].

Proof. It is clear that

$$M[h, F, I'] \leq M[f, F, I] \tag{4.203}$$

since $h(x) \leq f(x)$ for all $x \in F$ and $I' \subset I$ [Parvate (2009); Parvate and Gangal (2009, 2011)]. Now, by definition

$$m[h, F, I'] = \inf_{x \in F \cap I'} h(x) \tag{4.204}$$

where

$$h(x) = \min_{\{z \in F : S_F^\alpha(z) = S_F^\alpha(x)\}} f(z). \tag{4.205}$$

Thus

$$m[h, F, I'] = \inf_{x \in F \cap I'} \left[\min_{\{z \in F : S_F^\alpha(z) = S_F^\alpha(x)\}} f(z) \right]$$
$$= \inf_{\{x \in F : S_F^\alpha(q) \leq S_F^\alpha(z) \leq S_F^\alpha(r)\}} f(z), \tag{4.206}$$

since for a collection of set $\{A_\theta\}_{\theta \in \Theta}$,

$$\inf_{\theta \in \Theta} = \inf \left\{ \bigcup_{\theta \in \Theta} A_\theta \right\}. \tag{4.207}$$

Further, as S_F^α is monotonic increasing, the set $\{z \in F : S_F^\alpha(q) \leq S_F^\alpha(z) \leq S_F^\alpha(r)\}$ is a subset of $F \cap I = F \cap [p, s]$. Thus,

$$m[h, F, I'] \geq m[f, F, I]. \tag{4.208}$$

Thus we can finish the proof in view of Eqs.(4.203) and (4.208). $\square$

Example 4.10. Let $f = x\chi_C(x)$, where C is the middle-$\frac{1}{3}$ Cantor set, and let $I = [\frac{2}{3}, 1]$. Then, $\omega[f, C, I] = \frac{1}{3}$, but $\omega[h, C, I] = \frac{2}{3}$.

Theorem 4.20. *If $f \in \mathcal{L}(F)$, then $h = \eta[f] \in \tilde{\mathcal{L}}(F)$*

$$\int_a^b f(x)d_F^\alpha x = \int_a^b h(x)d_F^\alpha x, \tag{4.209}$$

which says that η preserves F^α-integrals.

Proof. Let $\epsilon > 0$ be arbitrary but fixed. As f is F^α-integrable, there exists a subdivision $P_{[a,b]} = \{x_0, \ldots, x_n\}$ such that

$$U^\alpha[f, F, P] - L^\alpha[f, F, P] < \frac{\epsilon}{2} \tag{4.210}$$

First consider the case $\omega[h, F, [a, b]] = 0$. Then h is a constant over $F \cap [a, b]$, and hence F^α-integrable over $[a, b]$. Now we consider the case $\omega[h, F, [a, b]] \neq 0$. Indicate

$$\mathcal{I} = \{i : 0 \leq i \leq n - 1; S_F^\alpha(x_{i+1}) \neq S_F^\alpha(x_i)\}. \tag{4.211}$$

For each $i \in \mathcal{I}$, let $y_i, z_i \in [x_i, x_{i+1}]$ be such that they satisfy the following three conditions

$$S_F^\alpha(x_i) < S_F^\alpha(y_i) < S_F^\alpha(z_i) < S_F^\alpha(x_{i+1}) \tag{4.212}$$

$$S_F^\alpha(y_i) - S_F^\alpha(x_i) < \frac{\epsilon}{4n\omega[h, F, [a, b]]} \tag{4.213}$$

and

$$S_F^\alpha(x_{i+1}) - S_F^\alpha(z_i) < \frac{\epsilon}{4n\omega[h, F, [a, b]]}. \tag{4.214}$$

let Q be a new subdivision of $[a, b]$ defined by $Q = P \cup \{y_i, z_i : i \in \mathcal{I}\}$. Moreover, if $i \notin \mathcal{I}$ then $S_F^\alpha(x_{i+1}) - S_F^\alpha(x_i) = 0$, allowing us to omit the corresponding terms in the sum in the following. Thus we have

$$U^\alpha[h, F, Q] - L^\alpha[h, F, Q]$$

$$= \sum_{i \in \mathcal{I}} \omega[h, F, [x_i, y_i]](S_F^\alpha(y_i) - S_F^\alpha(x_i))$$

$$+ \sum_{i \in \mathcal{I}} \omega[h, F, [y_i, z_i]](S_F^\alpha(z_i) - S_F^\alpha(y_i))$$

$$+ \sum_{i \in \mathcal{I}} \omega[h, F, [z_i, x_{i+1}]](S_F^\alpha(x_{i+1}) - S_F^\alpha(z_i))$$

$$\leq \omega[h, F, [a, b]] \sum_{i \in \mathcal{I}}$$

$$+ \sum_{i \in \mathcal{I}} \omega[f, F, [x_i, x_{i+1}]](S_F^\alpha(x_{i+1}) - S_F^\alpha(x_i)), \qquad (cf \ Lemma \ 4.13)$$

$$+ \omega[h, F, [a, b]] \sum_{i \in \mathcal{I}} (S_F^\alpha(x_{i+1}) - S_F^\alpha(z_i))$$

$$< \omega[h, F, [a, b]] \left[\frac{\epsilon}{4\omega[h, F, [a, b]]} \right] \qquad (cf \ Eq.(4.213))$$

$$+ (U^\alpha[f, F, P] - L^\alpha[f, F, P])$$

$$+ \omega[h, F, [a, b]] \left[\frac{\epsilon}{4\omega[h, F, [a, b]]} \right] \qquad (cf \ Eq.(4.214))$$

$$< \frac{\epsilon}{4} + \frac{\epsilon}{2} + \frac{\epsilon}{4} = \epsilon \qquad (cf \ Eq.(4.210))$$

which implies that h is F^α-integrable over $[a, b]$. Now the equality (4.209) of the respective integrals is proved using the F^α-integrability of h along with the inequality (4.199). $\qquad \square$

Remark 4.20. It is possible that a function f is not F^α-integrable but $h = \eta[f]$ is F^α-integrable.

Example 4.11. Consider the following function

$$f(x) = \begin{cases} 2, \; x \in C, \, (x - \delta, x) \cap C = \emptyset, \; \text{for some } \delta > 0 \\ 1, \; x \in C, \, (x - \delta, x) \cap C \neq \emptyset, \; \text{for any } \; \delta > 0 \\ 0, \; x \notin C \end{cases} \qquad (4.215)$$

where C is the middle-$\frac{1}{3}$ Cantor set. Function f is not F^α-integrable, while $\eta[f] = \chi_C$ which is F^α-integrable.

Theorem 4.21. *Given a function $v \in \mathcal{L}(F)$, v and $\eta[v]$ are $\mathcal{N}_p$-equivalent for all $p \geq 1$.*

Proof. As we know for any $x \in F$, $\eta[v(x)] \leq v(x)$. Thus

$$\int_a^b |v(x) - \eta[v(x)]| d_F^\alpha x$$
$$= \int_a^b (v(x) - \eta[v(x)]) d_F^\alpha x$$
$$= \int_a^b (v(x) d_F^\alpha x - \int_a^b \eta[v(x)]) d_F^\alpha x$$
$$= 0 \qquad (4.216)$$

which implies the required results for $p = 1$. This can be extended for $p > 1$ using Lemma 4.11. $\qquad \square$

Theorem 4.22. *Let $f \in \mathcal{L}(F)$, and let $g = \psi[f] = \phi \circ \eta[f]$. Then $g \in \mathcal{L}(K)$ is Riemann integrable over $K = [S_F^\alpha(a), S_F^\alpha(b)]$ and*

$$\int_a^b f(x) d_F^\alpha x = \int_{S_F^\alpha(a)}^{S_F^\alpha(b)} g(u) du. \qquad (4.217)$$

Definition 4.42. The map $\bar{\psi} : L'(F) \to L'(K)$ is such that if $v \in \bar{v} in L'(F)$, then $\bar{\psi}[\bar{v}]$ is defined to be the equivalence class $\bar{u} \in L'(K)$ containing $u = \psi[v]$.

Theorem 4.23. *The map $\bar{\psi}$ is a linear isometric isomorphism between the space $L'(F)$ and $L'(K)$.*

Proof. The proof follows from linearity of ϕ and Theorems 4.17, 4.19, 4.21 and 4.22. $\qquad \square$

Theorem 4.24. *The space $L_p(F)$ is separable.*

Proof. Let $\omega \in C_0^\infty(R)$ be such that $\omega(x) \geq 0$, $\forall x \in R$, $\text{supp}(\omega) = [-1, 1]$, and

$$\int_{-1}^{1} \omega(x)dx = 1 \qquad (4.218)$$

Here, for $u \in L'(K)$ a mollifier is defined by

$$R_\epsilon(u(x)) = \frac{1}{\epsilon} \int_{S_F^\alpha(a)}^{S_F^\alpha(b)} \omega(\frac{x-y}{\epsilon})u(y)dy \qquad (4.219)$$

i.e.

$$R_\epsilon(u(x)) = \int_{-1}^{1} u(x - \epsilon y)\omega(y)dy. \qquad (4.220)$$

Since $u \in L'(K)$ which belongs to the corresponding function space based on Lebesgue integral. Thus from Theorem 2.5.3 of [Kufner *et al.* (1977)] we have

$$R_\epsilon u \in C^\infty(R) \qquad (4.221)$$

and

$$\lim_{\epsilon \to 0+} \|R_\epsilon u - u\|_p = 0, \quad p > 1. \qquad (4.222)$$

Hence for every $u \in L'(K)$, there exists a sequence $\{u_n\}$ in $C^\infty(K)$ converging to u, implying that $C^\infty(K)$ is dense in $L'(K)$. Further, C^0 is also dense in $L'(K)$, since $C^\infty(K) \subset C^0(K)$. As $C^0(K)$ is separable, so is $L'(K)$. Now as $L'(K)$ is isomorphic to $L'(F)$ then it is separable. Therefore $L_p(F)$ being the completion of $L'(F)$ by definition, is also separable. □

Definition 4.43. The map $\bar{\bar{\psi}} : L_p \to L_p(K)$ is such that if $\{v_n\}$ is a Cauchy sequence in the equivalence class $\bar{\bar{v}} \in L_p(F)$, then $\bar{\bar{\psi}}[\bar{\bar{v}}]$ is defined to be the equivalence class $\bar{\bar{u}} \in L_p(K)$ containing the Cauchy sequence $\{\bar{\bar{u}}_n\} = \{\bar{\psi}[\bar{v}_n]\}$.

Remark 4.21. As $\bar{\psi}$ is an isometric isomorphism from $L'(F)$ to $L'(K)$, $\bar{\bar{\phi}}$ is an isometric isomorphism from $L_p(F)$ to $L_p(K)$.

Remark 4.22. For visualizing of the conjugacy, let $f \in \mathcal{L}(F)$ and $x \in [a, b]$, then the indefinite F^α-integral is

$$f_1(x) = \int_a^x f(x')d_F^\alpha x', \qquad (4.223)$$

belongs to $\tilde{C}(F) \subset \tilde{\mathcal{L}}(F)$. Let $g = \psi[f]$. Then $g \in \mathcal{L}(K)$. Further for $y \in [S_F^\alpha(a), S_F^\alpha(b)]$, the indefinite Riemann integral

$$g_1(y) = \int_{S_F^\alpha(a)}^{y} g(y')dy' \tag{4.224}$$

which belongs to $C(K) \subset \mathcal{L}(K)$. Now, Theorem 4.22 implies that

$$g_1(S_F^\alpha(x)) = f_1(x), \quad x \in [a, b]. \tag{4.225}$$

Definition 4.44. An operator $I_F^\alpha : \mathcal{L}(F) \to \tilde{C}(F)$ which is called the indefinite F^α-integral and defined by

$$I_F^\alpha[f(x)] = \int_a^x f(x')d_F^\alpha x' \tag{4.226}$$

and an operator $I : L(K) \to C(K)$ as the indefinite Riemann integral

$$I[g(y)] = \int_{S_F^\alpha(a)}^{y} g(y')dy' \tag{4.227}$$

We note that the indefinite F^α-integral of f can be calculated by applying ϕ^{-1} to the indefinite Riemann integral of $\psi[f]$ as follows:

$$I_F^\alpha = \phi^{-1}I\psi. \tag{4.228}$$

The relation between I_F^α and I is shown in the commutative diagram of Figure 4.1.

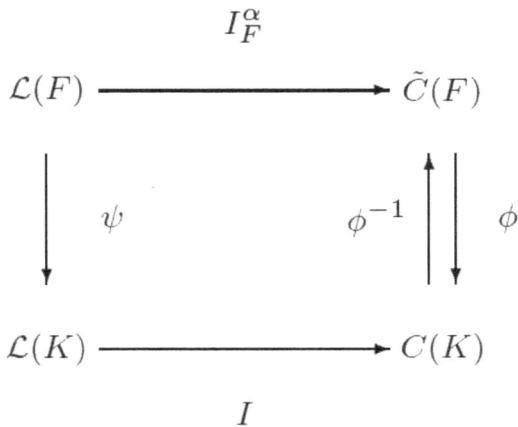

Fig. 4.1: The relation between F^α-integral and Riemann integral

4.11.3   *Conjugacy between fractal derivative on fractal sets and the ordinary derivative*

In this section we study the map ϕ which establishes the conjugacy between the F^{α}-derivative and the ordinary derivative under certain conditions. This consistent with fundamental theorem of both F^{α}-calculus and ordinary calculus [Parvate (2009); Parvate and Gangal (2009, 2011)].

Remark 4.23. The derivative of a function on $K = [S_F^{\alpha}(a), S_F^{\alpha}(b)]$ at the left end $S_F^{\alpha}(a)$ always means the right handed derivative, while one at the right end $S_F^{\alpha}(b)$ always means the left handed derivative.

Remark 4.24. As F is α-perfect , for any $u \in K$, the set $\{x \in F : S_F^{\alpha}(x) = u\}$ after this we denote $x \in (S_F^{\alpha})^{-1}(u)$, contains at the most two points [Parvate (2009); Parvate and Gangal (2009, 2011)].

Theorem 4.25. *Let h be a function in $\tilde{B}(F)$ such that the image $g = \phi[h]$ of h is ordinarily differentiable on K. Then $D_F^{\alpha}h(x)$ exists, belongs to $\tilde{B}(F)$, and*

$$D_F^{\alpha}h(x) = \frac{dg(t = S_F^{\alpha}(x))}{dt}, \qquad (4.229)$$

for all $x \in F$ [Parvate (2009); Parvate and Gangal (2009, 2011)].

Proof. Let $t \in K$, then by definition

$$\frac{dg}{dt} = \lim_{u \to t} \frac{g(u) - g(t)}{u - t}, \qquad (4.230)$$

i.e. for given $\epsilon_0 > 0$, there exists $\delta_0 > 0$ such that

$$u \in K \text{ and } 0 < |u - t| < \delta_0 \Rightarrow \left| \frac{dg}{dt} - \frac{g(u) - g(t)}{u - t} \right| < \epsilon_0. \qquad (4.231)$$

Let $x \in (S_F^{\alpha})^{-1}, y \in (S_F^{\alpha})^{-1}(u)$. Then $x, y \in F$, $h(y) = g(u)$ and $h(x) = g(t)$ [Parvate (2009); Parvate and Gangal (2009, 2011)]. Thus

$$0 < |S_F^{\alpha}(y) - S_F^{\alpha}(x)| < \delta_0 \Rightarrow \left| \frac{dg}{dt} - \frac{h(y) - h(x)}{S_F^{\alpha}(y) - S_F^{\alpha}(x)} \right| < \epsilon_0 \qquad (4.232)$$

whenever $S_F^{\alpha}(y) \neq S_F^{\alpha}(x)$.
As S_F^{α} is continuous, there exists $\delta_1 > 0$ such that for $y \in F$,

$$|y - x| < \delta_1 \Rightarrow |S_F^{\alpha}(y) - S_F^{\alpha}(x)| < \delta_0. \qquad (4.233)$$

For some $y_0 \in F$ such that $y_0 - x|\delta_1, y_0 \neq x$, it is possible that $S_F^{\alpha}(y_0) = S_F^{\alpha}(x)$ Then y_0 is the only such point. In that case we choose δ_2 such that

$0 < \delta_2 < |y_0 - x|$. Else we choose $\delta_2 = \delta_1$. Hence for $x \in (S_F^\alpha)^{-1}(t) \subset F$, there exists $\delta_2 > 0$ such that for $y \in F$,

$$0 < |y - x| < \delta_2 \Rightarrow \left| \frac{dg}{dt} - \frac{h(y) - h(x)}{S_F^\alpha(y_0) - S_F^\alpha(x)} \right| < \epsilon_0, \qquad (4.234)$$

In view of F-limit and D_F^α we have

$$D_F^\alpha h(x) = F - \lim_{y \to x} \frac{h(y) - h(x)}{S_F^\alpha(y) - S_F^\alpha(x)} = \frac{dg}{dt}(t = S_F^\alpha(x)). \qquad (4.235)$$

This holds for any $x \in (S_F^\alpha)^{-1}(t)$, we also see that $D_F^\alpha h(x) \in \tilde{B}(F)$. $\qquad \square$

Example 4.12. Let us consider the function $h(x) = (S_F^\alpha(x))^n, n = 1, 2, \ldots$ where $h \in \tilde{B}(F)$. Let $g = \phi[h]$. Then by Eq.(4.170), $g(t) = t^n$. Now according to Theorem 4.25, for $x \in F$

$$D_F^\alpha h(x) = \frac{dg(t = S_F^\alpha(x))}{dt} = nt^{n-1} \bigg|_{t = S_F^\alpha(x)} = n(S_F^\alpha(x))^{n-1}. \qquad (4.236)$$

Theorem 4.26. *Let $h \in \tilde{B}(F)$ be an F^α-differentiable function at all $x \in (S_F^\alpha)^{-1}(u)$ for some $u \in K$. Further, let $g = \phi[h]$*

(1) If $(S_F^\alpha)^{-1}(u) = \{y, x\}$ where $y < x$, then dg/dt_- and dg/dt_+ exists at $t = u$ and

$$\frac{dg(t = u)}{dt_-} = D_F^\alpha h(y), \qquad (4.237)$$

and

$$\frac{dg(t = u)}{dt_+} = D_F^\alpha h(x), \qquad (4.238)$$

where the subscripts $-$ and $+$ of dt denote the left and the right handed derivatives respectively.

(2) If $x \in F$ is the only point of $(S_F^\alpha)^{-1}(u)$, then dg/dt exists at $t = u$ and

$$\frac{dg(t = u}{dt} = D_F^\alpha h(x). \qquad (4.239)$$

Proof. As $g = \phi[h]$, we have $g(S_F^\alpha(x)) = h(x)$ for all $x \in F$ from Eq.(4.170) we have two cases as follows:

(1) By definition and substitution

$$\begin{aligned} D_F^\alpha h(x) &= F - \lim_{z \to x} \frac{h(z) - h(x)}{S_F^\alpha(z) - S_F^\alpha(x)} \\ &= F - \lim_{z \to x} \frac{g(S_F^\alpha(z)) - g(S_F^\alpha(x))}{S_F^\alpha(z) - S_F^\alpha(x)} \end{aligned} \qquad (4.240)$$

Thus, give $\epsilon_0 >$, there exists $\delta_0 > 0$ such that

$$z \in F, \quad |z-x| < \delta_0 \Rightarrow \left| \frac{g(S_F^\alpha(z)) - g(S_F^\alpha(x))}{S_F^\alpha(z) - S_F^\alpha(x)} - D_F^\alpha h(x) \right| < \epsilon_0. \quad (4.241)$$

Since S_F^α is constant on $[y, x]$ and F is α-perfect, $(x, x + \delta) \cap F$ is nonempaty for any $\delta > 0$. Let $z_0 \in F$ be such that $x < z_0 < x + \delta_0$. Thus $S_F^\alpha(z)) \neq S_F^\alpha(x))$ since F is α-perfect. Now let $t = S_F^\alpha(x)$ and $v_0 = S_F^\alpha(z_0)$. For $z \in (x, z_0)$, we can write $z - x < \delta_0$. Using monotonicity of S_F^α, we have $S_F^\alpha(z) - t \leq S_F^\alpha(z_0) - t (\equiv \delta_1)$. Let $v = S_F^\alpha(z)$. Further, $v = S_F^\alpha(z)$ takes all values between t and v_0 as z varies between x and z_0. Therefore, we arrive at for given ϵ_0, there exists δ_1 such that

$$0 < v - t < \delta_1 \Rightarrow \left| \frac{g(v) - g(t)}{v - t} - D_F^\alpha h(x) \right| < \epsilon_0 \quad (4.242)$$

which implies

$$\frac{dg(t = u)}{dt+} = D_F^\alpha h(x) \quad (4.243)$$

In the same manner we can see that

$$\frac{dg(t = u)}{dt-} = D_F^\alpha h(x) \quad (4.244)$$

(2) In this case, both $(x - \delta, x) \cap F$ and $(x, x + \delta) \cap F$ are nonempty for any $\delta > 0$. Thus the results can be obtained by considering two points $z_0, z_1 \in F$ such that $x - \delta_0 < z_0 < x < z_1 < x + \delta_0$ and doing same as case (1). $\qquad \square$

Remark 4.25. In a typical fractal set F, there are generally countably many points of type (1) and uncountably many of type (2) as the following for the middle-$\frac{1}{3}$ Cantor set.

(1) Points are $\frac{1}{3}, \frac{2}{3}, \frac{1}{9}$, etc.
(2) Points are $\frac{1}{4}, \frac{3}{4}$, etc.

Remark 4.26. We note that since the fractal F has fractured nature then F-limits at some points of F correspond only to one side limits in K. Thus at such points, only one-sided ordinary derivatives are guaranteed.

Example 4.13. Let us consider the following function

$$h(x) = \begin{cases} S_F^\alpha(x), & 0 \leq 0.5 \\ \Gamma(\alpha + 1) - S_F^\alpha(x), & 0.5 < x \leq 1. \end{cases} \quad (4.245)$$

where $F = C$ is the middle-$\frac{1}{3}$ Cantor set and $\alpha = \ln(2)/\ln(3)$ is its γ-dimension. If $g = \phi[h]$, then

$$g(t) = \begin{cases} t, & 0 \le t \le \frac{\Gamma(\alpha+1)}{2} \\ \Gamma(\alpha+1) - t, & \frac{\Gamma(\alpha+1)}{2} < t < \Gamma(\alpha+1). \end{cases} \qquad (4.246)$$

As $S_F^\alpha(\frac{1}{3}) = \Gamma(\alpha+1)/2$. Therefore by Theorem 4.26,

$$D_F^\alpha h(1/3) = \frac{dg(t = \frac{\Gamma(\alpha+1)}{2})}{dt-} = +1 \qquad (4.247)$$

whereas

$$D_F^\alpha h(2/3) = \frac{dg(t = \frac{\Gamma(\alpha+1)}{2})}{dt+} = -1 \qquad (4.248)$$

We see that $t = \Gamma(\alpha+1)/2$, the right and left derivatives of $g = \phi[h]$ are different even if h is F^α-differentiable everywhere on C [Parvate (2009); Parvate and Gangal (2009, 2011)].

Theorem 4.27. *Let $h \in \tilde{B}(F)$ be an F^α-differentiable function on F such that $D_F^\alpha h \in \tilde{B}(F)$. Let $g = \phi[h]$. Then for all $x \in F$,*

$$\frac{dg(t = S_F^\alpha(x))}{dt} = D_F^\alpha h(x). \qquad (4.249)$$

Proof. The function h satisfies the hypothesis of Theorem 4.26. If $x \in F$ satisfies the hypothesis of case(2) of Theorem 4.26, then the results is clear. But if x satisfies the assumption in case (1) of Theorem 4.26, i.e. there exists $y \in F$ such that $y < x$ and $S_F^\alpha(x) = S_F^\alpha(y) (\equiv u)$, then, for $t = u$,

$$\frac{dg}{dt-} = D_F^\alpha h(y) \quad and \quad \frac{dg}{dt+} = D_F^\alpha h(x). \qquad (4.250)$$

But as $D_F^\alpha h \in \tilde{B}(F)$, $D_F^\alpha h(y) = D_F^\alpha h(x)$. Thus

$$\frac{dg}{dt-} = \frac{dg}{dt+} \qquad (4.251)$$

which leads to Eq.(4.249), that completes the proof. $\qquad \square$

Remark 4.27. We can visualise of the conjugacy between the F^α-derivative and ordinary derivative. Let $\tilde{B}_1(F)$ be the subclass of function $h \in \tilde{B}(F)$ such that $D_F^\alpha h(x) \in \tilde{B}(F)$. Let $B_1(K)$ denote the subclass of function $g \in \tilde{B}(K)$ such that $dg/dt \in B(K)$. If D_F^α denotes the F^α-differentiation operator $(D_F^\alpha : \tilde{B}_1(F) \to \tilde{B}(F))$ and D denotes the ordinary differentiation operator $(D : (\tilde{B}_1(K) \to B(K))$, then the conjugacy between them is seen in Figure 4.2.

Thus the F^α-differentiation can be performed as follows:

$$D_F^\alpha = \phi^{-1} D \phi. \qquad (4.252)$$

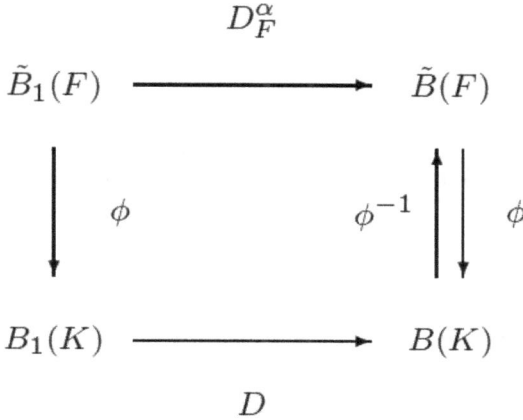

$$D_F^\alpha$$

$$\tilde{B}_1(F) \longrightarrow \tilde{B}(F)$$

$$\phi \qquad \phi^{-1} \qquad \phi$$

$$B_1(K) \longrightarrow B(K)$$

$$D$$

Fig. 4.2: The relation between F^α-derivative and ordinary derivative

4.11.4 *Conjugacy between the Sobolev spaces on fractal sets and ordinary Sobolev spaces*

We want to preset the conjugacy of the Sobolev spaces $\tilde{W}^{k,p}(F)$ and $\tilde{W}^{k,p}(K)$ which defined in Section 4.10[Parvate (2009); Parvate and Gangal (2009, 2011)]. To distinguish between the domain F and K, we use the notation $X(F)$ in place of X when each of the X_j in Eq.(4.151) is $L_p(F)$, and $X(K)$ when each of the X_j is $L_p(K)$. The map $\Psi : X(F) \to X(K)$ underlying the conjugacy is defined by:

$$\Psi(\mathbf{u}) = (\bar{\bar{\psi}}(u_{j_1}), \ldots, \bar{\bar{\psi}}(u_{j_m})), \quad \mathbf{u} \in X(F). \qquad (4.253)$$

The map Ψ is an isomorphism when the norm defined by Eq.(4.152) is considered. Using Theorem 4.27, if $u \in \tilde{C}^\infty(F)$, then $\bar{\bar{\psi}}u = \phi u \in C^\infty(K)$, and $\tilde{C}^\infty(F)$ is isometrically isomorphic to $C^\infty(K)$ under $\bar{\bar{\psi}}$ if we consider the $||.||_J$ norm defined by Eq.(4.153). It is obvious that

$$\Psi(J_J(u)) = I_J(\bar{\bar{\psi}}(u)), \quad u \in \tilde{C}^\infty(F) \qquad (4.254)$$

where the I_J operator on both sides act sides act upon the function on respective domain. The image of both I_J i.e. $[\tilde{Y}^{k,p}(F)]$ and $[\tilde{Y}^{k,p}(K)]$ are isometrically isomorphic under Ψ, implying that their closures $[\tilde{W}^{k,p}(F)]$ and $[\tilde{W}^{k,p}(K)]$ are also isometrically isomorphic, and so are $\tilde{W}^{k,p}(F)$ and $\tilde{W}^{k,p}(K)$. Also the abstract Sobolev derivatives are conjugate:

$$D_W^j(\bar{\bar{\psi}}(u)) = \bar{\bar{\psi}}((D_F^\alpha)_W^j u). \tag{4.255}$$

4.12   Calculus on fractal curves in R^n

Fractal curves because they are not smooth, so one can not use a ordinary calculus for them, such as path of a quantum mechanical particles, Brownian and Fractional Brownian trajectories , Koch curve and Weierstrass function . Fractal curves are continuous but non-differentiable. In this section, we adopt a Riemann-Stieltjes like approach for defining integrals and derivatives on fractal curves [Parvate *et al.* (2011)]. We present the framework as the follows.

Definition 4.45. The fractal curve $F \subset R^n$ is said to be continuously paramatrizable if there exists a function $w : [a_0, b_0] \to F \subset R^n$ which is continuous, one-to-one and onto F. In this section F stand for a fractal curve.

Example 4.14. The function $w : R \to R^2$ as follows

$$w(t) = (t, W_\lambda^s(t)) \tag{4.256}$$

where

$$W_\lambda^s(t) = \sum_{k=1}^{\infty} \lambda^{(s-2)k} \sin \lambda^k t, \quad \lambda > 1, \quad 1 < s < 2. \tag{4.257}$$

The graph of $W_\lambda^s(t)$ is known as a fractal curve with box-dimension s.

4.12.1   *Parameterizing fractal curves*

In this section, we want to explain how to parameterize self-similar fractal curve in two dimension [Parvate *et al.* (2011)].
Let $T_i, i = 0, \ldots, n-1$ be a linear operations which are composed of rotation and scaling. As T_i can be written as follows

$$T_i = s_i \begin{pmatrix} \cos\theta_i & -\sin\theta_i \\ \sin\theta_i & \cos\theta_i \end{pmatrix} \tag{4.258}$$

with the condition

$$\sum_{i=0}^{n-1} T_i(\mathbf{v}) = \mathbf{v} \tag{4.259}$$

where $\mathbf{v}$ is vector and $0 < s_i < 1$, for $i = 0, \ldots, n-1$. Then the fractal is defined by

$$S_j(\mathbf{v}) = \sum_{i=0}^{j-1} T_i(\mathbf{v_0}) + T_j(\mathbf{v}), \quad j = 0, \ldots, n-1, \tag{4.260}$$

where $\mathbf{v_0}$ is a fixed vector. The limit set of Eq.(4.260) will be in the form of a curve since the way S_j are constructed by from T_i. If $\lfloor nt \rfloor$ denotes the integer part of nt then we can define

$$\mathbf{w}(t) = \sum_{i=0}^{\lfloor nt \rfloor - 1} T_i(\mathbf{v_0}) + T_{\lfloor nt \rfloor}(\mathbf{w}(nt - \lfloor nt \rfloor)), \quad 0 \le t \le 1, \tag{4.261}$$

which parameterizes the above fractal curve.

Example 4.15. The Koch curve is parameterized by setting $s_i = 1/3$, $\theta_0 = \theta_3 = 0$, $\theta_1 = -\theta_2 = \pi/3$, and $v_0 = (1,0)$.

From now, a, b, c, etc. denote numbers in $[a_0, b_0]$ and θ, θ' etc. denote points of F.

Definition 4.46. For a set F and a subdivision $P_{[a,b]}, a < b, [a,b]$

$$\sigma^\alpha[F, P] = \sum_{i=0}^{n-1} \frac{|\mathbf{w}(t_{i+1}) - \mathbf{w}(t_i)|^\alpha}{\Gamma(\alpha + 1)}, \tag{4.262}$$

where $|.|$ denotes the Euclidean norm on R^n, $P_{[a,b]} = \{a = t_0, \ldots, t_n = b\}$.

Definition 4.47. For a given $\delta > 0$ and $a_0 \le a \le b \le b_0$ the coarse grained mass is defined by

$$\gamma_\delta^\alpha(F, a, b) = \inf_{\{P_{[a,b]} : |P| \le \delta\}} \sigma^\alpha[F, P], \tag{4.263}$$

where $|P| = \max_{0 \le i \le n-1}(t_{i+1} - t_i)$ for a subdivision P.

Lemma 4.14. *Let $\delta > 0$ and $a_0 \le a < b < c < b_0$. Then $\gamma_\delta^\alpha(F, a, b) \le \gamma_\delta^\alpha(F, a, c)$ and $\gamma_\delta^\alpha(F, b, c) \le \gamma_\delta^\alpha(F, a, c)$.*

Proof. Let $\epsilon > 0$. Using Definition 4.47 we can write

$$\sigma^\alpha[F, P] < \gamma_\delta^\alpha(F, a, c) + \epsilon \tag{4.264}$$

Let $Q_{[a,b]} = \{t \in P : t < b\} \cup \{b\}$. viz $Q_{[a,b]} = \{y_0, y_1, \ldots, y_m\}$ where $y_i = t_i$ if $t_i < b$ and $y_m = b$. $|Q_{[a,b]}| \le |P_{[a,c]}| \le \delta$ since $[y_{m-1}, y_m] \subset [t_{m-1}, t_m]$. Therefore,

$$\sigma^\alpha[F, Q_{[a,b]}] \le \sigma^\alpha[F, P_{[a,b]}] < \gamma_\delta^\alpha(F, a, c) + \epsilon \tag{4.265}$$

Since $\gamma_\delta^\alpha(F, a, c) \leq sigma^\alpha[F, Q]$ and ϵ is arbitrary, so that

$$\gamma_\delta^\alpha(F, a, b) \leq \gamma_\delta^\alpha(F, a, c) \tag{4.266}$$

which completes the proof of the first part. By a similar argument one can proof the second part. $\square$

Theorem 4.28. *For $a_0 \leq a \leq b \leq b_0$, $\gamma_\delta^\alpha(F, a, b)$ is continuous in b and a.*

Proof. We want to proof continuity $\gamma_\delta^\alpha(F, a, b)$ in b. Let δ, α, and a be fixed. Since $\mathbf{w}(t)$ is a continuous function then for given $\epsilon > 0$, there exists $\Delta' > 0$ such that

$$|c - b| < \Delta' \Rightarrow |\mathbf{w}(c) - \mathbf{w}(b)| < (\epsilon\Gamma(\alpha + 1))^{1/\alpha} \tag{4.267}$$

Let $\Delta = \min(\Delta', \delta)$. For $\epsilon_1 > 0$, there exists a subdivision $P_{[a,b]}$ such that $|P| \leq \delta$ and

$$\sigma^\alpha[F, P] < \gamma_\delta^\alpha(F, a, b) + \epsilon_1. \tag{4.268}$$

Let $Q = P \cup \{b + \Delta\}$ is a subdivision of $[a, b + \Delta]$ such that $|Q| \leq \delta$. Therefore

$$\gamma_\delta^\alpha(F, a, b + \Delta) \leq \sigma^\alpha[F, Q] \tag{4.269}$$

$$= \sigma^\alpha[F, P] + \frac{|\mathbf{w}(b + \Delta) - \mathbf{w}(b)|^\alpha}{\Gamma(\alpha + 1)} \tag{4.270}$$

$$\leq \sigma^\alpha[F, P] + \epsilon \tag{4.271}$$

$$< \gamma_\delta^\alpha(F, a, b) + \epsilon_1 + \epsilon. \tag{4.272}$$

Since ϵ_1 is arbitrary so that we have

$$\gamma_\delta^\alpha(F, a, b + \Delta) \leq \gamma_\delta^\alpha(F, a, b) + \epsilon. \tag{4.273}$$

As $\gamma_\delta^\alpha(F, a, b)$ is a nondecreasing function of b then we can write

$$\gamma_\delta^\alpha(F, a, b + t) \leq \gamma_\delta^\alpha(F, a, b) + \epsilon, \quad 0 < t < \Delta. \tag{4.274}$$

Thus for given $\epsilon > 0$, there exists a $\Delta > 0$ such that

$$0 < c - b < \Delta \Rightarrow \gamma_\delta^\alpha(F, a, c) - \gamma_\delta^\alpha(F, a, b) < \epsilon \tag{4.275}$$

which shows that $\gamma_\delta^\alpha(F, a, b)$ is continuous in b from right. For proofing continuity from left one need to replace b by $b - \Delta$ and $b + \Delta$ by b in the above proof. $\square$

Definition 4.48. For $a_0 \leq a < b < b_0$, the mass function $\gamma^\alpha(F, a, b)$ is defined by

$$\gamma^\alpha(F, a, b) = \lim_{\delta \to 0} \gamma^\alpha_\delta(F, a, b). \tag{4.276}$$

Remark 4.28. As $\gamma^\alpha(F, a, b)$ is a monotonic function of δ, then limit exists but could be finite or $+\infty$.

Theorem 4.29. *Let $a_0 \leq a < b < c < b_0$ and $\gamma^\alpha(F, a, c) < \infty$. Then we have*

$$\gamma^\alpha(F, a, c) = \gamma^\alpha(F, a, b) + \gamma^\alpha(F, b, c) \tag{4.277}$$

Proof. Let $\delta' > 0$. There exists a $\delta > 0$ such that

$$|t - t'| < \delta \Rightarrow |\mathbf{w}(t) - \mathbf{w}(t')| < \delta, \tag{4.278}$$

since $\mathbf{w}(t)$ is continuous. Suppose P_1 is any subdivision of $[a, b]$ and P_2 is any subdivision of $[b, c]$, such that $|P_1|\delta$ and $|P_2| < \delta$. Thus $P_1 \cup P_2$ is a subdivision of $[a, c], |P_1 \cup P_2| \leq \delta$, and

$$\sigma^\alpha[F, P_1 \cup P_2] = \sigma^\alpha[F, P_1] + \sigma^\alpha[F, P_2] \tag{4.279}$$

Taking the infimum of Eq.(4.279) over all P_1 and P_2 such that $|P_1| \leq \delta$ and $|P_2| \leq \delta$. We note all subdivision of $[a, c]$ can not be written in the form of $P_1 \cup P_2$. Then we have

$$\gamma^\alpha_\delta(F, a, c) \leq \gamma^\alpha_\delta(F, a, b) + \gamma^\alpha_\delta(F, b, c) \tag{4.280}$$

Suppose $0 < \delta_1 \leq \delta$. Now for every subdivision $P_{[a,c]}$, $|P| < \delta_1$, one can construct a subdivision $P' = P \cup b$ and $P' = P_1 \cup P_2$ where P_1 is a subdivision of $[a, c]$ and P_2 is a subdivision of $[b, c]$. Suppose $\{t_0, t_1, \ldots, t_n\}$ are points of P. If $b \in P$, then $P = P'$ and $\sigma^\alpha[F, P] = \sigma^\alpha[F, P']$. Otherwise, suppose $[t_k, t_{k+1}]$ is the interval which contains b. In that case

$$\sigma^\alpha[F, P'] - \sigma^\alpha[F, P] = \frac{|\mathbf{w}(b) - \mathbf{w}(t_k)|^\alpha}{\Gamma(\alpha + 1)}$$
$$+ \frac{|\mathbf{w}(t_{k+1}) - \mathbf{w}(b)|^\alpha}{\Gamma(\alpha + 1)} - \frac{|\mathbf{w}(t_{k+1}) - \mathbf{w}(t_k)|^\alpha}{\Gamma(\alpha + 1)}. \tag{4.281}$$

Hence

$$\sigma^\alpha[F, P'] - \sigma^\alpha[F, P] \leq \frac{3\delta'}{\Gamma(\alpha + 1)} \tag{4.282}$$

we see that

$$\sigma^\alpha[F, P] + \frac{3\delta'}{\Gamma(\alpha + 1)} \geq \sigma^\alpha[F, P']$$
$$= \sigma^\alpha[F, P_1] + \sigma^\alpha[F, P_2]$$
$$\geq \gamma^\alpha_{\delta_1}(F, a, b) + \gamma^\alpha_{\delta_1}(F, b, c) \tag{4.283}$$

for all such that $P < \delta_1$. Hence, by taking infimum over all subdivision P such that $|P| \leq \delta_1$, we get

$$\gamma_{\delta_1}^{\alpha}(F, a, c) + \frac{3\delta'}{\Gamma(\alpha + 1)} \geq \gamma_{\delta_1}^{\alpha}(F, a, b) + \gamma_{\delta_1}^{\alpha}(F, b, c). \qquad (4.284)$$

Eq.(4.284) holds for all δ_1 such that $0 < \delta_1 \leq \delta$. Taking limit as $\delta_1 \to 0$, we have

$$\gamma^{\alpha}(F, a, c) + \frac{3\delta'}{\Gamma(\alpha + 1)} \geq \gamma^{\alpha}(F, a, b) + \gamma^{\alpha}(F, b, c) \qquad (4.285)$$

Since δ' is arbitrary,

$$\gamma^{\alpha}(F, a, c) \geq \gamma^{\alpha}(F, a, b) + \gamma^{\alpha}(F, b, c) \qquad (4.286)$$

Taking into account Eq.(4.280) and Eq.(4.286) we complete the proof. $\qquad\square$

Theorem 4.30. *Let $a < b$ and $\gamma^{\alpha}(F, a, b) \neq 0$ be finite. Suppose y is such that $0 < y < \gamma^{\alpha}(F, a, b)$. Then there exists $c \in (a, b)$ where $a_0 \leq a < c < b \leq b_0$ such that $\gamma^{\alpha}(F, a, c) = y$.*

Proof. Let $z = \gamma^{\alpha}(F, a, b) - y$. For a given $\delta > 0$ consider the set of all points x of $[a, b]$ such that $\gamma_{\delta}^{\alpha}(F, x, b) \leq z$. This set is an interval of the form $[s_{\delta}, b]$ for some s_{δ}, $a \leq s_{\delta} < b$, because $\gamma_{\delta}^{\alpha}(F, x, b)$ is continuous and decreasing in x. Since $\gamma_{\delta}^{\alpha}(F, x, b)$ increasing as δ decreases, s_{δ} increasing as δ decreases. Likewise the set of all points x of $[a, b]$ such that $\gamma_{\delta}^{\alpha}(F, a, x) \leq y$ is an interval of the form $[a, t_{\delta}]$, $a < t_{\delta} \leq b$, and t_{δ} decreases as δ decreases. If $x \in (a, b)$, then by Theorem 4.29 we can write

$$\gamma^{\alpha}(F, a, b) = \gamma^{\alpha}(F, a, x) + \gamma^{\alpha}(F, x, b) \geq \gamma_{\delta}^{\alpha}(F, a, x) + \gamma_{\delta}^{\alpha}(F, x, b). \quad (4.287)$$

As $y, z < \gamma^{\alpha}(F, a, b)$, there exists a $\delta_0 > 0$ such that $\delta < \delta_0$ implies that $y, z < \gamma^{\alpha}(F, a, b)$. As $\gamma^{\alpha}(F, a, b) > y$ and $\gamma^{\alpha}(F, a, u)$ is continuous and increasing in u, there exists an $x \in (a, b)$ such that $\gamma^{\alpha}(F, a, x) = y$. This leads to $x \in [a, t_{\delta}]$. By Eq.(4.287) we obtain

$$z = \gamma^{\alpha}(F, a, b) - y = \gamma^{\alpha}(F, a, b) - \gamma^{\alpha}(F, a, x) \geq \gamma_{\delta}^{\alpha}(F, x, b) \qquad (4.288)$$

implies that x also belongs to $[s_{\delta}, b]$. This can happen only $s_{\delta} \leq t_{\delta}$. Therefore for each δ there exists an interval $[s_{\delta}, s_{\delta}]$ such that

$$x \in [s_{\delta}, s_{\delta}] \Rightarrow \gamma_{\delta}^{\alpha}(F, x, b) \leq z, \text{and } \gamma_{\delta}^{\alpha}(F, a, x) \leq y \leq \qquad (4.289)$$

Suppose

$$s = \sup_{0 < \delta < \delta_0} s_{\delta}, \qquad t = \inf_{0 < \delta < \delta_0} t_{\delta}. \qquad (4.290)$$

Since s_δ increase and t_δ decreases as δ goes to zero, but $s_\delta \leq t_\delta$ for any δ. Thus $s \leq t$ and

$$[s,t] = \bigcap_{0<\delta<\delta_0} [s_\delta, t_\delta]. \tag{4.291}$$

As a result $x \in [s,t]$ implies $\gamma_\delta^\alpha(F,x,b) \leq z$ and $\gamma_\delta^\alpha(F,a,x) \leq y$ for any δ. Thus it follows that

$$x \in [s,t] \Rightarrow \gamma^\alpha(F,x,b) \leq z, \text{ and } \gamma_\delta^\alpha(F,a,x) \leq y. \tag{4.292}$$

Since $\gamma^\alpha(F,a,x) + \gamma^\alpha(F,x,b) = \gamma^\alpha(F,a,b) = y + z$, so that inequalities in (4.292) must be equalities. In summery for given y, $0 < y < \gamma^\alpha(F,a,b)$ there exists a set $[s,t] \subset [a,b]$ such that

$$x \in [s,t] \Rightarrow \gamma^\alpha(F,a,x) \tag{4.293}$$

which completes the proof. $\qquad\qquad\qquad\qquad\qquad\qquad\qquad\qquad\square$

4.12.2 *Properties of F^α-calculus on fractal curves*

Let $F \subset R^n$ be parameterizable fractal curve and λ be a positive real number, $\mathbf{v} \in R^n$, and T denotes a rotation operator. Then we have the following properties in this framework [Parvate *et al.* (2011)].

(1) $F + \mathbf{v} = \{\mathbf{w}(t) + \mathbf{v} : t \in [a_0, b_0]\}$
(2) $\lambda F = \{\lambda \mathbf{w}(t) : t \in [a_0, b_0]\}$
(3) $TF = \{T\mathbf{w}(t) : t \in [a_0, b_0]\}$
(4) $\gamma^\alpha(F + \mathbf{v}, a, b) = \gamma^\alpha(F, a, b)$
(5) $\gamma^\alpha(\lambda F, a, b) = \lambda^\alpha \gamma^\alpha(F, a, b)$
(6) $\gamma^\alpha(TF, a, b) = \gamma^\alpha(F, a, b)$

Definition 4.49. Let $p_0 \in [a_0, b_0]$ be arbitrary but fixed. The staircase function $S_F^\alpha : [a_0, b_0] \to R$ of order α for a set F is given by

$$S_F^\alpha(t) = \begin{cases} \gamma^\alpha(F, p_0, t), & t \geq p_0; \\ \gamma^\alpha(F, p_0, t), & t < p_0, \end{cases} \tag{4.294}$$

where $t \in [a_0, b_0]$. From now set $p_0 = a_0$ unless stated otherwise. In Figure 4.3 we have plotted the staircase function a fractal curve [Golmankhaneh (2017)].

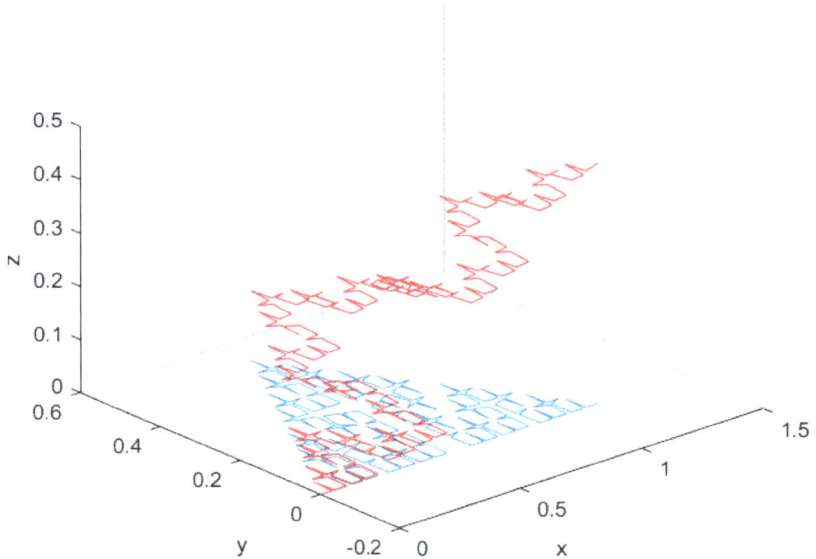

Fig. 4.3: The staircase function of a fractal curve for $\alpha = 1.7$

Assume that $1 \leq \alpha \leq n$, and $\gamma^\alpha(F, a_0, b_0)$ is finite, then for all $t_0, t_1 \in (a_0, b_0)$ such that $t_0 < t_1$, the following declarations hold:

(1) $S_F^\alpha(t)$ is increasing in t
(2) $S_F^\alpha(t_1) - S_F^\alpha(t_0) = \gamma^\alpha(F, t_0, t_1)$
(3) $S_F^\alpha(t)$ is continuous on (a_0, b_0)

Definition 4.50. The staircase function is defined by

$$J(\theta) = S_F^\alpha(\mathbf{w}^{-1}(\theta)), \quad \theta \in F. \tag{4.295}$$

As S_F^α is strictly increasing function then it is invertible and it is also one-to-one.

Definition 4.51. The mass function $\gamma^\alpha(F, a, b)$ of fractals gives the most useful information such as dimension. If $1 \leq \alpha < \beta \leq n$ then by Definition 4.46 we have

$$\sigma^\beta[F, P] = \sum_{i=0}^{n-1} \frac{|\mathbf{w}(t_{i+1}) - \mathbf{w}(t_i)|^\beta}{\Gamma(\beta + 1)} \tag{4.296}$$

Let $\delta' > 0$. There exists a $\delta > 0$ such that

$$|t - t'| \leq \delta \Rightarrow |\mathbf{w}(t) - \mathbf{w}(t)| \leq \delta', \qquad (4.297)$$

so $\mathbf{w}(t)$ is called uniformly continuous. Suppose $|P| < \delta$, then we can write

$$\sigma^\beta[F, P] \leq \delta'^{\beta - \alpha} \sum_{i=0}^{n-1} \frac{|\mathbf{w}(t_{i+1}) - \mathbf{w}(t_i)|^\alpha}{\Gamma(\alpha + 1)} \frac{\Gamma(\alpha + 1)}{\Gamma(\beta + 1)}$$

$$= \delta'^{\beta - \alpha} \sigma^\beta[F, P] \frac{\Gamma(\alpha + 1)}{\Gamma(\beta + 1)} \qquad (4.298)$$

Let $\delta' < \delta$ and taking infimum of Eq.(4.298) over all subdivision $|P| < \delta'$ we obtain

$$\gamma^\alpha_{\delta_1}(F, a, b) \leq \delta'^{\beta - \alpha} \gamma^\alpha_{\delta_1}(F, a, b) \frac{\Gamma(\alpha + 1)}{\Gamma(\beta + 1)}. \qquad (4.299)$$

In the limit $\delta_1 \to 0$, Eq.(4.299) turn to

$$\gamma^\alpha(F, a, b) \leq \delta'^{\beta - \alpha} \gamma^\alpha(F, a, b) \frac{\Gamma(\alpha + 1)}{\Gamma(\beta + 1)} \qquad (4.300)$$

Since δ' is arbitrary

$$\gamma^\alpha(F, a, b) = 0, \quad \text{for} \quad \gamma^\alpha(F, a, b) < \infty, \quad \text{and} \quad \beta > \alpha. \qquad (4.301)$$

Also, $\gamma^\alpha(F, a, b)$ is infinite up to certain value of α, say α_0, and jumps down to zero for $\alpha > \alpha_0$. By this preface, we can write the following definition. The γ-dimension of F, denoted by $\dim_\gamma(F)$, and defined by

$$\dim_\gamma(F) = \inf\{\alpha : \gamma^\alpha(F, a, b) = 0\}$$

$$= \sup\{\alpha : \gamma^\alpha(F, a, b) = \infty\} \qquad (4.302)$$

Example 4.16. Let α denote the γ-dimension of self similar curve, which is made up m copies of itself, scaled by factor of $\frac{1}{n}$ and rotated and translated appropriately. Thus we can write the mass function as follows

$$\gamma^\alpha(F, a_0, b_0) = m\gamma^\alpha\left(\frac{1}{n} F, a_0, b_0\right). \qquad (4.303)$$

It follows that

$$\gamma^\alpha(F, a_0, b_0) = m\left(\frac{1}{n}\right)^\alpha \gamma^\alpha(F, a_0, b_0). \qquad (4.304)$$

Then we have

$$\alpha = \frac{\log m}{\log n} \qquad (4.305)$$

This is same as the Hausdorff dimension of self-similar curves.

Definition 4.52. Let $F \subset R^n$ be a fractal curve, and let $f : R \to R$. Let $\theta \in F$. A number l is said to be the limit of f through points of F, if for a given $\epsilon > 0$ there exists $\delta > 0$ such that

$$\theta' \in F \quad and \quad |\theta' - \theta| < \delta \Rightarrow |f(\theta') - l|\epsilon. \qquad (4.306)$$

If such a number exists it is denoted by

$$l = F - \lim_{\theta' - \theta} f(\theta) \qquad (4.307)$$

Definition 4.53. A function $f : F \to R$ is said to be F-continuous at $\theta \in F$ if

$$f(\theta) = F - \lim_{\theta' \to \theta} f(\theta'). \qquad (4.308)$$

Definition 4.54. A function $f : F \to R$ is called to be uniformly continuous on $E \subset F$ if for any $\epsilon > 0$ there exists $\delta > 0$ such that for any $\theta \in F$ and $\theta' \in F$

$$|\theta' - \theta| < \delta \Rightarrow |f(\theta') - f(\theta)| < \epsilon. \qquad (4.309)$$

Definition 4.55. For $t_1, t_2 \in [a_0, b_0]$, $t_1 \leq t_2$ a section segment $C(t_1, t_2)$ of the curve is defined by

$$C(t_1, t_2) = \{\mathbf{w}(t) : t' \in [t_1, t_2]\} \qquad (4.310)$$

Definition 4.56. Let $f : F \to R$ and $t_1, t_2 \in [a_0, b_0]$, $t_1 \leq t_2$. Then

$$M[f, C(t_1, t_2)] = \sup_{\theta \in C(t_1, t_2)} f(\theta) \qquad (4.311)$$

and

$$m[f, C(t_1, t_2)] = \inf_{\theta \in C(t_1, t_2)} f(\theta) \qquad (4.312)$$

Definition 4.57. Let $S_F^\alpha(t)$ be finite for $t \in [a, b] \subset [a_0, b_0]$. Let P be the subdivision of $[a, b]$ with points $\{t_0, \ldots, t_n\}$. The upper and lower F^α-sum for the function f over the subdivision P are given respectively by:

$$U^\alpha[f, F, P] = \sum_{i=0}^{n-1} M[f, C(t_i, t_{i+1})][S_F^\alpha(t_{i+1}) - S_F^\alpha(t_i)] \qquad (4.313)$$

$$L^\alpha[f, F, P] = \sum_{i=0}^{n-1} m[f, C(t_i, t_{i+1})][S_F^\alpha(t_{i+1}) - S_F^\alpha(t_i)] \qquad (4.314)$$

From the Eqs(4.313) and (4.314) it is clear that

$$U^\alpha[f, F, P] \geq L^\alpha[f, F, P] \qquad (4.315)$$

Lemma 4.15. *Let $f \in B(f)$. If Q is a refinement of a subdivision P, then*

$$U^\alpha[f, F, Q] \le U^\alpha[f, F, P], \tag{4.316}$$

and

$$L^\alpha[f, F, Q] \ge L^\alpha[f, F, P]. \tag{4.317}$$

Proof. Let $P = \{t_0, t_1, \ldots, t_n\}$ and $Q = \{P \cup t'\}$ where $t' \in (t_i, t_{i+1})$. Then $M[f, F, [t_i, t']] \le M[f, F, [t_i, t_{i+1}]]$ and $M[f, F, [t', t_{i+1}]] \le M[f, F, [t_i, t_{i+1}]]$. Hence, $U^\alpha[f, F, P] \ge U^\alpha[f, F, Q]$. This results can be extended for any refinement of P. Likewise, we can prove

$$L^\alpha[f, F, Q] \ge L^\alpha[f, F, P] \tag{4.318}$$

□

Lemma 4.16. *If P and Q are any two subdivisions of $[a, b]$, then*

$$L^\alpha[f, F, P] \ge L^\alpha[f, F, Q]. \tag{4.319}$$

Proof. As $P \cup Q$ is a refinement of both P and Q, then in view of Lemma 4.15 and Eq.(4.318) we have

$$U^\alpha[f, F, Q] \ge U^\alpha[f, F, Q \cup F] \ge L^\alpha[f, F, P \cup Q] \ge L^\alpha[f, F, Q] \tag{4.320}$$

□

Definition 4.58. Let F be such that S_F^α is finite on $[a, b]$. For $f \in B(F)$, the lower and upper F^α-integral of the function f respectively, on the section $C(a, b)$ are

$$\underline{\int_{C(a,b)}} f(\theta) d_F^\alpha \theta = \sup_{P_{[a,b]}} L^\alpha[f, F, P], \tag{4.321}$$

and

$$\overline{\int_{C(a,b)}} f(\theta) d_F^\alpha \theta = \inf_{P_{[a,b]}} U^\alpha[f, F, P]. \tag{4.322}$$

Definition 4.59. If $f \in B(f)$, we say that f is F^α-integrable on $C(a, b)$ if

$$\underline{\int_{C(a,b)}} f(\theta) d_F^\alpha \theta = \overline{\int_{C(a,b)}} f(\theta) d_F^\alpha \theta, \tag{4.323}$$

and the common value is called the F^α-integral

$$\int_{C(a,b)} f(\theta) d_F^\alpha \theta. \tag{4.324}$$

Lemma 4.17. *Let $f \in B(F)$. Then f is F^α-integrable on $C(a,b)$ if and only if, for any $\epsilon > 0$, there exists a subdivision P of $[a,b]$ such that*

$$U^\alpha[f,F,P] < L^\alpha[f,F,P] + \epsilon. \qquad (4.325)$$

Theorem 4.31. *Let f be an F^α-integrable function on $C(a,b)$ and a $\leq c \leq b$. Then f is F^α-integrable on $C(a,c)$ and $C(c,b)$. Then we can write*

$$\int_{C(a,b)} f(\theta)d_F^\alpha\theta = \int_{C(a,c)} f(\theta)d_F^\alpha\theta + \int_{C(c,b)} f(\theta)d_F^\alpha\theta \qquad (4.326)$$

Proof. The proof is similar to Riemann integral. □

Lemma 4.18. *F^α-integration is a linear operation.*

Lemma 4.19. *Let $\gamma^\alpha(F,a,b)$ be finite, and $f(\theta) = 1$. Then we have*

$$\int_{C(a,b)} f(\theta)d_F^\alpha\theta = \int_{C(a,b)} d_F^\alpha\theta = S_F^\alpha(b) - S_F^\alpha(a) = J(\boldsymbol{w}(b)) - J(\boldsymbol{w}(a)). \qquad (4.327)$$

Proof. Let $I = C(a,b)$, $M[f,I] = m[f,I] = 1$. Thus $U^\alpha[f,F,P] = L^\alpha[f,F,P] = S_F^\alpha(b) - S_F^\alpha(a)$ for any subdivision P of $[a,b]$. □

Definition 4.60. Let F be a fractal curve. Then the F^α-derivative of function f at $\theta \in F$ is defined by:

$$D_F^\alpha f(\theta) = F - \lim_{\theta' \to \theta} \frac{f(\theta') - f(\theta)}{J(\theta') - J(\theta)} \qquad (4.328)$$

if the limit exists.

Theorem 4.32. *If $D_F^\alpha f(\theta)$ exists for all $\theta \in C(a,b)$, then f is F-continuous on $C(a,b)$.*

Lemma 4.20. *F^α-derivative is a linear operation.*

Example 4.17. Let $f : F \to R$, $f(\theta) = k \in R$, then we have

$$D_F^\alpha f(\theta) = 0. \qquad (4.329)$$

Example 4.18. Let $J : F \to R$ then its derivative is

$$D_F^\alpha J(\theta) = 1. \qquad (4.330)$$

Lemma 4.21. *Let f be a F-continuous function on the segment $C(a,b)$. If the maximum or minimum value for f is attained at $\boldsymbol{w}(c)$ where $a < c < b$ and if $D_F^\alpha(f(\boldsymbol{w}(c)))$ exists then $D_F^\alpha(f(\boldsymbol{w}(c))) = 0$.*

Proof. Let the contrary is true $D_F^\alpha(f(\mathbf{w}(c))) \neq 0$. If $D_F^\alpha(f(\mathbf{w}(c))) > 0$ then

$$F - \lim_{t \to c} \frac{f(\mathbf{w}(t)) - f(\mathbf{w}(c))}{S_F^\alpha(t) - S_F^\alpha(c)} > 0 \quad \text{and so} \quad \frac{f(\mathbf{w}(t)) - f(\mathbf{w}(c))}{S_F^\alpha(t) - S_F^\alpha(c)} > 0 \quad (4.331)$$

for $0 < |t - c| < \delta_1$ where δ_1 is a suitable positive number. If $t \in (c, c + \delta_1)$ then $S_F^\alpha(t) - S_F^\alpha(c) > 0$ and hence $f(\mathbf{w}(t) - \mathbf{w}(c)) > 0$. This contradicts the hypothesis that attains a maximum at $\mathbf{w}(c)$. If

$$\frac{f(\mathbf{w}(t) - \mathbf{w}(c))}{S_F^\alpha(t) - S_F^\alpha(c)} < 0 \quad (4.332)$$

for $0 < |t - c| < \delta_2$. If $t \in (c - \delta_2, c)$ then $S_F^\alpha(t) - S_F^\alpha(c) < 0$ and hence $f(\mathbf{w}(t) - \mathbf{w}(c)) > 0$, which is again a contradiction. Thus we have $D_F^\alpha(f(\mathbf{w}(c))) = 0$. □

Theorem 4.33. *Let $f : F \to R$ be a F-continuous function such that $D_F^\alpha f(\theta)$ is defined on $C(a, b)$ and $f(\mathbf{w}(a)) = f(\mathbf{w}(b)) = 0$. Then there is some point $c \in (a, b)$ where $D_F^\alpha f(\mathbf{w}(c)) = 0$.*

Theorem 4.34. *Let $f : F \to R$ be a F-continuous function such that $D_F^\alpha f(\mathbf{w}(t))$ exists on $C(a, b), a < b$. Then there exists a point $c \in [a, b]$ such that*

$$D_F^\alpha(\mathbf{w}(c)) = \frac{f(\mathbf{w}(b)) - f(\mathbf{w}(a))}{S_F^\alpha(b) - S_F^\alpha(a)}. \quad (4.333)$$

Proof. By using Theorem 4.33 to the function h

$$h(\mathbf{w}(t)) = f(\mathbf{w}(t)) - f(\mathbf{w}(a)) - \frac{f(\mathbf{w}(b)) - f(\mathbf{w}(a))}{S_F^\alpha(b) - S_F^\alpha(a)} S_F^\alpha(t) - S_F^\alpha(a) \quad (4.334)$$

for $a \le t \le b$. □

Corollary 4.6. *Let $f : F \to R$ be a F-continuous function such that $D_F^\alpha f = 0$. Then $f = k$ where k on $C(a, b)$.*

Proof. Let f be not a constant, then there exist y and z, $a \le y < z \le b$, such that $f(\mathbf{w}(y)) \neq f(\mathbf{w}(z))$. This implies that either $f(\mathbf{w}(y)) < f(\mathbf{w}(z))$ or $f(\mathbf{w}(y)) > f(\mathbf{w}(z))$. In both cases there exists $c \in (y, z)$ such that $D_F^\alpha f(\mathbf{w}(c)) \neq 0$ by Theorem 4.34, which is a contradiction. □

4.12.3 *Fundamental theorem of F^α-calculus on fractal curves*

In this subsection, we give relation of F^α-integration and F^α-differentiation [Parvate *et al.* (2011)].

Theorem 4.35. *Let $f \in B(F)$ is an F-continuous function on $C(a,b)$, and let $g : f \to R$ be defined as*

$$g(\boldsymbol{w}(t)) = \int_{C(a,t)} f(\theta)d_F^\alpha\theta \qquad (4.335)$$

for all $t \in [a,b]$. Then we have

$$D_F^\alpha g(\theta) = f(\theta). \qquad (4.336)$$

Proof. From Theorem 4.31, for $t' \in (t,b]$, we get

$$g(\mathbf{w}(t')) - g(\mathbf{w}(t)) = \int_{C(t,t')} f(\theta)d_F^\alpha\theta, \qquad (4.337)$$

Using Definition 4.60, we obtain

$$D_F^\alpha g(\theta) = F - \lim_{\theta'-\theta} \frac{g(\theta') - g(\theta)}{J(\theta') - J(\theta)} \qquad (4.338)$$

viz

$$D_F^\alpha g(\mathbf{w}(t)) = F - \lim_{t'\to t} \frac{\int_{C(t,t')} f(\theta)d_F^\alpha\theta}{S_F^\alpha(t') - S_F^\alpha(t)}, \quad \text{when} \quad \theta = \mathbf{w}(t), \ \theta' = \mathbf{w}(t). \qquad (4.339)$$

Now,

$$m[f,C(t,t')]\int_{C(t,t')} d_F^\alpha\theta \le \int_{C(t,t')} f(\theta)d_F^\alpha\theta \le M[f,C(t,t')]\int_{C(t,t')} f(\theta)d_F^\alpha\theta, \qquad (4.340)$$

and as we know by Lemma 4.19

$$\int_{C(t,t')} f(\theta)d_F^\alpha\theta = S_F^\alpha(t') - S_F^\alpha(t) \qquad (4.341)$$

so that

$$m[f,C(t,t')] \le \frac{\int_{C(t,t')} f(\theta)d_F^\alpha\theta}{S_F^\alpha(t') - S_F^\alpha(t)} \le M[f,C(t,t')] \qquad (4.342)$$

As f is continuous and $\mathbf{w}$ is continuous, namely,

$$\lim_{t'\to t+} m[f,C(t,t')] = \lim_{t'\to t+} M[f,C(t,t')] = f(\mathbf{w}(t)) \qquad (4.343)$$

Likewise

$$\lim_{t'\to t-} m[f,C(t,t')] = \lim_{t'\to t-} M[f,C(t,t')] = f(\mathbf{w}(t)) \qquad (4.344)$$

In view of Eqs.(4.339),(4.342),(4.343) and (4.344), we complete the proof. $\square$

Theorem 4.36. *Let $f : F \to R$ be F^α-differentiable function and $h : F \to R$ be F-continuous , such that $h(\theta) = D_F^\alpha f(\theta)$. Then*

$$\int_{C(t,t')} f(\theta)h(\theta)d_F^\alpha \theta = f(\boldsymbol{w}(b)) - f(\boldsymbol{w}(a)) \qquad (4.345)$$

Proof. If

$$g(\theta) = \int_{C(t,t')} f(\theta)h(\theta)d_F^\alpha \theta \qquad (4.346)$$

Thus using Theorem 4.35 we get $D_F^\alpha g(\theta) = h(\theta)$. Therefore $D_F^\alpha(g-f)(\theta) = 0$ for all $\theta \in C(a,b)$. Then Corollary 4.6 implies that $g - f = k$, a constant, or $g = f + k$. Thus

$$f(\boldsymbol{w}(b)) - f(\boldsymbol{w}(a)) = g(\boldsymbol{w}(b)) - g(\boldsymbol{w}(a)) = \int_{C(a,b)} h(\theta)d_F^\alpha \theta \qquad (4.347)$$

$\square$

4.13 Conjugacy of F^α-calculus on fractal curves and ordinary calculus

In this section, we present a map from F^α-integrable functions space to the Riemann integrable functions space [Parvate *et al.* (2011)].

Definition 4.61. We introduce the following definitions:

(1) $B(F)$: class of bounded functions $f : F \to R$
(2) $B([c,d])$: class of bounded functions $f : [c,d] \to R$
(3) $\mathcal{L}(F)$: set of all functions which are F^α-integrable on $C(a_0, b_0)$.
(4) The image of F under S_F^α is denoted by K, i.e. $K = [S_F^\alpha(a_0), S_F^\alpha(b_0)]$, and $B(K)$ denotes the class of functions bounded on K.
(5) $\mathcal{L}(K)$ denotes the class of functions in $B(K)$ which are Riemann integrable over the interval $K = [S_F^\alpha(a_0), S_F^\alpha(b_0)]$.

Remark 4.29. We give a brief review of Riemann integral to fix the notation. If $g \in B([c,d])$ and $I \subset [c,d]$ is a closed interval, then we denote $M'[g, I] = \sup_{x \in I} g(x)$ and $m'[g, I] = \inf_{x \in I} g(x)$. More, the upper and lower sum over a subdivision $P_{[c,d]} = \{y_0, \ldots, y_n\}$ are given by $U'[g, P] = \sum_{i=0}^{n-1} M'[g, [y_i, y_{i+1}]]$ and $L'[g, P] = \sum_{i=0}^{n-1} m'[g, [y_i, y_{i+1}]]$. If the upper and lower integral which are given by $\inf_P U'[g, P]$ and $\sup_P L'[g, P]$ are equal, therefore g is called to be Riemann integrable and denoted by

$$\int_c^d g(y)dy \qquad (4.348)$$

which is common value of both the upper and lower integral.

Definition 4.62. The map $\phi : B(F) \to B([S_F^\alpha(a_0), S_F^\alpha(b_0)])$ takes $f \in B(F)$ to $\phi[f] \in B([S_F^\alpha(a_0), S_F^\alpha(b_0)])$ such that for each $t \in [a_0, b_0]$,

$$\phi[f](S_F^\alpha(t)) = f(\mathbf{w}(t)) \tag{4.349}$$

Lemma 4.22. *The map* $\phi : B(F) \to B(K)$ *is one to one and onto.*

Theorem 4.37. *A function* $f \in B(F)$ *is* F^α*-integrable over* $C(a, b)$ *if and only if* $g = \phi[f]$ *is Riemann integrable over* $[S_F^\alpha(a_0), S_F^\alpha(b_0)]$, *viz*

$$f \in B(F) \quad and \quad f \in \mathcal{L}(F) \Leftrightarrow g \in \mathcal{L}(K), \tag{4.350}$$

then we can write

$$\int_{C(a,b)} f(\theta) d_F^\alpha \theta = \int_{S_F^\alpha(a)}^{S_F^\alpha(b)} g(u) du \tag{4.351}$$

Proof. Suppose $f : F \to R$ is F^α-integrable. Then there exists a subdivision $P_{[a,b]} = \{t_0, t_1, \ldots, t_n\}$ such that

$$U^\alpha[f, F, P] - L^\alpha[f, F, P] < \epsilon \tag{4.352}$$

for any $\epsilon > 0$. Set $y_i = S_F^\alpha(t_i)$. Then $Q = \{y_i : 0 \le i \le n\}$ is a subdivision of $[S_F^\alpha(a), S_F^\alpha(b)]$. Thus for any component $[t_i, t_{i+1}]$

$$\begin{aligned}
M[f, C(t_i, t_{i+1})] &= \sup_{w \in C(t_i, t_{i+1})} f(w) \\
&= \sup_{t \in [t_i, t_{i+1}]} f(\mathbf{w}(t)) \\
&= \sup_{t \in [t_i, t_{i+1}]} g(S_F^\alpha(t)) \\
&= \sup_{y \in [y_i, y_{i+1}]} g(y) \\
&= M'[g, [y_i, y_{i+1}] \tag{4.353}
\end{aligned}$$

Thus we have

$$\begin{aligned}
U^\alpha[f, F, P] &= \sum_{i=0}^{n-1} M[f, C(t_i, t_{i+1})][S_F^\alpha(t_{i+1}) - S_F^\alpha(t_i)] \\
&= \sum_{i=0}^{n-1} M[f, C(t_i, t_{i+1})][y_{i+1} - y_i] \\
&= \sum_{i=0}^{n-1} M'[g, [y_i, y_{i+1}]][y_{i+1} - y_i] \\
&= U'[g, Q] \tag{4.354}
\end{aligned}$$

Likewise we can show

$$L^\alpha[f, F, P] = L'[g, Q] \tag{4.355}$$

By Eqs.(4.352),(4.354), and (4.355) we get

$$U'[g, Q] - L'[g, Q] < \epsilon \tag{4.356}$$

which implies that g is Riemann integrable over $[S_F^\alpha(a), S_F^\alpha(b)]$ and

$$\int_{S_F^\alpha(a)}^{S_F^\alpha(b)} g(u)d(u) = \int_\theta f(\theta)d_F^\alpha\theta \tag{4.357}$$

Conversely if g is Riemann integrable then for given $\epsilon >$ there exists a subdivision $Q' = \{v_0, \ldots, v_m\}$ of $[S_F^\alpha(a), S_F^\alpha(b)]$ such that $U'[g, Q'] - L'[g, Q'] < \epsilon$. Then the converse can be proved by the following the above steps in the reverse order. Let f_1 denote the indefinite F^α-integral viz. $f_1(\mathbf{w}(t)) = \int_{C(a,t)} f(\theta)d_F^\alpha\theta$ and let g_1 denote the ordinary indefinite Riemann integral viz. $g_1(y) = \int_{S_F^\alpha(a)}^y g(y')dy'$. If the indefinite F^α-integral operator denote by I_F^α and indefinite Riemann integral operator by I, then we can rewrite Theorem 4.37 as follows:

$$I_F^\alpha = \phi^{-1}I\phi \tag{4.358}$$

which is shown in the commutative diagram of Figure 4.4. $\quad\square$

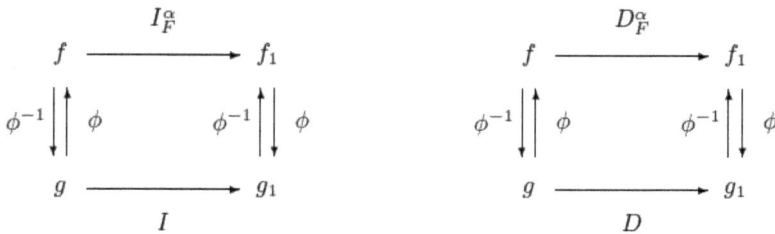

Fig. 4.4: The relation between F^α-integral and Riemann integral and also, between F^α-derivatives and ordinary derivatives

Theorem 4.38. *Let h be a function in $B(F)$ such that $g = \phi[h]$ is ordinary differentiable on $K = range of S_F^\alpha$. Then $D_F^\alpha h(\mathbf{w}(t))$ exists for all $t \in (a_0, b_0)$ and*

$$D_F^\alpha h(\mathbf{w}(t)) = \frac{dg(v)}{dv}\bigg|_{v=S_F^\alpha(t)} \tag{4.359}$$

Proof. If $v \in K$, then by definition of ordinary derivative we have

$$\frac{dg}{dv} = \lim_{u \to v} \frac{g(u) - g(v)}{u - v} \tag{4.360}$$

e.g. given $\epsilon_0 > 0$, there exists $\delta_0 > 0$ such that

$$|u - v| < \delta_0 \Rightarrow \left| \frac{dg}{dv} - \frac{g(u) - g(v)}{u - v} \right| < \epsilon_0 \tag{4.361}$$

As S_F^α is monotonically increasing and one-to-one. If $t = (S_F^\alpha)^{-1}(v)$, $t' = (S_F^\alpha)^{-1}(u)$, then

$$h(\mathbf{w}(t')) = g(u)$$
$$h(\mathbf{w}(t)) = g(v) \tag{4.362}$$

where $t', t \in [a_0, b_0]$. It follows that

$$|S_F^\alpha(t') - S_F^\alpha(t)| < \delta_0 \Rightarrow \left| \frac{dg}{dv} - \frac{h(\mathbf{w}(t')) - h(\mathbf{w}(t))}{S_F^\alpha(t') - S_F^\alpha(t)} \right| < \epsilon_0 \tag{4.363}$$

As $(\mathbf{w})^{-1}$ and S_F^α are continuous, hence their composition $S_F^\alpha \circ (\mathbf{w})^{-1}$. Thus there exists $\delta_1 > 0$ such that

$$|\mathbf{w}(t') - \mathbf{w}(t)| < \delta_1 \Rightarrow |S_F^\alpha(t') - S_F^\alpha(t)| < \delta_0$$
$$\Rightarrow \left| \frac{dg}{dv} - \frac{h(\mathbf{w}(t')) - h(\mathbf{w}(t))}{S_F^\alpha(t') - S_F^\alpha(t)} \right| < \epsilon_0. \tag{4.364}$$

By definition of F-limit and D_F^α we can write

$$D_F^\alpha h(\mathbf{w}(t)) = F - \lim_{\mathbf{w}(t') \to \mathbf{w}(t)} \frac{h(\mathbf{w}(t')) - h(\mathbf{w}(t))}{S_F^\alpha(t') - S_F^\alpha(t)} = \frac{dg}{dv}\bigg|_{v = S_F^\alpha(t)} \tag{4.365}$$

which completes the proof. $\qquad\qquad\qquad\qquad\qquad\qquad\qquad\quad\square$

Theorem 4.39. *Suppose $h \in B(F)$ is an F^α-differentiable function at all $\mathbf{w} \in F$. Further, let $g = \phi[h]$. Thus dg/dv exists at $v = S_F^\alpha(t)$ and*

$$\frac{dg(v)}{dv}\bigg|_{v = S_F^\alpha(t)} = D_F^\alpha h(\mathbf{w}(t)). \tag{4.366}$$

Proof. Since $g = \phi[h]$, we have $g(S_F^\alpha(t)) = h(\mathbf{w}(t))$ for all $t \in [a_0, b_0]$. By Definition 4.60 and substitution we have

$$D_F^\alpha h(\mathbf{w}(t)) = F - \lim_{\mathbf{w}(t') \to \mathbf{w}(t)} \frac{h(\mathbf{w}(t')) - h(\mathbf{w}(t))}{S_F^\alpha(t') - S_F^\alpha(t)}$$
$$= F - \lim_{\mathbf{w}(t') \to \mathbf{w}(t)} \frac{g(S_F^\alpha(t')) - g(S_F^\alpha(t))}{S_F^\alpha(t') - S_F^\alpha(t)}. \tag{4.367}$$

Thus given $\epsilon_0 > 0$ there exists $\delta_0' > 0$ such that

$$|\mathbf{w}(t') - \mathbf{w}(t)|\delta_0' \Rightarrow \left| \frac{g(S_F^\alpha(t')) - g(S_F^\alpha(t))}{S_F^\alpha(t') - S_F^\alpha(t)} - D_F^\alpha h(\mathbf{w}(t)) \right| < \epsilon_0. \quad (4.368)$$

Let $v = S_F^\alpha(t)$ and $u = S_F^\alpha(t')$. Thus $\mathbf{w} \circ (S_F^\alpha)^{-1}(v) = \mathbf{w}(t)$ and $\mathbf{w} \circ (S_F^\alpha)^{-1}(u) = \mathbf{w}(t')$. Further, since $\mathbf{w} \circ (S_F^\alpha)^{-1}$ is continuous there exists $\delta > 0$ such that

$$|u - v| < \delta \Rightarrow |\mathbf{w}(t') - \mathbf{w}(t)| < \delta_0' \Rightarrow \left| \frac{g(u) - g(v)}{u - v} - D_F^\alpha h(\mathbf{w}(t)) \right| < \epsilon_0.$$
$$(4.369)$$

By definition of ordinary derivative we can write

$$\left. \frac{dg}{dv} \right|_{v = S_F^\alpha(t)} = \lim_{u \to v} \frac{g(u) - g(v)}{u - v} = D_F^\alpha h(\mathbf{w}(t)). \quad (4.370)$$

The proof is now easily completed. $\qquad\square$

Remark 4.30. The above conjugacy can also be expressed as $D_F^\alpha = \phi^{-1} D \phi$ and also it is shown in the commutative diagram of Figure 4.4.

4.14   Function space on fractal curves

In this section we give function spaces on fractal curves [Parvate *et al.* (2011)].

4.14.1   *Spaces of F^α-differentiable functions on fractal curves*

Definition 4.63. We give the definition the following spaces:

(1) $C^k[c, d]$: Set of all functions k-times continuously differentiable on $[c, d]$ (in the ordinary sense of differentiation)
(2) $C^0(F)$ or $C(F)$: Set of all functions which are F-continuous.
(3) $C^k(F)$, $k \in N$: Set of all functions $f : F \to R$ such that

$$(D_F^\alpha)^n f \in C^0(F) \quad \text{for all} \quad n \le k. \quad (4.371)$$

which is set of all functions that have F-continuous F^α-derivatives upto order k.

Definition 4.64. The norm on $C^k(F)$ is defined by

$$\|f\| = \sum_{0 \le n \le k} \sup_{\theta \in F} |(D_F^\alpha)^n f(\theta)|, \quad f \in C^k(F). \quad (4.372)$$

Remark 4.31. The spaces $C^k(F)$ are complete with respect to this norm.

Remark 4.32. The class of function $C^k(F)$ is mapped one-to-one onto $C^k[c,d]$, where $c = S_F^\alpha(a_0)$, $d = S_F^\alpha(b_0)$ by ϕ. Due to this mapping, many results related to $C^k[c,d]$ is translated to analogous results for $C^k(F)$. For example, $C^k(F)$ is separable since $C^k[c,d]$ is separable.

4.14.2 F^α-integrable functions on fractal curves

In this section we introduce F^α-integrable function and their completion [Parvate *et al.* (2011)].

Definition 4.65. We denote F^α-integrable function by $\mathcal{L}(F)$ and it is obviously a vector space with usual operations of addition and scalar multiplication.

Definition 4.66. A norm on $f \in \mathcal{L}(F)$ is defined by

$$\mathcal{N}_p(f) = ||f||_p = \left[\int_{C(a,b)} |f(\theta)|^p d_F^\alpha \theta\right]^{1/p}, \quad 1 \le p < \infty. \tag{4.373}$$

It is satisfies the homogeniety property

$$||\lambda f||_p = |\lambda|\,||f||_p, \quad \lambda \in R. \tag{4.374}$$

Theorem 4.40. *For $f, g \in \mathcal{L}(F)$ and $p \in (1, \infty)$, we have*

$$\int_{C(a,b)} |f(\theta)g(\theta)| d_F^\alpha \theta \le \mathcal{N}_p(f)\mathcal{N}_{p'}(g) \tag{4.375}$$

where p, p' are related by Eq.(4.141).

Proof. If either $\mathcal{N}_p(f)$ or $\mathcal{N}_{p'}(g)$ is zero, the result is trivial. Else using Eq.(4.141) with

$$a = \frac{|f(\theta)|}{\mathcal{N}_p(f)}, \quad b = \frac{|g(\theta)|}{\mathcal{N}_p(g)}, \tag{4.376}$$

we get

$$\frac{|f(\theta)|}{\mathcal{N}_p(f)}\frac{|g(\theta)|}{\mathcal{N}_p(g)} \le \frac{1}{p}\frac{|f(\theta)|^p}{\mathcal{N}_p^p(f)} + \frac{1}{p'}\frac{|g(\theta)|^{p'}}{\mathcal{N}_{p'}^{p'}(g)} \tag{4.377}$$

for all $\theta \in F$. F^α-integrating Eq.(4.377) and applying Eq.(4.141), we complete the proof. $\square$

Theorem 4.41. *For $1 \leq p < \infty$ and $f, g \in \mathcal{L}(F)$ we have*

$$\mathcal{N}_p(f + g) \leq \mathcal{N}_p(f) + \mathcal{N}_p(g). \tag{4.378}$$

Proof. The case $p = 1$ is trivial. For $p > 1$

$$\mathcal{N}_p(f + g) \leq \int_{C(a,b)} |f(\theta)| |f(\theta) + g(\theta)|^{p-1} d_F^\alpha \theta +$$

$$\int_{C(a,b)} |g(\theta)| |f(\theta) + g(\theta)|^{p-1} d_F^\alpha \theta \tag{4.379}$$

Using Eq.(4.375) we have

$$\int_{C(a,b)} |f(\theta)| |f(\theta) + g(\theta)|^{p-1} d_F^\alpha \theta \leq \mathcal{N}_p(f) \mathcal{N}_{p'}(|f + g|^{p-1})$$

$$= \mathcal{N}_p(f) \left[\int_{C(a,b)} |f(\theta) + g(\theta)|^{(p-1)p'} d_F^\alpha \theta \right]^{1/p'}$$

$$= \mathcal{N}_p(f) \left[\int_{C(a,b)} |f(\theta) + g(\theta)|^{p} d_F^\alpha \theta \right]^{(p-1)/p}$$

$$= \mathcal{N}_p(f) \mathcal{N}_p^{p-1}(f + g) \tag{4.380}$$

In the same manner we can see that

$$\int_{C(a,b)} |g(\theta)| |f(\theta) + g(\theta)|^{p-1} d_F^\alpha \theta = \mathcal{N}_p(g) \mathcal{N}_p^{p-1}(f + g) \tag{4.381}$$

which completes the proof. $\qquad\square$

Remark 4.33. The $\mathcal{N}_p$ is a seminorm since it satisfies the triangle inequality.

Lemma 4.23. *For two functions $f, g \in \mathcal{L}(F)$, $\mathcal{N}_p(f - g) = 0$ for $p > 1$, if and only if $\mathcal{N}_1(f - g) = 0$.*

Proof. The proof is straightforward. $\qquad\square$

Definition 4.67. Two functions $f, g \in \mathcal{L}(F)$, are $\mathcal{N}_p$-equivalent if $\mathcal{N}_p(f - g) = 0$.

Remark 4.34. This equivalence relation partitions $\mathcal{L}(F)$ into equivalence classes of functions.

Definition 4.68. The space $L'_p(F)$ is a vector space with addition and scalar multiplication. The function $||.||_p = \mathcal{N}_p$ acts as a norm on $L'_p(F)$. Recalling Lemma 4.23 for any $p, q \in [1, \infty)$, $L'_p(F) = L'_q(F)$. Thus one can remove the subscript p if irrelevant and denote the space by $L'(F)$.

Definition 4.69. Two Cauchy sequences $\{f_n\}$, $\{g_n\}$ are $\mathcal{N}_p$-equivalent if

$$\lim_{n\to\infty} ||f_n - g_n||_p = 0. \tag{4.382}$$

Remark 4.35. This equivalence relation partition the set of sequences in $L'_p(F)$ into equivalence classes. The set of the equivalence classes of sequences in $L'_p(F)$ is denoted by $L_p(F)$. The $L_p(F)$ is complete by definition and thus is a Banach space. Constructions of $L'_p(K), L_p(K), \mathcal{N}_p$-norm can be made in analogy with $L'_p(F)$ $L_p(F)$, $\mathcal{N}_p$-norm respectively using Riemann integral.

Theorem 4.42. *If $v_1, v_2 \in \mathcal{L}(F)$ are $\mathcal{N}_p$-equivalent, then $\phi[v_1]$ and $\phi[v_2]$ are $\mathcal{N}_p$-equivalent in $\mathcal{L}(K)$.*

Proof. We see that

(1) $\phi[v_1 - v_2] = \phi[v_1] - \phi[v_2]$
(2) $\phi[|v|] = |\phi[v]|$
(3) $\phi[|v|^p] = (\phi[|v|])^p$

The proof follows from the above properties and Theorem 4.37. $\quad\square$

Definition 4.70. The map $\bar{\phi} : L'(F) \to L'(K)$ is defined, such that if $v \in \bar{v} \in L'(F)$, then $\bar{\phi}[\bar{v}]$ is the equivalence class $u \in L'(K)$ containing $u = \phi[v]$.

Theorem 4.43. *The map $\bar{\phi}$ is a linear isometric isomorphism between the spaces $L'(F)$ and $L'(K)$.*

Proof. The proof follows from linearity of ϕ and Theorems 4.37, 4.22 and 4.42. $\quad\square$

Theorem 4.44. *The spaces $L'_p(F)$ and $L_p(F)$ are separable.*

Proof. Let $\omega \in C_0^\infty(R)$ be such that $\omega(x) \geq 0$ for all $x \in R$, $supp(\omega) = [-1, 1]$,and

$$\int_{-1}^{1} \omega(x)dx = 1. \tag{4.383}$$

where $C_0^\infty(R)$ is the space of all functions in $C^\infty(R)$ with compact support. $\quad\square$

Definition 4.71. A mollifier is defined by

$$R_\epsilon(u(x)) = \frac{1}{\epsilon} \int_{S_F^\alpha(a)}^{S_F^\alpha(b)} \omega(\frac{x-y}{\epsilon})u(y)dy \qquad (4.384)$$

where $u \in L'(K)$. Then Eq.(4.384) turn to

$$R_\epsilon(u(x)) = \int_{-1}^{1} u(x - \epsilon y)\omega(y)dy. \qquad (4.385)$$

As $u \in L'(K)$, it also belongs to the corresponding function space based on Lebesgue integral. Then one can write

$$R_\epsilon u \in C^\infty(R), \qquad (4.386)$$

and

$$\lim_{\epsilon \to 0+} ||R_\epsilon u - u||_p = 0, p \geq 1. \qquad (4.387)$$

Thus for every $u \in L'(K)$, there exists a sequence $\{u_n\}$ in $C^\infty(K)$ converging to u. This indicate that $C^\infty(K)$ is dense in $L'(K)$. Since $C^\infty(K) \subset C^0(K)$ thus $C^0(K)$ is also dense in $L'(K)$. As $C^0(K)$ is separable $L'(K)$ is separable. Further $L'(K)$ is isomorphic to $L'(F)$ by Theorem 4.43, hence it is separable. Therefore $L_p(F)$, being the completion of $L_p'(F)$ by definition is also separable.

4.15 Analogues of abstract Sobolev spaces on fractal curves

Recalling section 4.10 we take $X_j = L_p(F)$ for each $j \in J$, where $p \in [1, \infty)$ is fixed. Also, $C^\infty \subset L_p(F)$, and if $u \in C^\infty(F)$, then $D_F^\alpha u \in C^\infty(F) \subset L_p(F)$ [Parvate *et al.* (2011)].

Definition 4.72. A norm on $C^\infty(F)$ is defined by

$$||u||_J = \sum_{j \in J} ||(D_F^\alpha)^j u||_p, \qquad (4.388)$$

where $u \in C^\infty(F)$. In general, the space is not complete under this norm.

Definition 4.73. A mapping $I_J : C^\infty(F) \to X$ is defined by

$$I_J(u) = ((D_F^\alpha)^{j_1}u, \ldots, (D_F^\alpha)^{j_m}u) \qquad (4.389)$$

A projection operator $P_n : X \to X_n$, $n \in J$ is defined by

$$P_n(\mathbf{u} = (u_{j_1}, \ldots, u_{j_m})) = u_n. \qquad (4.390)$$

Remark 4.36. The mapping I_J is linear and isometric, i.e. for $u \in C^\infty(F)$

$$||I_J(u)||_X = ||u||_J. \tag{4.391}$$

The image $I_J(C^\infty(F))$ is denoted by $[Y^{k,p}(F)]$. Then I_J is isometric isomorphism between $C^\infty(F)$ and $[Y^{k,p}(F)]$. The mapping P_n is continuous linear mapping from X to X_n. The closure of $[Y^{k,p}(F)]$ in the topology of X by $[W^{k,p}(F)]$. Since $[W^{k,p}(F)]$ is closed in X, it is a Banach space. Further since X is separable, so $[W^{k,p}(F)]$ is also separable.

Definition 4.74. An abstract Sobolev space is defined by

$$W^{k,p}(F) = P_0([W^{k,p}(F)]) \tag{4.392}$$

It is easy to see, this is a Banach space , and is separable since $L_p(F)$ is separable .

Definition 4.75. The abstract (j^{th}) Sobolev F^α-derivative of $u \in W^{k,p}$ is defined by

$$(D_F^\alpha)_W^j u = P_j(P_0^{-1}(u)). \tag{4.393}$$

Chapter 5

Local Fractal Differential Equations

In this chapter, we study the fractal differential equations and their solutions. Furthermore, we present Laplace and Fourier transform to solve the fractal differential equations.

5.1 Local differential equations with α-order on fractal sets and curves

Before start this chapter we summarize fractal calculus as follows:
The totally discontinuous thin Cantor set which is support of function f. To calculate integral or derivative of f we do the following steps:

(1) Calculate the staircase function of corresponding thin Cantor set.
(2) Replace x by $S_F^\alpha(x)$
(3) Use the standard rules of integration and derivation

In Figures 5.1, 5.2, 5.3, and 5.4, we have plotted ternary Cantor set, the staircase function, its dimension, and characteristic function on triadic Cantor set Cantor set [Golmankhaneh (2019b)].

Example 5.1. Consider a function $f(x) : F \to R$ with the fractal set support as follows [Golmankhaneh *et al.* (2018)]:

$$f(x) = \sin(2\pi x \chi_F) \tag{5.1}$$

Hence, its fractal derivative is

$$D_F^\alpha f(x) = \frac{2\pi}{\Gamma(\alpha+1)} \cos(2\pi x \chi_F), \tag{5.2}$$

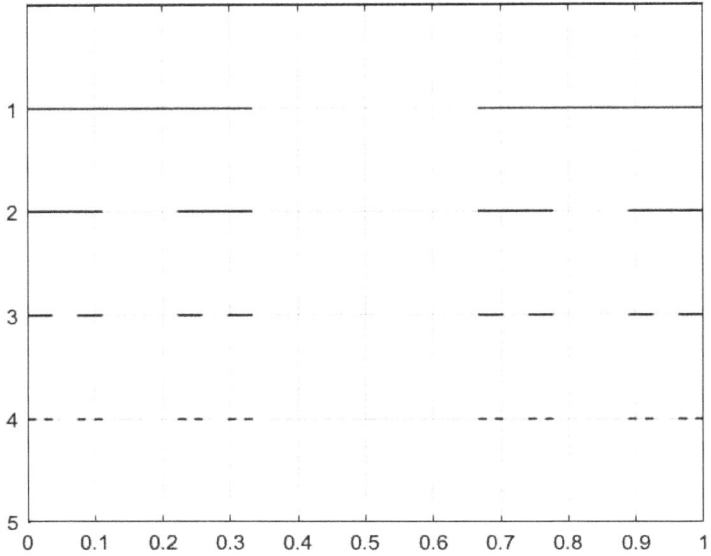

Fig. 5.1: Graph of triadic Cantor set for 4-iteration

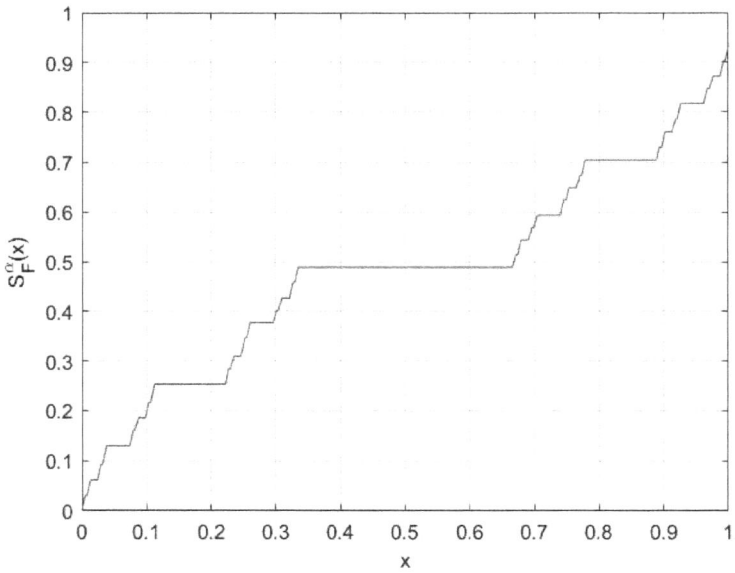

Fig. 5.2: Graph of staircase function

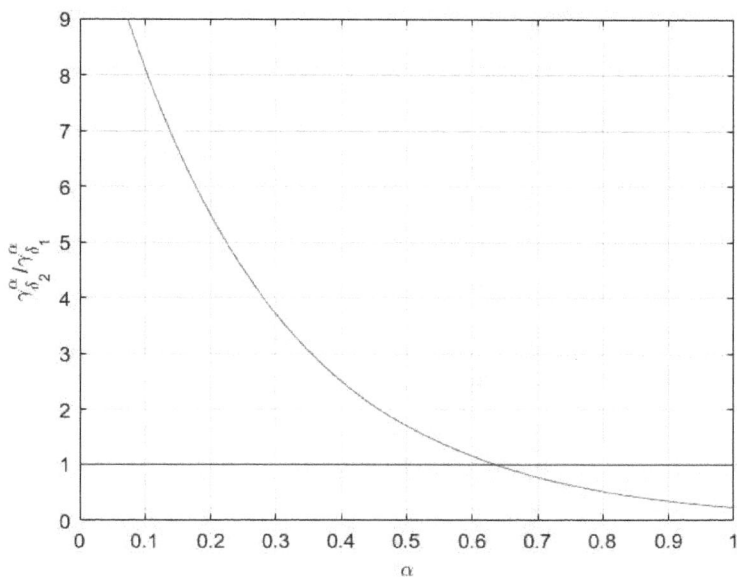

Fig. 5.3: Graph of $\gamma_{\delta_2}^{\alpha}(F, a, b)/\gamma_{\delta_1}^{\alpha}(F, a, b)$ gives γ-dimension=0.63

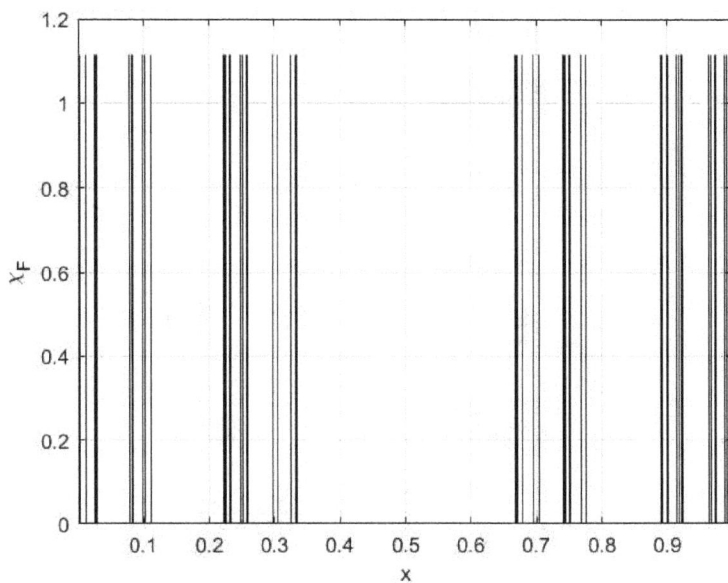

Fig. 5.4: Graph of characteristic function

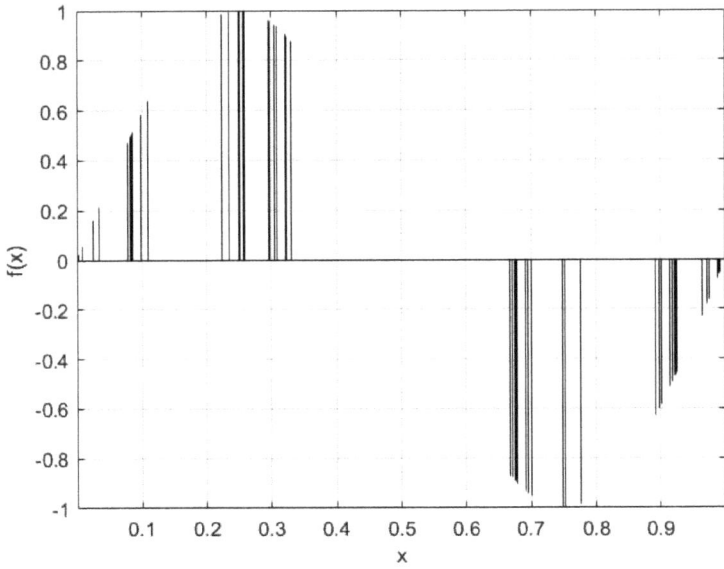

Fig. 5.5: Graph of f on the Cantor set

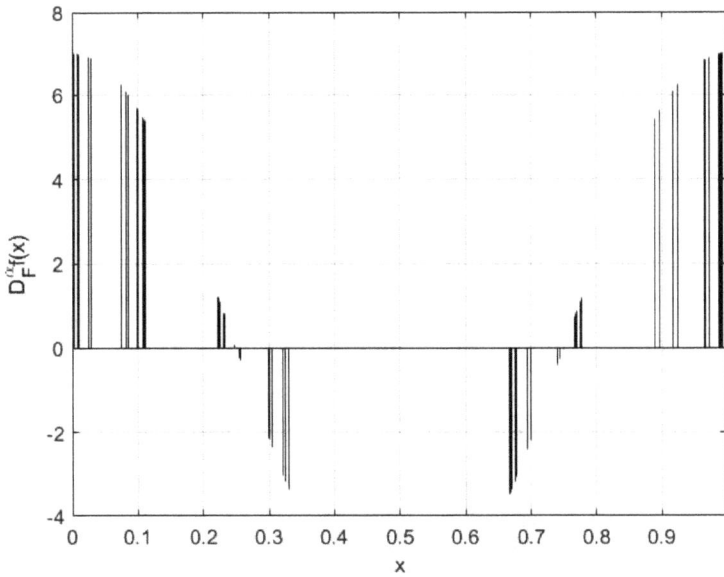

Fig. 5.6: Graph of fractal derivative of f on the Cantor set

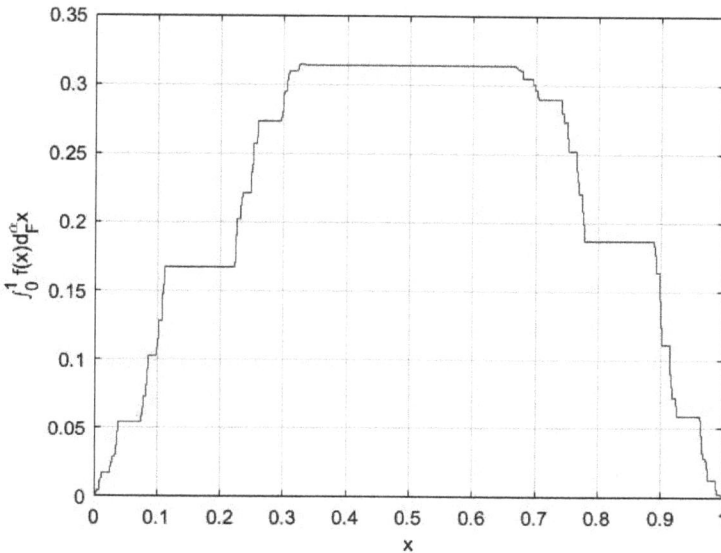

Fig. 5.7: Graph of fractal integral of f on the Cantor set

and its fractal integral will be

$$\int_0^1 \sin(2\pi x \chi_F) d_F^\alpha x = \frac{-\Gamma(1+\alpha)}{2\pi}\left[\cos\left(2\pi\frac{S_F^\alpha(x)}{\Gamma(1+\alpha)}\right)\right] = 0, \qquad (5.3)$$

where $\alpha = 0.63$ is for the middle-1/3 Cantor set.

Example 5.2. Consider the Weierstrass function $w : R \to R^2$ that is defined by $w(t) = (t, W_\lambda^s(t))$ where $W_\lambda^s(t)$ is

$$W_\lambda^s(t) = \sum_{k=1}^\infty \lambda^{(s-2)k}\sin(\lambda^k t) \qquad (5.4)$$

and $\lambda > 1$ and $1 < s < 2$ which is the box dimension of the Weierstrass function [Gowrisankar *et al.* (2021)].
The fractal derivative of the Weierstrass functions $D_F^\alpha W_\lambda^s(t)$ is given by:

$$D_F^\alpha W_\lambda^s(t) = 1.7411\cos(2S_F^\alpha(t)) + 3.03143\cos(4S_F^\alpha(t)) + 5.27803\cos(8S_F^\alpha(t))$$
$$+ 9.18959\cos(16S_F^\alpha(t)) + 16.\cos(32S_F^\alpha(t)) + 27.8576\cos(64S_F^\alpha(t))$$
$$+ 48.5029\cos(28S_F^\alpha(t)) + 84.4485\cos(256S_F^\alpha(t)) + 147.033\cos(512S_F^\alpha(t))$$
$$+ 256\cos(1024S_F^\alpha(t)) + \cdots \qquad (5.5)$$

The fractal integral of the Weierstrass function $\int W_\lambda^s(t)d_F^\alpha(t)$ is given by

$$\int W_\lambda^s(t)d_F^\alpha(t) = -0.870551\cos(S_F^\alpha(t))^2 - 0.189465\cos(4S_F^\alpha(t))$$

$$- 0.0824692\cos(8S_F^\alpha(t)) - 0.0358968\cos(16S_F^\alpha(t))0.015625\cos(32S_F^\alpha(t))$$

$$- 0.00680118\cos(64S_F^\alpha(t)) - 0.00296038\cos(128S_F^\alpha(t))$$

$$- 0.00128858\cos(256S_F^\alpha(t)) - 0.000560888\cos(512S_F^\alpha(t))$$

$$- 0.000244141\cos(1024S_F^\alpha(t)) - \cdots \tag{5.6}$$

Figure 5.8 is graph of the Weierstrass function, and Figures 5.9 and 5.10 illustrate the fractal derivative and fractal integral of the Weierstrass function, respectively.

Next, we preset the fractal differential equations.

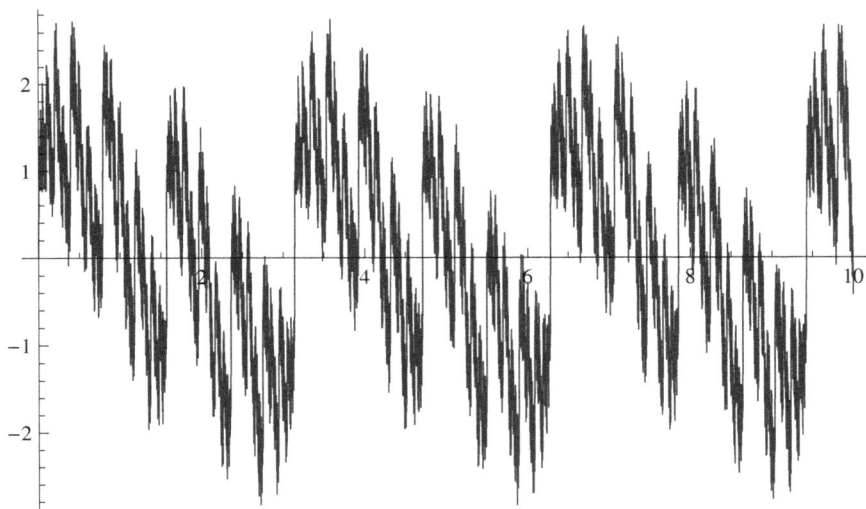

Fig. 5.8: Graph of the Weierstrass function for $\lambda = 2, s = 1.8$.

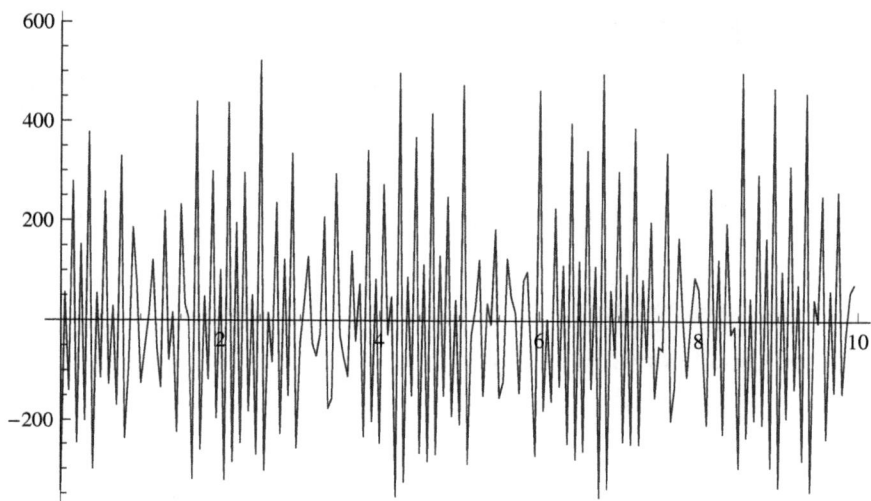

Fig. 5.9: Graph of fractal derivative of order $\alpha = 1.8$ of the Weierstrass function

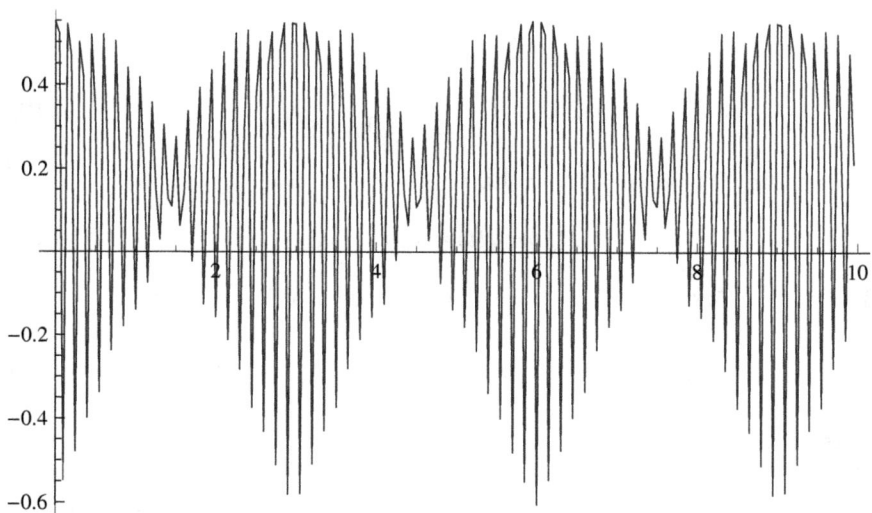

Fig. 5.10: Graph of fractal integral of order $\alpha = 1.8$ of the Weierstrass function

5.2 The local Fourier transforms on fractal sets

We give some basic definition to give the local fractal Fourier transforms [Golmankhaneh and Baleanu (2016a); Satin and Gangal (2016); Golmankhaneh *et al.* (2021a)].

Definition 5.1. The Dirac delta function on a thin Cantor set F is defined by

$$\delta_F(t - t_0) = 0, \quad \text{if} \quad t \neq t_0, \ t \in F \tag{5.7}$$

and its fractal integral is given by

$$\int_{a-\epsilon}^{a+\epsilon} f(t)\delta_F(t - t_0)d_F^\alpha t = \Gamma(\alpha + 1)f(t_0). \tag{5.8}$$

The fractal step function $u_{F,t_0}(t)$ relates the fractal Dirac delta function as the following

$$\int_{-\infty}^{t} \delta_F(t - t_0)d_F^\alpha t = u_{F,t_0}(t) \tag{5.9}$$

where

$$u_{F,t_0}(t) = \begin{cases} 0, & t < t_0 \\ 1/\Gamma(\alpha + 1), & t \leq t_0. \end{cases} \tag{5.10}$$

By recalling Theorem 4.12, we have

$$D_F^\alpha u_{F,t_0}(t) = \delta_F(t - t_0). \tag{5.11}$$

Definition 5.2. The fractal convolution is defined by

$$f(t) * g(t) = \int_0^t f(t - \tau)g(\tau)d_F^\alpha \tau = \int_0^t f(\tau)g(t - \tau)d_F^\alpha \tau, \tag{5.12}$$

Definition 5.3. Let f be a function defined on a thin Cantor-like set F, then f is a piecewise F-continuous if and only if there exists a finite subdivision $[a_0, a_1], \dots, [a_{n-1}, a_n]$ of $F \subset [a_0, a_n]$ such that f is a F-continuous on all (a_{i-1}, a_i).

Definition 5.4. Let f be a function defined on a thin Cantor-like set F, then f is called absolutely F^α-integrable if and only if its absolute value $|f|$ has the F^α-integral.

Definition 5.5. If f is a F-continuous and fractal piecewise smooth, and absolutely F^α-integrable function, then the fractal Fourier transform of f is denoted by $\mathbf{F}(\omega)$ and is given by:

$$\mathcal{F}_F^\alpha\{f(t)\} = \mathbf{F}(\omega) = \frac{1}{\sqrt{2\pi}} \int_{-\infty}^{\infty} \exp(iS_F^\alpha(\omega)S_F^\alpha(t))f(t)\,d_F^\alpha t, \qquad (5.13)$$

where ω is called the fractal Fourier transform variable and $\exp(-iS_F^\alpha(\omega)S_F^\alpha(t))$ is called the fractal kernel of the transform.

Definition 5.6. The inverse local fractal Fourier transform of $\mathbf{F}(\omega)$ is defined by

$$(\mathcal{F}_F^\alpha)^{-1}\{\mathbf{F}(\omega)\} = f(t) = \frac{1}{\sqrt{2\pi}} \int_{-\infty}^{\infty} \exp(-iS_F^\alpha(\omega)S_F^\alpha(t))\mathbf{F}(\omega)\,d_F^\alpha \omega,$$

$$(5.14)$$

Theorem 5.1. *The local fractal Fourier transformation of f is linear operator.*

Proof. As we have

$$\mathcal{F}_F^\alpha\{f(t)\} = \frac{1}{\sqrt{2\pi}} \int_{-\infty}^{\infty} \exp(iS_F^\alpha(\omega)S_F^\alpha(t))f(t)\,d_F^\alpha t. \qquad (5.15)$$

Then, for any constants γ and β, we can write

$$\mathcal{F}_F^\alpha\{\gamma f(t) + \beta g(t)\}$$

$$= \frac{1}{\sqrt{2\pi}} \int_{-\infty}^{\infty} [\gamma f(t) + \beta g(t)]\exp(iS_F^\alpha(\omega)S_F^\alpha(t))d_F^\alpha t,$$

$$= \frac{\gamma}{\sqrt{2\pi}} \int_{-\infty}^{\infty} f(t)\exp(iS_F^\alpha(\omega)S_F^\alpha(t))d_F^\alpha t \qquad (5.16)$$

$$+ \frac{\beta}{\sqrt{2\pi}} \int_{-\infty}^{\infty} g(t)\exp(iS_F^\alpha(\omega)S_F^\alpha(t))d_F^\alpha t,$$

$$= \gamma \mathcal{F}_F^\alpha\{f(t)\} + \beta \mathcal{F}_F^\alpha\{g(t)\},$$

which completes the proof. $\square$

Theorem 5.2. *Let* $\mathcal{F}_F^\alpha\{f(t)\}$ *be a local fractal Fourier transform of* $f(t)$. *Then*

$$\mathcal{F}_F^\alpha\left[f\left(t-a\right)\right] = \exp\left(iaS_F^\alpha\left(t\right)\right)\mathcal{F}_F^\alpha\left(f\left(t\right)\right), \tag{5.17}$$

where a is a real constant.

Proof. According to the definition, we have, for $a > 0$,

$$
\begin{aligned}
\mathcal{F}_F^\alpha\left[f\left(t-a\right)\right] &= \frac{1}{\sqrt{2\pi}}\int_{-\infty}^{\infty}\exp\left(iS_F^\alpha\left(\omega\right)S_F^\alpha\left(t\right)\right)f\left(t-a\right)d_F^\alpha t,\\
&= \frac{1}{\sqrt{2\pi}}\int_{-\infty}^{\infty}\exp\left(iS_F^\alpha\left(\omega\right)S_F^\alpha\left(t\right)\right)f\left(\eta\right)d_F^\alpha\eta, \qquad \eta = t - a \\
&= \exp\left(-iS_F^\alpha\left(\omega\right)S_F^\alpha\left(a\right)\right)\mathcal{F}_F^\alpha\left\{f\left(t\right)\right\}.
\end{aligned}
\tag{5.18}
$$

$\square$

Theorem 5.3. *If $\mathcal{F}_F^\alpha$ is a local fractal Fourier transform of f, then*

$$\mathcal{F}_F^\alpha\left\{f\left(S_F^\alpha\left(\lambda t\right)\right)\right\} = \frac{1}{|\lambda^\alpha|}\mathcal{F}_F^\alpha\left(\frac{S_F^\alpha\left(\omega\right)}{\lambda^\alpha}\right), \tag{5.19}$$

where λ is a real nonzero constant.

Proof. For $\lambda \neq 0$, we can write

$$\mathcal{F}_F^\alpha\left\{f\left(S_F^\alpha\left(\lambda t\right)\right)\right\} = \frac{1}{\sqrt{2\pi}}\int_{-\infty}^{\infty}\exp\left(iS_F^\alpha\left(\omega\right)S_F^\alpha\left(\lambda t\right)\right)f\left(S_F^\alpha\left(\lambda t\right)\right)d_F^\alpha(\lambda t). \tag{5.20}$$

If we let $\eta = \lambda^\alpha S_F^\alpha\left(t\right)$, then

$$
\begin{aligned}
\mathcal{F}_F^\alpha\left\{f\left(S_F^\alpha\left(\lambda t\right)\right)\right\} &= \frac{1}{|\lambda^\alpha|}\frac{1}{\sqrt{2\pi}}\int_{-\infty}^{\infty}\exp\left(i\left(\frac{S_F^\alpha\left(\omega\right)}{\lambda^\alpha}\right)\eta\right)f\left(\eta\right)d_F^\alpha\eta \\
&= \frac{1}{|\lambda^\alpha|}\mathcal{F}_F^\alpha\left(\frac{S_F^\alpha\left(\omega\right)}{\lambda^\alpha}\right).
\end{aligned}
\tag{5.21}
$$

$\square$

Example 5.3. Let F be the middle-$\frac{1}{3}$ thin Cantor set with $\alpha = \frac{\ln 2}{\ln 3}$ and $\lambda = \frac{1}{3^n}$ for any positive integer number n.

Theorem 5.4. *Let f be F^α-continuous and fractal piecewise smooth in $(-\infty, \infty)$. Suppose $f(t)$ approach zero as $|t| \to \infty$. If f and $D_F^\alpha f$ are fractal absolutely integrable, then*

$$\mathcal{F}_F^\alpha \{ D_{F,t}^\alpha f \} = -iS_F^\alpha(\omega)\mathcal{F}_F^\alpha \{ f(t) \} = -iS_F^\alpha(\omega)\, \boldsymbol{F}(\omega). \qquad (5.22)$$

Proof. Let

$$\mathcal{F}_F^\alpha \{ D_{F,t}^\alpha f(t) \} = \frac{1}{\sqrt{2\pi}} \int\limits_{-\infty}^{\infty} D_{F,t}^\alpha f(t) \exp\left(iS_F^\alpha(\omega)S_F^\alpha(t) \right) d_F^\alpha t$$

$$= \frac{1}{\sqrt{2\pi}} \left[f(t) \exp\left(iS_F^\alpha(\omega)S_F^\alpha(t) \right)\big|_{-\infty}^{\infty} - \int\limits_{-\infty}^{\infty} f(t) \exp\left(iS_F^\alpha(\omega)S_F^\alpha(t) \right) d_F^\alpha t \right],$$

$$= -iS_F^\alpha(\omega)\mathcal{F}_F^\alpha \{ f(t) \} = -iS_F^\alpha(\omega)\, \boldsymbol{F}(\omega). $$

$$(5.23)$$

where $D_{F,t}^\alpha$ indicate the fractal derivative respect to t. $\qquad \square$

Remark 5.1. In general, if f and its first $(n-1)$ derivatives are F^α-continuous, and if its αth derivatives is fractal piecewise continues, then the local fractal Fourier transform of order $n - 1 < \alpha \le n$, $n \in \mathbb{N}$ is

$$\mathcal{F}_F^\alpha \{ (D_{F,t}^\alpha)^n f(t) \} = (-iS_F^\alpha(\omega))^n \mathcal{F}_F^\alpha \{ f(t) \} = (-iS_F^\alpha(\omega))^n \boldsymbol{F}(\omega). \quad (5.24)$$

Provided f and its derivatives are absolutely F^α-integrable. Furthermore, we suppose that f and its first $(n-1)$ derivatives tend to zero as $|t|$ approaches to infinity.

Theorem 5.5. *If $\boldsymbol{F}(\omega)$ and $\boldsymbol{G}(\omega)$ are the local fractal Fourier transforms of $f(t)$ and $g(t)$ respectively, then the local fractal Fourier transform of the convolution $(f * g)$ is the product $\boldsymbol{F}(\omega)\,\boldsymbol{G}(\omega)$. That is,*

$$\mathcal{F}_F^\alpha \{ f(t) * g(t) \} = \boldsymbol{F}(\omega)\,\boldsymbol{G}(\omega). \qquad (5.25)$$

Or, equivalently,

$$\mathcal{F}_F^{\alpha,-1} \{ \boldsymbol{F}(\omega)\,\boldsymbol{G}(\omega) \} = f(t) * g(t). \qquad (5.26)$$

More explicitly,

$$\frac{1}{\sqrt{2\pi}} \int\limits_{-\infty}^{\infty} \boldsymbol{F}(\omega)\,\boldsymbol{G}(\omega) \exp\left(iS_F^\alpha(\omega)S_F^\alpha(t) \right) d_F^\alpha t = (f * g)(t)$$

$$(5.27)$$

$$= \frac{1}{\sqrt{2\pi}} \int\limits_{-\infty}^{\infty} f(t - \xi)g(\xi)\, d_F^\alpha \xi.$$

Proof. By definition, we have

$$\mathcal{F}_F^\alpha \{(f * g)(t)\} = \frac{1}{2\pi} \int\limits_{-\infty}^{\infty} \exp\left(iS_F^\alpha(\omega) S_F^\alpha(t)\right) d_F^\alpha t \int\limits_{-\infty}^{\infty} f(t - \xi) g(\xi) d_F^\alpha \xi,$$

$$= \frac{1}{2\pi} \int\limits_{-\infty}^{\infty} g(\xi) \exp\left(iS_F^\alpha(\omega) S_F^\alpha(\xi)\right) d_F^\alpha \xi$$

$$* \int\limits_{-\infty}^{\infty} f(t - \xi) \exp\left(iS_F^\alpha(\omega) \left(S_F^\alpha(t) - S_F^\alpha(\xi)\right)\right) d_F^\alpha t. \tag{5.28}$$

With the change of variable $S_F^\alpha(\eta) = S_F^\alpha(t) - S_F^\alpha(\xi)$, we get

$$\mathcal{F}_F^\alpha \{(f * g)(t)\} = \frac{1}{\sqrt{2\pi}} \int\limits_{-\infty}^{\infty} g(\xi) \exp\left(iS_F^\alpha(\omega) S_F^\alpha(\xi)\right) d_F^\alpha \xi$$

$$* \frac{1}{\sqrt{2\pi}} \int\limits_{-\infty}^{\infty} f(\eta) \exp\left(iS_F^\alpha(\omega) S_F^\alpha(\eta)\right) d_F^\alpha \eta, \tag{5.29}$$

$$= \mathbf{F}(\omega) \mathbf{G}(\omega).$$

$\square$

In Table 5.1, we have presented the fractal Fourier transform of some function.

Theorem 5.6. *The fractal n-order derivative of the Dirac delta function is*

$$S_F^\alpha(t)^n D_F^{\alpha n} \delta_F(t) = (-1)^n n! \delta_F(t). \tag{5.30}$$

Proof. Using the fractal integration by parts, we have

$$\int_{-\infty}^{\infty} f(t) D_F^{n\alpha} \delta_F(t) d_F^\alpha t = f(t) D_F^{(n-1)\alpha} \delta_F(t) \Big|_{-\infty}^{\infty}$$

$$- \int_{-\infty}^{\infty} D_F^\alpha f(t) D_F^{(n-1)\alpha} \delta_F(t) d_F^\alpha t. \tag{5.31}$$

Table 5.1: Some analogies of the Fourier transform of both F^α-calculus and the ordinary calculus

Function	Fourier transform	Local fractal Fourier transform
$f(t)$	$f(\omega)$	$\mathbf{F}(\omega)$
$f(t) = \begin{cases} 1 & |t| < b \\ 0 & |t| > b \end{cases}$	$\dfrac{2\sin\omega}{\omega}$	$\dfrac{2\sin S_F^\alpha(\omega)}{S_F^\alpha(\omega)}$
$\dfrac{1}{t^2 + b^2}$	$\dfrac{\pi\exp(-b\omega)}{b}$	$\dfrac{\pi\exp(-bS_F^\alpha(\omega))}{b}$
$\dfrac{t}{t^2 + b^2}$	$-i\pi\exp(b\omega)$	$-i\pi\exp(bS_F^\alpha(\omega))$
$f^{(n)}(t)$	$(i\omega)^n f(\omega)$	$(-iS_F^\alpha(\omega))^n \mathbf{F}(\omega)$
$t^n f(t)$	$i^n \dfrac{d^n f}{d\omega^n}$	$i^n D_{F,\omega}^{n\alpha}(\mathbf{F}(\omega))$
$f(bt)\exp(ixt)$	$\dfrac{1}{b}f\left(\dfrac{w-x}{b}\right)$	$\dfrac{1}{b}\mathbf{F}\left(\dfrac{S_F^\alpha(\omega) - S_F^\alpha(x)}{b}\right)$

According to Eq.(5.7) we have

$$f(t)D_F^{\alpha(n-1)}\delta_F(t)\bigg|_{-\infty}^{\infty} = 0. \qquad (5.32)$$

Then, Eq.(5.31) reduces to the following form

$$\int_{-\infty}^{\infty} f(t)D_F^{n\alpha}\delta_F(t)d_F^\alpha t = -\int_{-\infty}^{\infty} D_F^\alpha f(t)D_F^{(n-1)\alpha}\delta_F(t)d_F^\alpha t, \qquad (5.33)$$

Letting $f(t) = S_F^\alpha(t) g(t)$ and $n = 1$, then Eq.(5.33) reduces to the following form

$$\int_{-\infty}^{\infty} S_F^\alpha(t) g(t) D_F^\alpha \delta_F(t) d_F^\alpha t$$

$$= -\int_{-\infty}^{\infty} \delta_F(t) D_F^\alpha [S_F^\alpha(t) g(t)] d_F^\alpha t$$

$$= -\int_{-\infty}^{\infty} \delta_F(t) (g(t) \chi_F(\alpha,t) + S_F^\alpha(t) D_F^\alpha g(t)) d_F^\alpha t$$

$$= -\int_{-\infty}^{\infty} \delta_F(t) g(t) \chi_F(\alpha,t) d_F^\alpha t, \tag{5.34}$$

since $\int_{-\infty}^{\infty} \delta_F(t) S_F^\alpha(t) D_F^\alpha g(t) d_F^\alpha t = 0$, then Eq.(5.34) takes the form

$$S_F^\alpha(t) D_F^\alpha \delta_F(t) = -\delta_F(t) \chi_F(\alpha,t). \tag{5.35}$$

In general, the same procedure gives

$$\int_{-\infty}^{\infty} [S_F^\alpha(t)^n g(t)] D_F^{\alpha n} \delta_F(t) d_F^\alpha t = (-1)^n \int_{-\infty}^{\infty} \delta_F(t) D_F^{\alpha n} (S_F^\alpha(t)^n g(t)) d_F^\alpha t, \tag{5.36}$$

but because of any power of $S_F^\alpha(t)$ times $\delta_F(t)$ integrates to it implies that only the constant term contributes. So, all terms multiplied by derivative of $g(t)$ disappear, leaving $n! g(t)$, so

$$\int_{-\infty}^{\infty} [S_F^\alpha(t)^n g(t)] D_F^{\alpha n} \delta_F(t) d_F^\alpha t = (-1)^n n! \int_{-\infty}^{\infty} g(t) \delta_F(t) d_F^\alpha t, \tag{5.37}$$

which means that

$$S_F^\alpha(t)^n D_F^{\alpha n} \delta_F(t) = (-1)^n n! \delta_F(t). \tag{5.38}$$

which complete the proof.   □

Remark 5.2. We note for convenience, in about we set $D_{F,t}^\alpha = D_F^\alpha$.

Example 5.4. The Dirac delta function is given as Local fractal Fourier transform as

$$\mathcal{F}_F^\alpha \{1\} = \delta_F(t) = \int_{-\infty}^{\infty} \exp(-2\pi i S_F^\alpha(\omega) S_F^\alpha(t)) d_F^\alpha \omega. \tag{5.39}$$

Similarly

$$\mathcal{F}_F^{\alpha,-1} \{\delta_F(t)\} = \int_{-\infty}^{\infty} \delta_F(t) \exp(2\pi i S_F^\alpha(t) S_F^\alpha(\omega)) d_F^\alpha \omega = 1. \tag{5.40}$$

The local fractal Fourier transform of the fractal Dirac delta is given by

$$\mathcal{F}_F^\alpha \{\delta_F (t - t_0)\} = \int_{-\infty}^{\infty} \delta_F (t - t_0) \exp(-2\pi i S_F^\alpha(\omega) S_F^\alpha(t)) d_F^\alpha t$$

$$= \exp\left(-2\pi i S_F^\alpha(\omega) S_F^\alpha(t_0)\right). \tag{5.41}$$

5.2.1 *Fourier transform on fractal curves*

In this section we preset the Fourier transform on fractal curves [Satin and Gangal (2016)].

Definition 5.7. First, let us recall the ordinary Fourier transform on real line for a function $g(v)$ as

$$g(v) = \int_{-\infty}^{\infty} \tilde{g}(y) \exp(-ivy) dy \tag{5.42}$$

Definition 5.8. The inverse Fourier Transform is

$$\tilde{g}(y) = \frac{1}{2\pi} \int_{-\infty}^{\infty} g(v) \exp(ivy) dv \tag{5.43}$$

From the Definition 4.349 of conjugacy we have

$$\phi[f(S_F^\alpha(t)] = f(\mathbf{w}(t)) \tag{5.44}$$

The Fourier transform of the LHS of Eq.(5.44) gives

$$\tilde{\phi}[f(v = J(\psi))] = \int_{-\infty}^{\infty} \phi[f(y = J(\theta))] \exp(-iyv) dy \tag{5.45}$$

$$= \int_{C(-\infty,\infty)} f(\theta) \exp(-iJ(\theta)v) d_F^\alpha \theta \tag{5.46}$$

where let $\tilde{g} = \phi[\tilde{f}]$, $g = \phi[f]$ also $v = S_F^\alpha(k)$, $y = S_F^\alpha(t)$, and we set $J(\theta) = S_F^\alpha(t)$, $J(\psi) = S_F^\alpha(k)$, where $\theta = \mathbf{w}(t)$ and $\psi = \mathbf{w}(k)$,

$$C(-\infty, \infty) = \lim_{a \to -\infty, b \to \infty} C(a, b) \tag{5.47}$$

and

$$\phi[f(v = J(\psi))] = \phi\left[\int_{-\infty}^{\infty} f(y = J(\theta)) \exp(-iyv) dy\right]$$

$$= \int_{C(-\infty,\infty)} f(\theta) \exp(-iJ(\theta)v) d_F^\alpha \theta \tag{5.48}$$

Comparing Eqs.(5.45) and (5.48), we can write

$$\phi[f] = \pi[\tilde{f}].$$ (5.49)

Also we can define the action of ϕ in Fourier space as

$$\phi[\tilde{f}](v = J(\psi)) = \tilde{f}(\psi).$$ (5.50)

Now using conjugacy we can rewrite Eq.(5.45) as

$$\tilde{f}(\psi) = \int_{C(-\infty,\infty)} f(\theta) \exp(-iJ(\theta)J(\psi)) d_F^\alpha \theta.$$ (5.51)

which is called Fourier Transform on fractal curves. Likewise, we can obtain from Eq.(5.43) the inverse Fourier Transform on fractal curves as

$$f(\theta) = \frac{1}{2\pi} \int_{C(-\infty,\infty)} \tilde{f}(\psi) \exp(iJ(\theta)J(\psi)) d_F^\alpha \psi.$$ (5.52)

5.2.2 *Solving fractal differential equation by fractal Fourier transform*

In this section, we solve fractal differential equation by fractal Fourier transform [Golmankhaneh *et al.* (2021a)].

Example 5.5. Let us consider fractal differential equation as

$$aD_F^\alpha x(t) + bx(t) = \cos(t),$$ (5.53)

where a, b are constant. For solving Eq.(5.53), we apply the local fractal Fourier transform, we obtain

$$(iS_F^\alpha(\omega)a + b)X(\omega) = \sqrt{\frac{\pi}{2}} (\delta_F(-1 + S_F^\alpha(\omega)) + \delta_F(1 + S_F^\alpha(\omega))).$$ (5.54)

Or,

$$X(\omega) = \frac{\sqrt{\frac{\pi}{2}} (\delta_F(-1 + S_F^\alpha(\omega)) + \delta_F(1 + S_F^\alpha(\omega)))}{(iS_F^\alpha(\omega)a + b)}.$$ (5.55)

Taking the inverse local fractal Fourier transform to both sides of Eq.(5.55), we get the following solution

$$x(t) = \frac{a\cos(S_F^\alpha(t)) + b\sin(S_F^\alpha(t))}{b^2 + a^2}.$$ (5.56)

By $(a_1 t^\alpha \le S_F^\alpha(t) \le a_2 t^\alpha)$, we have

$$x(t) \approx \frac{a\cos(t^\alpha) + b\sin(t^\alpha)}{b^2 + a^2}.$$ (5.57)

Example 5.6. Consider fractal differential equation as follows:

$$D_F^{2\alpha}u(t) + b^2 u = f(t), \ -\infty < t < \infty. \tag{5.58}$$

where b is constant. Applying the local fractal Fourier transform method gives

$$U(w) = \frac{\mathbf{F}\left(S_F^\alpha(w)\right)}{S_F^\alpha(w)^2 + b^2}. \tag{5.59}$$

Using the convolution Theorem 5.5, we get the following solution

$$u(t) = \frac{1}{2a} \int_{-\infty}^{\infty} f(\xi) \exp\left(-b\left(|S_F^\alpha(t) - S_F^\alpha(\xi)|\right)\right) d_F^\alpha \xi. \tag{5.60}$$

Example 5.7. Let us consider the fractal differential equation as

$$aD_F^{4\alpha}u(t) + bu(t) = t^2 \sin(t), \ \ -\infty < t < \infty \tag{5.61}$$

where a, b are constants. Rewritten Eq.(5.61) gives

$$D_F^{4\alpha}u(t) + c^4 u(t) = t^2 \sin(t), \tag{5.62}$$

where $c^4 = \frac{b}{a}$. Applying the local fractal Fourier transform to both sides of Eq.(5.62), we get

$$U(w) = \frac{i\sqrt{\frac{\pi}{2}}D_F^{2\alpha}\delta_F\left(-1 + S_F^\alpha(w)\right) + i\sqrt{\frac{\pi}{2}}D_F^{2\alpha}\delta_F\left(1 + S_F^\alpha(w)\right)}{S_F^\alpha(w)^4 + c^4}. \tag{5.63}$$

Taking the inverse local fractal Fourier transform to both sides of Eq.(5.63), we have

$$u(t) = \frac{8(1+c^4)S_F^\alpha(t)\cos(S_F^\alpha(t)) + (4(-5+3c^4) + (1+c^4)^2 S_F^\alpha(t)^2)\sin(S_F^\alpha(t))}{(1+c^4)^3}. \tag{5.64}$$

In view of $(a_1 t^\alpha \le S_F^\alpha(t) \le a_2 t^\alpha)$, we have

$$u(t) \approx \frac{8\left(1+c^4\right)t^\alpha\cos\left(t^\alpha\right) + \left(4\left(-5+3c^4\right) + \left(1+c^4\right)^2 t^{2\alpha}\right)\sin\left(t^\alpha\right)}{(1+c^4)^3}. \tag{5.65}$$

5.3 The fractal Laplace transform

In this section, we present the fractal Laplace transform [Golmankhaneh and Tunç (2019a); Kamal Ali et al. (2022)].

Definition 5.9. Let $f(t), t \in F$, then its fractal Laplace transform is defined by

$$\mathcal{B}(\omega) = \mathcal{L}_F^\alpha(f) = \int_0^\infty \exp\left(-S_F^\alpha(t) S_F^\alpha(\omega)\right) f(t) d_F^\alpha t, \qquad (5.66)$$

where $\mathcal{B}(\omega)$ and $\mathcal{L}_F^\alpha(f) : f(t) \to \mathcal{B}(\omega)$ are called the fractal Laplace transform of $f(t)$. If $\exists\, M > 0$, $\exists\, T > 0$, and $at^\alpha < S_F^\alpha(t) < bt^\alpha$, $a,\ b \in R$

$$e^{-\varepsilon S_F^\alpha(t)} |f(t)| < e^{-\varepsilon t^\alpha} |f(t)| \leq M, \qquad \forall\, t > T, \qquad (5.67)$$

then the fractal integral Eq.(5.66) exists.

Definition 5.10. The inverse fractal Laplace transform is given by

$$f(t) = (\mathcal{L}_F^\alpha)^{-1}(\mathcal{B}(\omega)). \qquad (5.68)$$

Theorem 5.7. *Let $f(t)$ and $g(t)$ are on fractal set. Then we have*

$$\mathcal{L}_F^\alpha(f(t) * g(t)) = \mathcal{B}(\omega)\mathcal{G}(\omega), \qquad (5.69)$$

where $\mathcal{G}(\omega) = \mathcal{L}_F^\alpha(g)$.

Theorem 5.8. *The fractal Laplace transform of the local fractal derivative is given by*

$$\mathcal{L}_F^\alpha((D_F^\alpha)^n f(t)) = S_F^\alpha(\omega)^n \mathcal{L}_F^\alpha(f(t)) - \sum_{k=0}^{n-1} S_F^\alpha(\omega)^{n-k-1} (D_F^\alpha)^n f|_{t=0}. \qquad (5.70)$$

Example 5.8. Consider a function $f(t) = \exp(aS_F^\alpha(t))$, where a is constant, then the fractal Laplace transform is obtained as follows

$$\mathcal{B}(\omega) = \mathcal{L}_F^\alpha(f)$$

$$= \int_0^\infty \exp\left(-S_F^\alpha(t)S_F^\alpha(\omega) + aS_F^\alpha(t)\right) g(t) d_F^\alpha t$$

$$= \text{F.} \lim_{R \to \infty} \int_0^R \exp\left(-(S_F^\alpha(\omega) - a)S_F^\alpha(t)\right)$$

$$= \text{F.} \lim_{R \to \infty} \frac{1 - \exp\left(-(S_F^\alpha(\omega) - a)R\right)}{S_F^\alpha(\omega) - a}$$

$$= \frac{1}{S_F^\alpha(\omega) - a}, \quad \text{if } S_F^\alpha(\omega) \approx \omega^\alpha > a. \qquad (5.71)$$

Example 5.9. Let $f(t) = \cos(at)$, where a is a constant. By recalling Example 5.8, the fractal Laplace transform is

$$\mathcal{L}_F^\alpha(e^{aS_F^\alpha(t)}) = \frac{1}{S_F^\alpha(\omega) - a} \tag{5.72}$$

then replacing a by ai where $i = \sqrt{-1}$ we obtain

$$\mathcal{L}_F^\alpha(e^{iaS_F^\alpha(t)}) = \frac{1}{S_F^\alpha(\omega) - ia}$$

$$\mathcal{L}_F^\alpha(\cos(at) + i\sin(at)) = \frac{S_F^\alpha(\omega) + ai}{S_F^\alpha(\omega)^2 + a^2}$$

$$= \frac{S_F^\alpha(\omega)}{S_F^\alpha(\omega)^2 + a^2} + i\frac{a}{S_F^\alpha(\omega)^2 + a^2}. \tag{5.73}$$

It follows easily that

$$\mathcal{B}(\omega) = \mathcal{L}_F^\alpha(f) = \mathcal{L}_F^\alpha(\cos(at)) = \frac{S_F^\alpha(\omega)}{S_F^\alpha(\omega)^2 + a^2}. \tag{5.74}$$

Theorem 5.9. *Suppose* $\mathcal{B}(\omega) = \mathcal{L}_F^\alpha(h)$, *Then the fractal Laplace transform of the fractal derivative is given by*

$$\mathcal{L}_F^\alpha(D_F^\alpha h(t)) = S_F^\alpha(\omega)\mathcal{L}_F^\alpha(h(t)) - h(0). \tag{5.75}$$

Proof. Using Eq.(5.66) one can write

$$\mathcal{L}_F^\alpha(D_F^\alpha f(t)) = \int_0^\infty \exp(S_F^\alpha(\omega)S_F^\alpha(t))D_F^\alpha f(t)d_F^\alpha t$$

$$= \text{F-}\lim_{R\to\infty}\{\exp(S_F^\alpha(\omega)S_F^\alpha(t))f(t)|_0^R + S_F^\alpha(\omega)\int_0^R \exp(S_F^\alpha(\omega)S_F^\alpha(t))d_F^\alpha t\}$$

$$= S_F^\alpha(\omega)\int_0^\infty \exp(S_F^\alpha(\omega)S_F^\alpha(t))d_F^\alpha t - h(0)$$

$$= S_F^\alpha(\omega)\mathcal{L}_F^\alpha(h(t)) - h(0), \tag{5.76}$$

if we assume that

$$\text{F-}\lim_{R\to\infty}\exp(-S_F^\alpha(\omega)S_F^\alpha(R))f(R) = 0, \tag{5.77}$$

which completes the proof. $\qquad\square$

In Table 5.2 we give some important formulas for the fractal Laplace transform.

Table 5.2: Some analogies of the Laplace transform of both F^α-calculus and the ordinary calculus

Operators in the fractal t-domain	Corresponding operators in the fractal w-domain	Remarks
a	$\dfrac{a}{S_F^\alpha(\omega)}$	$S_F^\alpha(\omega) > 0$
$\exp(aS_F^\alpha(t))$	$\dfrac{1}{S_F^\alpha(\omega) - a}$	$S_F^\alpha(\omega) > a$
$\sin(aS_F^\alpha(t))$	$\dfrac{a}{S_F^\alpha(\omega)^2 + a^2}$	$S_F^\alpha(\omega) > 0$
$S_F^\alpha(t)^m, \ m = 1, 2, 3, \ldots$	$\dfrac{m!}{S_F^\alpha(\omega)^{m+1}}$	$S_F^\alpha(\omega) > 0$
$\cos(aS_F^\alpha(t))$	$\dfrac{S_F^\alpha(\omega)}{S_F^\alpha(\omega)^2 + a^2}$	$S_F^\alpha(\omega) > 0$
$D_F^\alpha h(t)$	$S_F^\alpha(\omega)B(\omega) - h(0)$	Differentiation
$(D_F^\alpha)^2 h(t)$	$S_F^\alpha(\omega)^2 \mathcal{B}(\omega) - S_F^\alpha(\omega)h(0)$ $- D_F^\alpha h(t)|_{t=0}$	
$\int_0^t h(t')d_F^\alpha t'$	$\dfrac{\mathcal{B}(\omega)}{S_F^\alpha(\omega)}$	Integration

5.3.1 *Solving fractal differential equation by fractal Laplace transform*

In this section, we solve fractal differential equation by fractal Laplace transform [Kamal *et al.* (2021); Golmankhaneh *et al.* (2021c,b)].

Example 5.10. Let us consider fractal differential equation as the following form

$$D_F^\alpha x(t) + ax(t) = b, \qquad (5.78)$$

with the initial condition $x(0) = c$.
Applying the local fractal Laplace transform to both sides of Eq.(5.78), we get

$$P(\omega) = \frac{b}{S_F^\alpha(\omega)(S_F^\alpha(\omega) + a)} + \frac{c}{(S_F^\alpha(\omega) + a)}. \qquad (5.79)$$

by taking the inverse Laplace transform of Eq.(5.79), we get the following solution

$$x(t) = \frac{b}{a} + \left(c - \frac{b}{a}\right)\exp(-bS_F^\alpha(t)), \qquad (5.80)$$

or its smooth version as

$$x(t) \approx \frac{b}{a} + \left(c - \frac{b}{a}\right)\exp(-bt^\alpha). \qquad (5.81)$$

5.4 Sumudu transform on fractal set

In this section, we define the fractal Sumudu transform of a function on fractal set F [Golmankhaneh and Tunç (2019a)].

Definition 5.11. Suppose $g(t) : F \to R$. Hence, one can write $g(t)$ as a convergent infinite series which is called the analogue of the Maclaurin series as follows:

$$g(t) = c_0 + c_1 S_F^\alpha(t) + c_2 (S_F^\alpha(t))^2 + \cdots$$
$$= \sum_{m=0}^{\infty} \frac{(D_F^\alpha)^m f(t)|_{t=0}}{m!} (S_F^\alpha(t))^m, \qquad (5.82)$$

where $g(t)$ is n^{th}-order fractal differentiable at $t \in [a_1, a_2]$ and

$$c_m = \frac{(D_F^\alpha)^m f(t)|_{t=0}}{m!} \qquad (5.83)$$

are constants.

Definition 5.12. The analogue of the Sumudu transform on fractal sets is defined by

$$G(v) = c_0 + c_1 v + 2! c_2 v^2 + 3! c_3 v^3 + \cdots$$

$$= \sum_{m=0}^{\infty} (D_{F,v}^{\alpha})^m f(v) \bigg|_{v=0} (S_F^{\alpha}(v))^m. \tag{5.84}$$

If we consider finite terms of Eq.(5.84), hence, one can define the fractal Sumudu transform and denote by $S(v) = G^*(v)$. On the other hand, we can find $0 \le v < N$ such that $G^*(v)$ is convergent and it is called the fractal Sumudu transform of $f(t)$.

Example 5.11. The fractal Sumudu transform of the characteristic function $g(t)$ is defined by

$$g(t) = \begin{cases} 0, & t < 0; \\ 1/\Gamma(\alpha+1), & t \ge 0. \end{cases} \tag{5.85}$$

Using Eq.(5.84) we get the fractal Sumudu transform of $g(t)$ as follows:

$$S(v) = 1/\Gamma(\alpha+1), \quad t \ge 0. \tag{5.86}$$

Example 5.12. Let us consider a function as follows:

$$f(t) = \exp(c S_F^{\alpha}(t)), \tag{5.87}$$

where c is constant, then the fractal Sumudu transform of Eq.(5.87), using the fractal power series expansion is

$$e^{(c S_F^{\alpha}(t))} = 1 + (c S_F^{\alpha}(t)) + \frac{(c S_F^{\alpha}(t))^2}{2!} + \frac{(c S_F^{\alpha}(t))^3}{3!} + \cdots . \tag{5.88}$$

By virtue of Eq.(5.84) we easily get

$$S(v) = 1 + (c S_F^{\alpha}(v)) + (c S_F^{\alpha}(v))^2 + (c S_F^{\alpha}(v))^3 + \cdots . \tag{5.89}$$

This series is convergent for the case $|c S_F^{\alpha}(v))| < |c v^{\alpha}| < 1$. Thus, we can take $N = 1/|c|^{\alpha}$. We next claim

$$S(v) = \frac{1}{1 - c S_F^{\alpha}(v)}, \quad \text{for} \quad 0 \le u < N = 1/|c|^{\alpha}. \tag{5.90}$$

Example 5.13. Let us consider a function as follows:

$$h(t) = \sin(\chi_F \omega t). \tag{5.91}$$

In order to calculate the fractal Sumudu transform of $h(t)$, we first write

$$e^{j\omega S_F^{\alpha}(t)} = \cos(\omega S_F^{\alpha}(t)) + j \sin(\omega S_F^{\alpha}(t)), \quad \text{where,} \quad j = (-1)^{1/2}. \tag{5.92}$$

Next, in view of Example 5.11, we thus obtain

$$\mathcal{H}[e^{j\omega S_F^\alpha(t)}] = \frac{1}{1 - j\omega S_F^\alpha(v)} \tag{5.93}$$

It is fairly easy to see that

$$\mathcal{H}[e^{j\omega S_F^\alpha(t)}] = \frac{1 + j\omega S_F^\alpha(v)}{1 + \omega^2 S_F^\alpha(v)^2}. \tag{5.94}$$

Then we have

$$\mathcal{H}[\cos(j\omega S_F^\alpha(t))] = \frac{1}{1 + \omega^2 S_F^\alpha(v)^2}, \tag{5.95}$$

and

$$\mathcal{H}[\sin(j\omega S_F^\alpha(t))] = \frac{\omega S_F^\alpha(v)}{1 + \omega^2 S_F^\alpha(v)^2}. \tag{5.96}$$

Definition 5.13. The fractal Sumudu transform of a function $g(t)$ involving F^α-integral is defined by

$$S(v) = \mathcal{H}(g) = \frac{1}{S_F^\alpha(v)} \int_0^\infty \exp\left(-\frac{S_F^\alpha(t)}{S_F^\alpha(v)}\right) g(t) d_F^\alpha t, \tag{5.97}$$

where $S(v)$ is called the fractal Sumudu transform if the integral exists.

Example 5.14. Consider a delta function with on fractal set as follows:

$$g(t) = \delta(S_F^\alpha(t)), \quad t \in F. \tag{5.98}$$

Then the fractal Sumudu transform of $g(t)$ is given by

$$\begin{aligned}
S(v) &= \frac{1}{S_F^\alpha(v)} \int_0^\infty \exp\left(-\frac{S_F^\alpha(t)}{S_F^\alpha(v)} \delta(S_F^\alpha(t))\right) d_F^\alpha t \\
&= \frac{1}{S_F^\alpha(v)} \exp\left(-\frac{S_F^\alpha(t)}{S_F^\alpha(v)}\right)\Big|_{t=0} \\
&= \frac{1}{S_F^\alpha(v)}.
\end{aligned} \tag{5.99}$$

Theorem 5.10. *If $S(v) = \mathcal{H}(h(t))$, then we have*

$$\frac{S(v) - S(0)}{S_F^\alpha(v)} = \frac{S(v) - h(0)}{S_F^\alpha(v)}. \tag{5.100}$$

Proof. We first expand $h(t)$ using the Taylor expansion formula, accordingly, we have

$$h(t) = c_0 + c_1 S_F^\alpha(t) + c_2 S_F^\alpha(t)^2 + \cdots + c_n S_F^\alpha(t)^n + \cdots. \tag{5.101}$$

Applying of the definition of the fractal Sumudu transform we get

$$S(v) = \mathcal{H}[h(t)] = c_0 + c_1 S_F^\alpha(v) + 2! c_2 S_F^\alpha(t)^2 + \cdots + n! c_n S_F^\alpha(t)^n + \cdots .$$
(5.102)

The fractal derivative of Eq.(5.101) yields

$$D_F^\alpha h(t) = c_1 + 2c_2 S_F^\alpha(t) + 3c_2 S_F^\alpha(t)^2 + \cdots + nc_n S_F^\alpha(t)^{n-1} + \cdots ,$$
(5.103)

and its fractal Sumudu transform leads to

$$\mathcal{H}[D_F^\alpha h(t)] = c_1 + 2c_2 S_F^\alpha(v) + 3! c_2 S_F^\alpha(t)^2 + \cdots + n! c_n S_F^\alpha(t)^{n-1} + \cdots$$
$$= \frac{S(v) - c_0}{S_F^\alpha(v)} = \frac{S(v) - S(0)}{S_F^\alpha(v)} = \frac{S(v) - h(0)}{S_F^\alpha(v)}.$$
(5.104)

If we choose $h(0) = 0$, then we find the differential in fractal t-domain which corresponds to

$$\frac{S(v)}{S_F^\alpha(v)}$$
(5.105)

in the fractal v-domain.   □

Theorem 5.11. *The integral of $h(t)$ in fractal t-domain is*

$$\int_0^t h(t) d_F^\alpha t,$$
(5.106)

then its fractal Sumudu transform is the following

$$S_F^\alpha(v) S(v),$$
(5.107)

in the v-space.

Proof. Suppose the series expansion of the function $h(t)$ as follows:

$$h(t) = c_0 + c_1 S_F^\alpha(t) + c_2 S_F^\alpha(t)^2 + \cdots + c_n S_F^\alpha(t)^n + \cdots .$$
(5.108)

Then by fractal integrating both sides of Eq.(5.108) with respect to t, we obtain

$$\int_0^t h(t) d_F^\alpha t = c_0 S_F^\alpha(t) + \frac{1}{2} c_1 S_F^\alpha(t)^2$$
$$+ \frac{1}{3} c_2 S_F^\alpha(t)^3 + \cdots + \frac{1}{n+1} c_n S_F^\alpha(t)^{n+1} + \cdots .$$
(5.109)

Subsequently, the fractal Sumudu transform of Eq.(5.109) gives

$$
\mathcal{H}\left[\int_0^t h(t)d_F^\alpha t\right] = c_0 S_F^\alpha(v) + c_1 S_F^\alpha(v)^2 + 2!c_2 S_F^\alpha(v)^3 + \cdots + n!c_n S_F^\alpha(v)^{n+1}
$$

$$
= S_F^\alpha(v)S(v), \qquad (5.110)
$$

which finishes the proof. □

Theorem 5.12. *Let $h(t)$ be a function on fractal set, then the fractal Sumudu transform of the second order of the fractal derivative is given by*

$$
\mathcal{H}[(D_F^\alpha)^2 h(t)] = \frac{1}{S_F^\alpha(v)^2}\left[S(v) - h(0) - S_F^\alpha(v)D_F^\alpha h(t)\Big|_{t=0}\right]. \qquad (5.111)
$$

Proof. Let $k(t) = D_F^\alpha h(t)$ be the fractal derivative of $h(t)$ respect to t. Then by Eq.(5.104), we can write

$$
\mathcal{H}[D_F^\alpha k(t)] = \frac{1}{S_F^\alpha(v)}[\mathcal{H}[k(t)] - k(0)]. \qquad (5.112)
$$

A trivial verification shows that,

$$
\mathcal{H}[(D_F^\alpha)^2 h(t)] = \frac{1}{S_F^\alpha(v)}\left[\frac{1}{S_F^\alpha(v)}(\mathcal{H}[h(t)] - h(0)) - k(0)\right]
$$

$$
= \frac{1}{S_F^\alpha(v)^2}\mathcal{H}[h(t)] - \frac{1}{S_F^\alpha(v)^2}h(0) - \frac{1}{S_F^\alpha(v)}D_F^\alpha h(t)|_{t=0}. \qquad (5.113)
$$

Consequently, Theorem 5.11 holds. One can proceed by induction in the same manner given above to derive the following general formula of the fractal Sumudu transform for n^{th}-order derivative

$$
\mathcal{H}[(D_F^\alpha)^n h(t)] = \frac{1}{S_F^\alpha(v)^m}\left[S(v) - \sum_{j=0}^{m-1} S_F^\alpha(v)^j (D_F^\alpha)^j h(t)|_{t=0}\right]. \qquad (5.114)
$$

□

Some properties of the fractal Sumudu transform which are listed in Table 5.3 can be proved alike manner that mentioned above.

Table 5.3: Some properties of the Sumudu transform

Operators in the fractal fractal t-domain	Corresponding operators in the fractal v-domian	Remarks
$a_1 h(t) + a_2 g(t)$ a_1, a_2 are constants.	$a_1 S(v) + a_2 S(v)$	Linearity
$\dfrac{S_F^\alpha(t)^{n-1}}{(n-1)!}$, $n = 1,2,\dots$	$S_F^\alpha(v)^{n-1}$	
$\exp(aS_F^\alpha(t))$	$\dfrac{1}{1-(aS_F^\alpha(v))^2}$	
$\sin(aS_F^\alpha(t))$	$\dfrac{aS_F^\alpha(v)}{1-(aS_F^\alpha(v))^2}$	
$\cos(aS_F^\alpha(t))$	$\dfrac{1}{1-(aS_F^\alpha(v))^2}$	
$D_{F,t}^\alpha h(t)$	$\dfrac{S_F^\alpha(v) - h(0)}{S_F^\alpha(v)}$	Differentiation
$\int_0^t h(t')d_F^\alpha t'$	$vS_F^\alpha(v)$	Integration
$h(S_F^\alpha(\lambda t)) = h(\lambda^\alpha S_F^\alpha(t))$	$S(\lambda^\alpha S_F^\alpha(v))$	λ is constant.
$\int_0^t h(S_F^\alpha(\vartheta))g(S_F^\alpha(t) - S_F^\alpha(\vartheta))d_F^\alpha\vartheta$	$S_F^\alpha(v)S(S_F^\alpha(v))G(S_F^\alpha(v))$	Convolution

5.4.1 Solving the fractal differential equation via the fractal Sumudu transform

In this section, the fractal linear differential equations are solved using the fractal Sumudu transform [Golmankhaneh and Tunç (2019a)].

Example 5.15. Consider a fractal differential equation on Cantor-like set as,

$$D_F^\alpha h(t) + h = g(t), \tag{5.115}$$

with initial condition,

$$h(0) = 0, \tag{5.116}$$

where $g(t)$ is the step function as

$$g(t) = \begin{cases} 0, & t < 0; \\ 1/\Gamma(\alpha+1), & t \geq 0. \end{cases} \tag{5.117}$$

Applying the fractal Sumudu transform to Eq.(5.115), we obtain

$$\frac{S(v) - h(0)}{S_F^\alpha(v)} + S(v) = \frac{1}{\Gamma(\alpha+1)}. \tag{5.118}$$

Solving Eq.(5.118) for $S(v)$ we obtain

$$S(v) = \frac{1/\Gamma(\alpha+1)}{(1/S_F^\alpha(v)) + 1} + \frac{h(0)}{S_F^\alpha(v) + 1}. \tag{5.119}$$

Substituting $h(0) = 0$ we get

$$S(v) = \frac{1/\Gamma(\alpha+1)}{(1/S_F^\alpha(v)) + 1} = 1 - \frac{\Gamma(\alpha+1)}{1 + S_F^\alpha(v)}. \tag{5.120}$$

Using series expansion we get

$$S(v) = 1 - (1 - S_F^\alpha(v) + S_F^\alpha(v)^2 - S_F^\alpha(v)^3 + \cdots). \tag{5.121}$$

Inverse of the fractal Sumudu transform gives

$$h(t) = 1 - (1 - S_F^\alpha(t) + \frac{S_F^\alpha(t)^2}{2!} + \frac{S_F^\alpha(t)^3}{3!} + \cdots)$$
$$= 1 - \exp(-S_F^\alpha(t)). \tag{5.122}$$

Example 5.16. Consider a second order fractal differential equation as follows:

$$(D_F^\alpha)^2 f(t) + f(t) = S_F^\alpha(t), \tag{5.123}$$

with the initial condition

$$f(0) = 0, \quad D_F^\alpha f(t)|_0 = 2. \tag{5.124}$$

Taking the fractal Sumudu transform of Eq.(5.123) we arrive at

$$\mathcal{H}[(D_F^\alpha)^2 f(t) + f(t)] = \mathcal{H}[S_F^\alpha(t)]. \tag{5.125}$$

Then by Theorem 5.11 we obtain

$$\frac{1}{S_F^\alpha(v)^2}\mathcal{H}[f(t)] - \frac{1}{S_F^\alpha(v)^2}f(0) - \frac{1}{S_F^\alpha(v)}D_{F,t}^\alpha f(t)|_{t=0} + \mathcal{H}[f(t)] = S_F^\alpha(v) \tag{5.126}$$

Solving Eq.(5.126) for $\mathcal{H}[f(t)]$ using Eq.(5.124), we have

$$\mathcal{H}[f(t)] = S_F^\alpha(v)\left[\frac{S_F^\alpha(v)^2 + 2}{(S_F^\alpha(v)^2 + 1)}\right]$$

$$= S_F^\alpha(v) + \frac{S_F^\alpha(v)}{S_F^\alpha(v)^2 + 1}. \tag{5.127}$$

Applying inverse fractal Laplace transform we find

$$f(t) = \mathcal{H}^{-1}\{S_F^\alpha(v)\} + \mathcal{H}^{-1}\left\{\frac{S_F^\alpha(v)}{S_F^\alpha(v)^2 + 1}\right\}$$

$$= S_F^\alpha(t) + \sin(S_F^\alpha(t)). \tag{5.128}$$

Chapter 6

Stability of Fractal Differential Equations

In this section, we present stability of fractal differential equations [Tunç et al. (2020); Golmankhaneh et al. (2021d); Golmankhaneh and Tunç (2017)].

6.1 Lyapunov stability of fractal differential equations

Here, we generalize the Lyapunov stability definition for the functions on fractal set [Tunç et al. (2020)].

Definition 6.1. Let us consider the following fractal differential equation with an initial condition

$$D_{F,t}^{\alpha} h(t) = g(h(t)), \quad h(0) = h_0, \ 0 < \alpha \leq 1, \tag{6.1}$$

where $g(h(t)) : F \to R$, and g has an equilibrium point h_e, then $g(h_e) = 0$.

(1) The equilibrium point h_e is called fractal Lyapunov stabile if we have

$$\forall\, \epsilon > 0, \ \exists\, \delta > 0, \quad |h(0) - h_e| < \delta^{\alpha} \Rightarrow |h(t) - h_e| < \varepsilon^{\alpha}, \ \ t \geq 0. \tag{6.2}$$

(2) The stable equilibrium point h_e is said fractal asymptotically stable if

$$\forall\, \epsilon > 0, \ \exists\, \delta > 0, \quad |h(0) - h_e| < \delta^{\alpha} \Rightarrow \lim_{t \to \infty} |h(t) - h_e| = 0. \tag{6.3}$$

(3) The equilibrium point h_e is called fractal exponentially stable if

$$\exists\, \delta > 0, \quad |h(0) - h_e| < \delta^{\alpha} \Rightarrow |h(t) - h_e| \leq \kappa^{\alpha} |h(0) - h_e| e^{-\lambda \alpha t}, \tag{6.4}$$

where $t \geq 0$, $\kappa, \lambda \in R$, and $\kappa > 0$, $\lambda > 0$.

Fractal Lyapunov function of Eq.(6.1) is a function $L : R \rightarrow R^+$, $R^+ = [0, +\infty)$ which is F-continuous. Also, its α-order derivative is F-continuous. Thus the fractal derivative of L with respect to Eq.(6.1) is written as L^* and if it has following condition

$$L^* = \frac{\partial L}{\partial h} g < 0, \forall\, t \in F \setminus \{0\}, \tag{6.5}$$

then, the zero solution of Eq.(6.1) is fractal asymptotically stable.

Example 6.1. Consider the following fractal differential equation

$$D^\alpha_{F,t} z(t) = -\chi_F z, \quad z(0) = c. \tag{6.6}$$

where $0 < \alpha \leq 1$. The general the solution of Eq.(6.6) is

$$z(t) = c \exp(-S^\alpha_F(t)). \tag{6.7}$$

A fractal Lyapunov function for studying the stability of Eq.(6.6) is

$$L(z) = z^2. \tag{6.8}$$

Then, we have

$$L^* = \frac{dL}{dz}(z) = -2z^2 < 0, \quad (z \neq 0). \tag{6.9}$$

Thus, the zero solution of Eq.(6.6) is fractal asymptotically stable.

Example 6.2. Consider harmonic oscillator on the fractal time as follows:

$$(D^\alpha_{F,t})^2 y(t) + \mathfrak{C}_F y(t) = 0, \ t \in F, \ \mathfrak{C}_F > 0, \tag{6.10}$$

where $\mathfrak{C}_F$ is constant. The equivalent fractal system is

$$D^\alpha_{F,t} y = z,$$
$$D^\alpha_{F,t} z = -\mathfrak{C}_F y. \tag{6.11}$$

The fractal Lyapunov function correspond to Eq.(6.10) is

$$L(y, z) = \frac{1}{2} \mathfrak{C}_F y^2 + \frac{1}{2} z^2, \tag{6.12}$$

where $L(0, 0) = 0$ and $L(y, z) > 0$ for $(y, z) \in R^2 \setminus (0, 0)$.

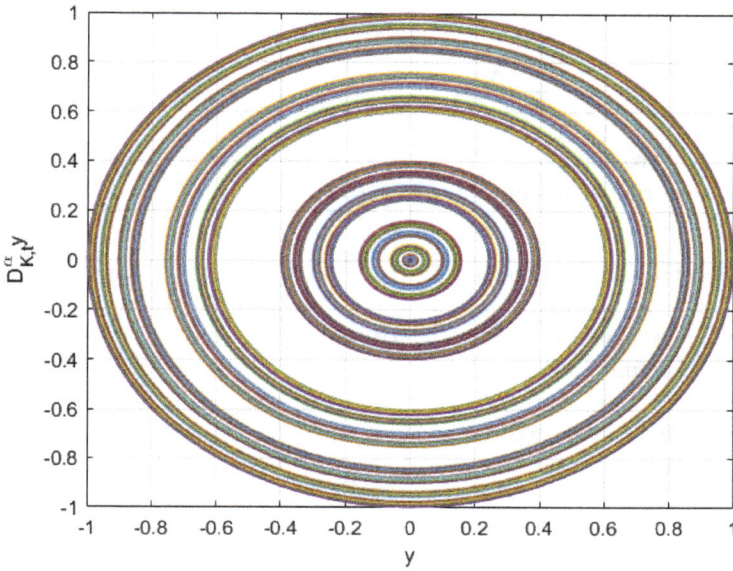

Fig. 6.1: Graph of solution of Eq.(6.10) with $\mu = 1/5$ and $\mathfrak{C}_F = 1$

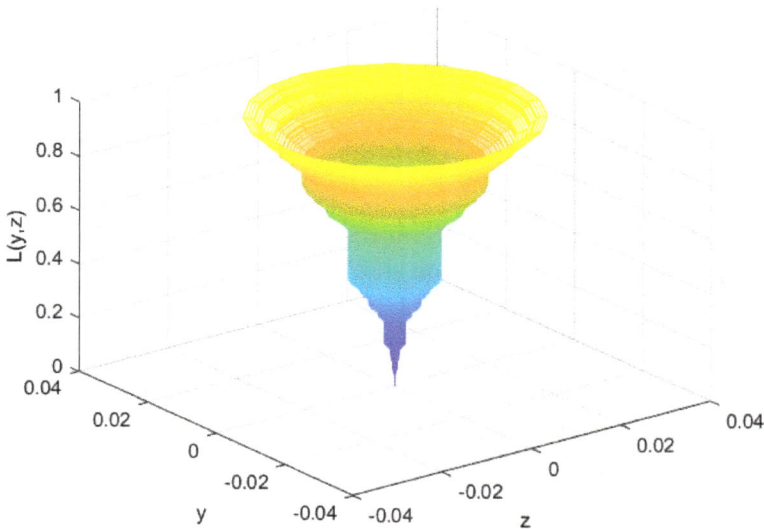

Fig. 6.2: Graph of fractal Lyapunov function Eq.(6.10) with $\mu = 1/5$ and $\mathfrak{C}_F = 1$

Then, it is obtain that

$$D^{\alpha}_{F,t}L(y,z) = \frac{\partial}{\partial y}L(y,z)D^{\alpha}_{F,t}y + \frac{\partial}{\partial y},L(y,z)D^{\alpha}_{F,t}z = 0. \tag{6.13}$$

Hence, the zero solution $(0,0)$ is a fractal stable point. In Figures 6.1 and 6.2, we have sketched solutions of Eq.(6.10) and Eq.(6.12).

6.2 Hyers-Ulam stability of the first order equation

In this subsection we review the Hyers-Ulam stability of the first order equation [Golmankhaneh *et al.* (2021d)].

Theorem 6.1. *Suppose function $g : I \to R$ and linear differential equation as the following form*

$$\frac{dg(x)}{dx} = g(x), \qquad \left|\frac{dg(x)}{dx} - g(x)\right| \le \epsilon, \quad \forall\, x \in I, \qquad \epsilon > 0 \tag{6.14}$$

where I is an open interval. If $f(x) : I \to R$, then

$$\frac{df(x)}{dx} = f(x), \qquad |g(x) - f(x)| \le 3\epsilon, \tag{6.15}$$

where $g(x)$ is the Hyers-Ulam stable.

6.2.1 Hyers-Ulam stability α-order fractal differential equation

In this section, we present the Hyers-Ulam stability α-order fractal differential equation/fractal Hyers-Ulam stability.

Definition 6.2. A function $h : F \to R$, if we have:

(1) $D^{\alpha}_F h(x) \ge 0 \Rightarrow h(x)$ is called a fractal increasing function.
(2) $D^{\alpha}_F h(x) < 0 \Rightarrow h(x)$ is called a fractal decreasing function.

Definition 6.3. A function $h : F \to R$ is called fractal Lipschitz /F-Lipschitz whenever

$$|h(x) - h(y)| \le F^{\alpha}|S^{\alpha}_F(x) - S^{\alpha}_F(y)|, \qquad \forall\, x, y \in F, \tag{6.16}$$

where $h(x)$ is the Jensen concave function.

Lemma 6.1. *Suppose $h : F \to R$ is a α-order fractal differentiable function. Then, one can write:*

(1)

$$h(x) \leq D_F^\alpha h(x) \Leftrightarrow h(x) = j(x)\exp(S_F^\alpha(x)), \quad \forall\, x \in F, \qquad (6.17)$$

where $j : F \to R$ is an arbitrary increasing α-order differentiable function.

(2)

$$D_F^\alpha h(x) \leq h(x) \Leftrightarrow h(x) = b(x)\exp(S_F^\alpha(x)), \quad \forall\, x \in F, \qquad (6.18)$$

where $b : F \to R$ is an arbitrary decreasing α-order differentiable function.

Proof. If $h(x) \leq D_F^\alpha h(x)$, then using Eq.(6.17) one can write

$$j(x) = h(x)\exp(-S_F^\alpha(x)). \qquad (6.19)$$

Since $j(x)$ is a fractal differentiable function, thus we have

$$\begin{aligned} D_F^\alpha j(x) &= D_F^\alpha h(x)\exp(-S_F^\alpha(x)) - \chi_F h(x)\exp(-S_F^\alpha(x)) \\ &= (D_F^\alpha h(x) - h(x))\exp(-S_F^\alpha(x)) \geq 0. \end{aligned} \qquad (6.20)$$

Then, according to Definition 6.2, $j(x)$ is the fractal increasing function. The converse can be easily proved by substituting $-h(x)$ into Eq.(6.17). $\square$

Theorem 6.2. *Let $h : F \to R$ be a α-order differentiable*

$$|D_F^\alpha h(x) - h(x)| \leq \epsilon^\alpha, \; \exists\, \epsilon, \; \forall\, x \in F, \qquad (6.21)$$

if and only if

$$h(x) = \epsilon^\alpha + \exp(S_F^\alpha(x))f(\exp(-S_F^\alpha(x))), \qquad (6.22)$$

where $f : \mathcal{H} \to R$, $\mathcal{H} = \{\exp(-S_F^\alpha(x))|x \in F\}$ is a α-order differentiable, decreasing and $2\epsilon^\alpha$-Lipschitz.

Proof. ($\Rightarrow$): Utilizing Eq.(6.21) we can write

$$h(x) - \epsilon^\alpha \leq D_F^\alpha h(x) \leq \epsilon^\alpha + h(x). \qquad (6.23)$$

Let $g(x) = h(x) - \epsilon^\alpha$ and considering part (i) of Lemma 6.1, so we can write

$$h(x) - \epsilon^\alpha = j(x)\exp(S_F^\alpha(x)). \qquad (6.24)$$

Therefore, $j(x)$ is α-order differentiable and increasing function. On the other hand, if $p(x) = \epsilon^{\alpha} + h(x)$, and in view of part (ii) of Lemma 6.1, one can write

$$h(x) + \epsilon^{\alpha} = b(x)\exp(S_F^{\alpha}(x)), \tag{6.25}$$

then $p(x)$ is a α-order differentiable and decreasing function. Using Eqs. (6.24) and (6.25) we have

$$b(x)\exp(S_F^{\alpha}(x)) - \epsilon^{\alpha} = j(x)\exp(S_F^{\alpha}(x)) + \epsilon^{\alpha}. \tag{6.26}$$

Taking the derivative of Eq.(6.26), we obtain

$$\begin{aligned}
D_K^{\alpha} j(x)\exp(S_F^{\alpha}(x)) &+ \chi_F j(x)\exp(S_F^{\alpha}(x)) \\
&= D_F^{\alpha} b(x)\exp(S_F^{\alpha}(x)) + \chi_F b(x)\exp(S_F^{\alpha}(x)) \\
&= D_K^{\alpha} b(x)\exp(S_F^{\alpha}(x)) + j(x)\exp(S_F^{\alpha}(x)) + 2\epsilon^{\alpha}.
\end{aligned} \tag{6.27}$$

By virtue of Eq.(6.27) and the decreasing property of $b(x)$, we arrive at

$$\begin{aligned}
D_F^{\alpha} b(x) &= \frac{D_F^{\alpha} j(x)\exp(S_F^{\alpha}(x)) - 2\epsilon^{\alpha}}{\exp(S_F^{\alpha}(x))} \\
&= D_F^{\alpha} j(x) - 2\epsilon^{\alpha}\exp(-S_F^{\alpha}(x)) \leq 0.
\end{aligned} \tag{6.28}$$

Subsequently, it follows that

$$0 \leq D_F^{\alpha} j(x) \leq 2\epsilon^{\alpha}\exp(-S_F^{\alpha}(x)). \tag{6.29}$$

By defining $f : \mathcal{H} \to R$, $\mathcal{H} = \{\exp(-S_F^{\alpha}(x))|x \in F\}$, and $f(y) = j(-\ln y)$, $y \in \mathcal{H}$, then we have

$$D_F^{\alpha} f(y) = \frac{-D_F^{\alpha} f(-\ln y)}{y} \leq 0, \quad y \in \mathcal{H}. \tag{6.30}$$

So $f(y)$ is decreasing. Now, we show f is $2\epsilon^{\alpha}$-Lipschitz as follows:

$$\begin{aligned}
|f(y_1) - f(y_2)| &= |D_F^{\alpha} f(y)|_{y_3} |y_1 - y_2| \\
&= \left| \frac{-D_F^{\alpha} f(-\ln y_3)}{y_3} \right| |y_1 - y_2| \\
&= D_F^{\alpha} f(-\ln y_3)\exp(-\ln y_3)|y_1 - y_2| \\
&\leq 2\epsilon^{\alpha}|y_1 - y_2|,
\end{aligned} \tag{6.31}$$

where

$$\exists\ y_3 \in (\min(y_1, y_2), \max(y_1, y_2)), \text{ for all } y_1,\ y_2.$$

Therefore, we can write

$$h(x) = \epsilon^\alpha + j(x)\exp(S_F^\alpha(x)) = \epsilon^\alpha + f(-\exp(S_F^\alpha(x)))\exp(S_F^\alpha(x)) \quad (6.32)$$

($\Leftarrow$): It is straightforward. $\qquad\qquad\qquad\qquad\qquad\qquad\qquad\qquad\qquad\square$

Theorem 6.3. *Consider a fractal α-order differential equation as follows:*

$$D_F^\alpha z(t) = z(t), \quad \forall\ t \in F \quad (6.33)$$

where $z : F \to R$,

$$|D_F^\alpha z(t) - z(t)| \le \epsilon^\alpha, \quad \epsilon > 0, \quad (6.34)$$

and

$$\exists\ u : F \to R \quad such\ that \quad D_F^\alpha u(t) = u(t), \quad (6.35)$$

implies

$$|z(t) - u(t)| \le 3\epsilon^\alpha. \quad (6.36)$$

Proof. We can write solution of Eq.(6.35) as the following form:

$$u(t) = c_1 \exp(S_F^\alpha(t)). \quad (6.37)$$

Since we have

$$\begin{aligned}
|z(t) - c_1\exp(S_F^\alpha(t))| &\le \epsilon^\alpha + \exp(S_F^\alpha(t))|f(\exp(-S_F^\alpha(t)) - c_1| \\
&\le \epsilon^\alpha + 2\epsilon^\alpha|\exp(-S_F^\alpha(t)) - M|\exp(S_F^\alpha(t)) \\
&= \epsilon^\alpha(1 + 2|1 - M\exp(S_F^\alpha(t))|), \quad (6.38)
\end{aligned}$$

where $M = \inf \mathcal{H}$. If we choose $\mathcal{H} = [0, \infty)$, then Eq.(6.38) turns into

$$|z(t) - c_1\exp(S_F^\alpha(t))| \le 3\epsilon^\alpha. \quad (6.39)$$

This completes the proof. $\qquad\qquad\qquad\qquad\qquad\qquad\qquad\qquad\qquad\square$

6.3 Existence and uniqueness theorems for α-order fractal linear differential equations

In this subsection, we present the existence and uniqueness theorems α-order fractal linear differential equations [Golmankhaneh and Tunç (2017)].

Theorem 6.4. *Consider the following linear fractal differential equation*

$$D_{F,x}^\alpha y(x) + \mathfrak{p}(x)y(x) = \mathfrak{g}(x), \quad y(x)|_{x=x_0} = y_0. \quad (6.40)$$

If $\mathfrak{p}(x)$ and $\mathfrak{g}(x)$ are F-continuous functions on an open interval and the interval contains x_0, therefor there exist a unique solution for Eq.(6.40).

Proof. First, we proof the existence of the solution. Since $\mathfrak{p}(x)$ is F-continuous using Theorem 4.13, we have

$$D^\alpha_{F,x} \int_{x_0}^x \mathfrak{p}(r)d^\alpha_F r = \mathfrak{p}(S^\alpha_F(x)). \quad (6.41)$$

Let us define

$$\mu(x) = \exp\left(\int_{x_0}^x \mathfrak{p}(r)d^\alpha_F r\right), \quad (6.42)$$

therefor Eq.(6.41) turns into

$$D^\alpha_{F,x}(\mu(x)y(x)) = \mu(x)\mathfrak{g}(x). \quad (6.43)$$

Integrating Eq.(6.43) we get

$$\mu(r)y(r)\Big|_{S^\alpha_F(x_0)}^{S^\alpha_F(x)} = \int_{x_0}^x \mu(r)\mathfrak{g}(r)d^\alpha_F r. \quad (6.44)$$

It follows that

$$\mu(S^\alpha_F(x))y(S^\alpha_F(x)) - \mu(S^\alpha_F(x_0))y(S^\alpha_F(x_0)) = \int_{x_0}^x \mu(r)\mathfrak{g}(r)d^\alpha_F r. \quad (6.45)$$

If

$$h_0 = \mu(S^\alpha_F(x_0))y(S^\alpha_F(x_0)), \quad (6.46)$$

then dividing Eq.(6.46) by $\mu(S^\alpha_F(x))$ we get

$$y(x) = \frac{1}{\exp\left(\int_{x_0}^x \mathfrak{p}(r)d^\alpha_F r\right)} \int_{x_0}^x \mu(r)\mathfrak{g}(r)d^\alpha_F r + \frac{h_0}{\exp\left(\int_{x_0}^x \mathfrak{p}(r)d^\alpha_F r\right)}. \quad (6.47)$$

Next, we show that uniqueness of the solution. Namely, we suppose that $y_1(S^\alpha_F(x))$ and $y_2(S^\alpha_F(x))$ are two solutions of Eq.(6.40). Then we define

$$\mathfrak{w}(x) = y_1(S^\alpha_F(x)) - y_2(S^\alpha_F(x)), \qquad S^\alpha_F(a) < S^\alpha_F(x) < S^\alpha_F(b). \quad (6.48)$$

For the proof of the uniqueness, we show that $\mathfrak{w}(x) = 0$. Since

$$D^\alpha_{F,x}\mathfrak{w}(x) + \mathfrak{p}(x)\mathfrak{w}(x) = 0 \quad (6.49)$$

Indeed, we get

$$(D^\alpha_{F,x}y_1(x)+\mathfrak{p}(x)y_1(x))-(D^\alpha_{F,x}y_2(x)+\mathfrak{p}(x)y_2(x)) = \mathfrak{g}(x)-\mathfrak{g}(x) = 0. \quad (6.50)$$

Multiply Eq.(6.49) by $\mu(x)$ we arrive at

$$D^\alpha_{F,x}\left(\exp\left(\int_{x_0}^x \mathfrak{p}(r)d^\alpha_F r\right)\mathfrak{w}(x)\right) = 0. \quad (6.51)$$

Now, by integrating of both sides of Eq.(6.51), it is easy to get

$$\mathfrak{w}(x) = D \, \exp\left(-\int_{x_0}^{x} \mathfrak{p}(r) d_F^\alpha r\right).$$ (6.52)

Due to the initial condition, we obtain

$$\mathfrak{w}(x_0) = y_1(x_0) - y_2(x_0) = y_0 - y_0 = 0.$$ (6.53)

which finishes the proof. □

Example 6.3. Consider the fractal linear differential equation with an initial condition as

$$D_{F,x}^\alpha y(x) + 2y(x) = 6, \qquad y(0) = 10.$$ (6.54)

Using Eq.(6.47) and in view of the initial condition we obtain the specific solution of Eq.(6.54) as

$$y(x) = 7 \exp(-2S_F^\alpha(x)) + 3.$$ (6.55)

Example 6.4. Consider the following linear fractal differential equation

$$D_{F,x}^\alpha y(x) + \frac{1}{S_F^\alpha(x)} y(x) = \exp(S_F^\alpha(x)), \quad y(1) = 1.$$ (6.56)

Applying Eq.(6.47) and initial condition we derive the solution of Eq.(6.56) as

$$y(x) = \frac{1}{S_F^\alpha(x)} - \frac{\exp(S_F^\alpha(x))}{S_F^\alpha(x)} + \exp(S_F^\alpha(x)).$$ (6.57)

6.4 The Lie method on fractal calculus

In this section, we study Sophus Lie method for solving linear and non-linear fractal differential equations [Golmankhaneh and Tunç (2019b)]. A Lie groups are functions with parameter η such that

$$L_\eta : (S_F^\alpha(t), x) \longmapsto (S_F^\alpha(t'), x'),$$ (6.58)

where $S_F^\alpha(t') = v(S_F^\alpha(t), x, \eta)$, $x' = w(S_F^\alpha(t), x, \eta)$ and have following properties:

(1) L_η onto and one-to-one.
(2) $L_{\eta_2} \circ L_{\eta_1} = L_{\eta_2 + \eta_1}$.
(3) $L_0 = I$.
(4) $\forall \eta_1 \in R, \exists \eta_2 = -\eta_1, \quad L_{\eta_2} \circ L_{\eta_1} = L_0$.

6.4.1 *Symmetry condition of fractal differential equations*

In this section we present symmetry condition of fractal differential equations

Definition 6.4. Let us consider fractal differential equations as follows:

$$D_{F,t}^{\alpha} x = h(S_F^{\alpha}(t), x(t)), \quad t \in F. \tag{6.59}$$

In view of symmetry one can write

$$D_{F,t'} \alpha x' = h(S_F^{\alpha}(t'), x'). \tag{6.60}$$

Since we can have

$$
\begin{aligned}
D_{F,t'}^{\alpha} x' &= \frac{D_{F,t}^{\alpha} x'}{D_{F,t} S_F^{\alpha}(t')} \\
&= \frac{D_{F,t}^{\alpha} x' + (D_{F,t} x)^{\alpha} D_{F,x}^{\alpha} x'}{D_{F,t}^{\alpha} S_F^{\alpha}(t') + (D_{F,t}^{\alpha} x) D_{F,x}^{\alpha} x'} = h(S_F^{\alpha}(t'), x') \\
&= \frac{D_{F,t}^{\alpha} x' + h(S_F^{\alpha}(t), x) x'}{D_{F,t}^{\alpha} S_F^{\alpha}(t') + h(S_F^{\alpha}(t), x) x'} = h(S_F^{\alpha}(t'), x'),
\end{aligned}
\tag{6.61}
$$

which is called symmetry condition .

Theorem 6.5. *If Consider following differential equation*

$$D_{F,t}^{\alpha} x(t) = \frac{1 - x^2}{S_F^{\alpha}(t)}. \tag{6.62}$$

Eq.(6.62), then has symmetry as follows:

$$(S_F^{\alpha}(t'), x') = (e^{\eta} S_F^{\alpha}(t), x). \tag{6.63}$$

Proof. By substituting Eq.(6.63) in the symmetry condition Eq.(6.61), we get

$$\frac{D_{F,t}^{\alpha} x' + \frac{1-x^2}{S_F^{\alpha}(t)} D_{F,x}^{\alpha} x'}{D_{F,t}^{\alpha} S_F^{\alpha}(t') + \frac{1-x^2}{S_F^{\alpha}(t)} D_{F,x}^{\alpha} S_F^{\alpha}(t')} = \frac{1 - x'^2}{S_F^{\alpha}(t')}, \tag{6.64}$$

where $D_{F,t}^{\alpha} x' = 0$, $D_{F,x}^{\alpha} S_F^{\alpha}(t') = 0$ and $D_{F,t}^{\alpha} S_F^{\alpha}(t') = e^{\eta}$.

$$\frac{1 - x^2}{e^{\eta} S_F^{\alpha}(t)} = \frac{1 - x'^2}{S_F^{\alpha}(t')}. \tag{6.65}$$

Therefore, symmetry condition is satisfied.   □

Definition 6.5. If H be a point on the solution of fractal differential equation. Applying the symmetry with different value η gives the orbit of point H. Let us consider the following fractal differential equation

$$D^{\alpha}_{F,t}x = \chi_F(x, \alpha), \quad 0 < t < 1, \quad t \in F. \tag{6.66}$$

Then orbit for the point on the fractal solution obtain under following symmetry

$$(S^{\alpha}_F(t'), x') = (S^{\alpha}_F(t), x + \eta). \tag{6.67}$$

For instance, under symmetry Eq.(6.67) orbit of $H = (1/4, 0)$ is denoted by $\mathcal{O}_H$

$$\mathcal{O}_H = \{(1/4, 0), (1/4, \eta), \eta = 1, 2, 3, \dots\}. \tag{6.68}$$

Definition 6.6. The analogue of the tangent vector for any orbit at a point $(S^{\alpha}_F(t'), x')$ are defined as follows:

$$\begin{aligned} D^{\alpha}_{F,\eta} S^{\alpha}_F(t') &= \phi(S^{\alpha}_F(t'), x'), \\ D^{\alpha}_{F,\eta} x' &= \psi(S^{\alpha}_F(t'), x'). \end{aligned} \tag{6.69}$$

At the initial point $(S^{\alpha}_F(t), x)$, we set $\eta = 0$, so that we have

$$\left(\frac{dS^{\alpha}_F(t')}{d\eta}\bigg|_{\eta=0}, \frac{dx'}{d\eta}\bigg|_{\eta=0} \right) = \left(\phi(S^{\alpha}_F(t'), x'), \psi(S^{\alpha}_F(t'), x') \right). \tag{6.70}$$

Invariant solution is obtained by using analogues tangent vectors, namely,

$$D^{\alpha}_{F,t}x(t) = h(S^{\alpha}_F(t), x) = \frac{\phi(S^{\alpha}_F(t), x)}{\psi(S^{\alpha}_F(t), x)}. \tag{6.71}$$

Definition 6.7. The fractal characteristic, $\mathcal{Q}$ is defined by

$$\mathcal{Q}\left(S^{\alpha}_F(t), x, D^{\alpha}_{F,t}x\right) = \phi(S^{\alpha}_F(t), x) - \left(D^{\alpha}_{F,t}x\right)\psi(S^{\alpha}_F(t), x). \tag{6.72}$$

Utilizing Eq.(6.71) we get reduced characteristic

$$\overline{\mathcal{Q}}\left(S^{\alpha}_F(t), x, D^{\alpha}_{F,t}x\right) = \phi(S^{\alpha}_F(t), x) - h(S^{\alpha}_F(t), x)\psi(S^{\alpha}_F(t), x) \tag{6.73}$$

Under given symmetry we conclude $\overline{\mathcal{Q}} = 0$.

Example 6.5. Consider the fractal analogue of the Riccati equation as follows:

$$D^{\alpha}_{F,t}x(t) = S^{\alpha}_F(t)x^2 - \frac{2x}{S^{\alpha}_F(t)} - \frac{1}{S^{\alpha}_F(t)^3}, \quad S^{\alpha}_F(t) \neq 0. \tag{6.74}$$

Then Eq.(6.74) has the following symmetry

$$(S^{\alpha}_F(t'), x') = (e^{\eta} S^{\alpha}_F(t) e^{-2\eta} x). \tag{6.75}$$

Corresponding analogous tangent vectors on Cantors sets are

$$\phi(S_F^\alpha(t), x) = S_F^\alpha(t), \tag{6.76}$$

and

$$\psi(S_F^\alpha(t), x) = -2x. \tag{6.77}$$

The fractal analogous reduced characteristic is

$$\overline{\mathcal{Q}}(S_F^\alpha(t), x) = -2x - \left(S_F^\alpha(t)x^2 - \frac{2x}{S_F^\alpha(t)} - \frac{1}{S_F^\alpha(t)^3} \right) S_F^\alpha(t)$$

$$= -S_F^\alpha(t)^2 x^2 + \frac{1}{S_F^\alpha(t)^2}. \tag{6.78}$$

Therefore, we get

$$\overline{\mathcal{Q}} = 0 \implies x(t) = \pm S_F^\alpha(t)^{-2}, \tag{6.79}$$

which is the solution Eq.(6.74) and it is invariant under symmetry.

Definition 6.8. Let us consider change of coordinate system for obtaining the symmetry as follows:

$$L_\eta : (v(S_F^\alpha(t), x), w(S_F^\alpha(t), x)) \longmapsto (v'(S_F^\alpha(t), x), w'(S_F^\alpha(t), x)). \tag{6.80}$$

If we suppose

$$v'(S_F^\alpha(t)) = v(S_F^\alpha(t)),$$
$$w'(S_F^\alpha(t), x) = w(S_F^\alpha(t), x) + \eta, \tag{6.81}$$

so we have

$$\frac{dv'}{d\eta}\bigg|_{\eta=0} = 0,$$

$$\frac{dw'}{d\eta}\bigg|_{\eta=0} = 1, \tag{6.82}$$

also we can write

$$\frac{dv'}{d\eta}\bigg|_{\eta=0} = \frac{\partial v'}{\partial S_F^\alpha(t)} \frac{\partial S_F^\alpha(t)}{\partial \eta}\bigg|_{\eta=0} + \frac{\partial v'}{\partial x(t)} \frac{\partial x(t)}{\partial \eta}\bigg|_{\eta=0}$$

$$= \frac{\partial v}{\partial S_F^\alpha(t)} \phi(S_F^\alpha(t), x) + \frac{\partial v}{\partial x(t)} \psi(S_F^\alpha(t), x) = 0,$$

$$\frac{dw'}{d\eta}\bigg|_{\eta=0} = \frac{\partial w'}{\partial S_F^\alpha(t)} \frac{\partial S_F^\alpha(t)}{\partial \eta}\bigg|_{\eta=0} + \frac{\partial w'}{\partial x(t)} \frac{\partial x(t)}{\partial \eta}\bigg|_{\eta=0}$$

$$= \frac{\partial w}{\partial S_F^\alpha(t)} \phi(S_F^\alpha(t), x) + \frac{\partial w}{\partial x(t)} \psi(S_F^\alpha(t), x) = 1. \tag{6.83}$$

We arrive at

$$D^\alpha_{F,t}x(t) = \frac{\psi(S^\alpha_F(t), x)}{\phi(S^\alpha_F(t), x)}, \tag{6.84}$$

which is called fractal symmetric equation.

Example 6.6. Suppose One parameter Lie group on fractal set as follows:

$$L_\eta : (S^\alpha_F(t'), x') \longmapsto (e^\eta S^\alpha_F(t), e^{k\eta}x), \ k > 0. \tag{6.85}$$

Analogues fractal tangent vector $\phi(S^\alpha_F(t), x)$, is gotten by taking derivative of $S^\alpha_F(t')$ with respect to η, namely

$$\phi(S^\alpha_F(t'), x') = e^\eta S^\alpha_F(t). \tag{6.86}$$

Evaluating Eq.(6.86) at $\eta = 0$ we get

$$\phi(S^\alpha_F(t), x) = S^\alpha_F(t). \tag{6.87}$$

Similarly to above following we find analogues tangent vector $\phi(S^\alpha_F(t), x)$ by taking derivative $x'(S^\alpha_F(t), x)$ with respect to η

$$\psi(S^\alpha_F(t'), x') = ke^{k\eta}x. \tag{6.88}$$

The evaluating Eq.(6.88) at $\eta = 0$ we obtain

$$\psi(S^\alpha_F(t), x) = kx. \tag{6.89}$$

So that one can get

$$D^\alpha_{F,t}x(t) = \frac{\psi(S^\alpha_F(t), x)}{\phi(S^\alpha_F(t), x)} = \frac{kx}{S^\alpha_F(t)}. \tag{6.90}$$

By using conjugacy of F^α-calculus with standard calculus we write

$$\int \frac{dx}{x} = k \int \frac{dt}{t}. \tag{6.91}$$

Integrating by Eq.(6.91) we get

$$x = c\, t^k. \tag{6.92}$$

In view of inverse conjugacy we have

$$x = c\, S^\alpha_F(t)^k. \tag{6.93}$$

If we work out to find c, we obtain

$$v(S^\alpha_F(t), x) = \frac{x}{S^\alpha_F(t)^k}. \tag{6.94}$$

Then we derive

$$w = \int \frac{d^\alpha_F t}{\psi(S^\alpha_F(t), x)} = \int \frac{d^\alpha_F t}{S^\alpha_F(t)}. \tag{6.95}$$

If $w = \ln S_F^\alpha(t)$ so the canonical coordinates are

$$(v, w) = \left(\frac{x}{S_F^\alpha(t)^k}, \ln S_F^\alpha(t) \right). \tag{6.96}$$

Example 6.7. Consider the analogous fractal Riccati equation as follows

$$D_{F,t}^\alpha x(t) = S_F^\alpha(t)x^2 - \frac{2x}{S_F^\alpha(t)} - \frac{1}{S_F^\alpha(t)^3}, \quad S_F^\alpha(t) \neq 0. \tag{6.97}$$

The corresponding symmetry of Eq.(6.97) is

$$(S_F^\alpha(t'), x') = (e^\eta S_F^\alpha(t), e^{-2\eta}x). \tag{6.98}$$

Analogues tangent vectors are given in the following

$$\phi(S_F^\alpha(t'), x') = e^\eta S_F^\alpha(t), \phi(S_F^\alpha(t), x) = S_F^\alpha(t), \tag{6.99}$$

and

$$\psi(S_F^\alpha(t'), x') = -2e^{-2\eta}x, \quad \psi(S_F^\alpha(t), x) = -2x. \tag{6.100}$$

The canonical coordinates is obtained as follows:

$$(v, w) = (S_F^\alpha(t)^2 x, \ln S_F^\alpha(t)). \tag{6.101}$$

Since we have

$$D_{F,v}^\alpha w = \frac{1}{v^2 - 1}, \tag{6.102}$$

so using conjugacy with ordinary calculus we can write

$$w = \frac{1}{2} \ln \left(\frac{r-1}{r+1} \right) + k. \tag{6.103}$$

By substituting and inverting conjugacy we get

$$x(t) = \frac{-k - S_F^\alpha(t)^2}{S_F^\alpha(t)^4 - kS_F^\alpha(t)^2}, \tag{6.104}$$

where by setting $k = 0$ it gives invariant solution.

Definition 6.9. Solving symmetry condition Eq.(6.61) is often very diffi-
cult or impossible. Then we linearize the equation by using Taylor series .
In order to do this we expand terms around η, namely:

$$S_F^\alpha(t') = S_F^\alpha(t) + \eta \; \phi(S_F^\alpha(t), x) + \mathcal{O}(\eta^2)$$
$$x' = x + \eta \; \psi(S_F^\alpha(t), x) + \mathcal{O}(\eta^2)$$
$$f(S_F^\alpha(t'), x') = f(S_F^\alpha(t), x)$$
$$+ \eta \left(D_{F,t}^\alpha f \phi(S_F^\alpha(t), x) + D_{F,x}^\alpha f \psi(S_F^\alpha(t), x) \right) + \mathcal{O}(\eta^2). \tag{6.105}$$

Here $\mathcal{O}(\eta^2)$ describes the error function of Taylor series expansions such that

$$\lim_{\eta \to 0} \frac{e(\eta)}{\eta^2} = c. \tag{6.106}$$

Substituting Eq.(6.105) into Eq.(6.61) and disregarding terms of η^2 or higher yields and simplify further we have linearized symmetry condition as follows:

$$D_{F,t}^{\alpha}\psi + (D_{F,x}^{\alpha}\psi - D_{F,t}^{\alpha}\phi)f - D_{F,x}^{\alpha}\phi f^2 = \phi D_{F,t}^{\alpha}f + \psi D_{F,x}^{\alpha}f. \tag{6.107}$$

Example 6.8. Consider the fractal differential equation as

$$D_{F,t}^{\alpha}x = \frac{x}{S_F^{\alpha}(t)} + S_F^{\alpha}(t). \tag{6.108}$$

Substitute Eq.(6.108) into the linearized symmetry condition Eq.(6.107) we get

$$D_{F,t}^{\alpha}\psi - D_{F,x}^{\alpha}\phi \left(\frac{x}{S_F^{\alpha}(t)} + S_F^{\alpha}(t) \right)^2 + (D_{F,x}^{\alpha}\psi - D_{F,t}^{\alpha}\phi) \left(\frac{x}{S_F^{\alpha}(t)} + S_F^{\alpha}(t) \right)$$
$$- \left(\phi(1 - \frac{x}{S_F^{\alpha}(t)^2}) + \psi(\frac{1}{S_F^{\alpha}(t)}) \right) = 0 \tag{6.109}$$

For solving Eq.(6.109) let $\phi = 0$ so we have

$$D_{F,t}^{\alpha}\psi(t) - \frac{\psi(t)}{S_F^{\alpha}(t)} = 0. \tag{6.110}$$

Using conjugacy of F^{α}-Calculus with standard calculus we get

$$\psi(t) = cS_F^{\alpha}(t). \tag{6.111}$$

Now, we can find analogues canonical coordinates (v, w). If $\phi(S_F^{\alpha}(t), x) = 0$, then $v = S_F^{\alpha}(t)$. To find w we can solve

$$w = \int \frac{d_F^{\alpha}x}{cS_F^{\alpha}(t)} = \frac{x}{cS_F^{\alpha}(t)}. \tag{6.112}$$

Now, if we set $c = 1$ we get

$$(v, w) = (S_F^{\alpha}(t), \frac{x}{S_F^{\alpha}(t)}). \tag{6.113}$$

Also, since we have

$$\frac{dw}{dv} = \frac{D_{F,t}^{\alpha}w + h(S_F^{\alpha}(t), x)D_{F,x}^{\alpha}w}{D_{F,t}^{\alpha}v + h(S_F^{\alpha}(t), x)D_{F,x}^{\alpha}r}, \tag{6.114}$$

therefore we can write

$$D^\alpha_{F,t} w = -\frac{x}{S^\alpha_F(t)^2}, \quad D^\alpha_{F,x} w = \frac{1}{S^\alpha_F(t)}. \tag{6.115}$$

Substituting Eq.(6.115) into Eq.(6.114) one obtain following

$$\frac{dw}{dv} = 1, w = v + k, \ k \text{ is constant} \tag{6.116}$$

Replacing $S^\alpha_F(t)$ and x back in Eq.(6.116) , we obtain

$$x(t) = S^\alpha_F(t)^2 + k S^\alpha_F(t). \tag{6.117}$$

Definition 6.10. In view of symmetry group of Eq.(6.58) infinitesimal generator on Lie group of fractal differential equations on fractal set is defined by

$$X_F = \phi D^\alpha_{F,t} + \psi D^\alpha_{F,x}. \tag{6.118}$$

Example 6.9. Consider the symmetry of the Lie group of fractal differential equation as follows:

$$(S^\alpha_F(t'), x') = (\frac{S^\alpha_F(t)}{1 - \eta x}, \frac{x}{1 - \eta x}). \tag{6.119}$$

Eq.(6.119) has analogous tangent vectors as

$$\phi(S^\alpha_F(t), x) = x S^\alpha_F(t), \tag{6.120}$$

and

$$\psi(S^\alpha_F(t), x) = x^2. \tag{6.121}$$

So the analogous fractal infinitesimal generator is

$$X_F = x S^\alpha_F(t) D^\alpha_{F,t} + x^2 D^\alpha_{F,x}. \tag{6.122}$$

Example 6.10. Consider the analogous fractal infinitesimal generator as

$$X_F = D^\alpha_{F,t} + x D^\alpha_{F,x}. \tag{6.123}$$

We can calculate the analogue of tangent vectors using Eq.(6.69) as follows:

$$\phi(S^\alpha_F(t), x) = 1, \tag{6.124}$$

and

$$\psi(S^\alpha_F(t), x) = x. \tag{6.125}$$

So we can say that

$$\phi(S_F^\alpha(t'), x') = 1, \quad \psi(S_F^\alpha(t'), x') = e^\eta x. \tag{6.126}$$

Also, we have

$$S_F^\alpha(t') = S_F^\alpha(t) + \eta,$$
$$x' = e^\eta x. \tag{6.127}$$

Corresponding symmetry to infinitesimals Eq.(6.123) is

$$(S_F^\alpha(t'), x') = (S_F^\alpha(t) + \eta, e^\eta x). \tag{6.128}$$

Chapter 7

Generalization of Fractal Calculus

In this chapter, we present some generalizations of fractal calculus.

7.1 Non-local fractal calculus

In this section, we present the non-local fractal calculus [Golmankhaneh and Baleanu (2016d,c)].

Suppose $f : F \to R$, $f(x) \in C^k(F)$ which is α-order fractal differentiable function on $F \subset [a, b]$.

Definition 7.1. The Gamma function on fractal is defined by

$$\Gamma^\alpha(x) = \int_0^\infty \exp(-S_F^\alpha(t)) S_F^\alpha(t)^{S_F^\alpha(x)-1} d_F^\alpha t, \qquad (7.1)$$

where

$$\exp(-S_F^\alpha(x)) = F - \lim_{n \to \infty} \left(1 - \frac{S_F^\alpha(x)}{n}\right)^n. \qquad (7.2)$$

In Figure 7.1 we have plotted the fractal Gamma function.

Definition 7.2. The fractal Beta function on the fractal set is defined as follows

$$B_F^\alpha(r, \omega) = \int_0^1 (S_F^\alpha(t))^{r-1}(1 - S_F^\alpha(t))^{\omega-1} d_F^\alpha t, \quad Re(r) > 0, \quad Re(\omega) > 0, \qquad (7.3)$$

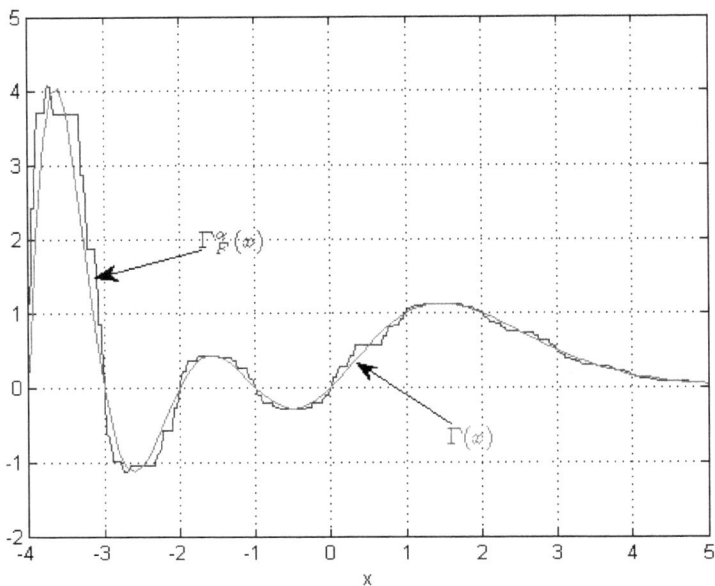

Fig. 7.1: Graph of the fractal Gamma function $\Gamma_F^\alpha(x)$ Eq.(7.1), and comparing with the standard case $\Gamma(x)$

which is called two-parameter r, ω fractal integral. Some properties of $B_F^\alpha(r, \omega)$ are as follows:

(1) $B_F^\alpha(r, \omega) = B_F^\alpha(\omega, r)$

(2)

$$B_F^\alpha(r, \omega) = 2 \int_0^{\pi/2} \sin^{2r-1}(S_F^\alpha(t)) \cos^{2\omega-1}(S_F^\alpha(t)) d_F^\alpha t. \qquad (7.4)$$

Theorem 7.1. *The fractal Beta function is related to the fractal Gamma function as*

$$B_F^\alpha(r, \omega) = \frac{\Gamma^\alpha(r)\Gamma_F^\alpha(\omega)}{\Gamma^\alpha(r + \omega)}. \qquad (7.5)$$

Definition 7.3. The fractal convolution of two functions is defined by

$$f(x) * g(x) = \int_0^x f(S_F^\alpha(x) - S_F^\alpha(\tau)) g(S_F^\alpha(\tau)) d_F^\alpha \tau, \qquad (7.6)$$

Remark 7.1. Some important formulas of the local fractal calculus are given below:

(1)

$$\mathcal{L}^\alpha[S_F^\alpha(x)^n] = \frac{\Gamma^\alpha(n+1)}{S_F^\alpha(\omega)^{n+1}} \tag{7.7}$$

(2)

$$\mathcal{L}^\alpha\left[\int_0^x f(S_F^\alpha(t))d_F^\alpha t\right] = \frac{\mathbf{B}(\omega)}{S_F^\alpha(\omega)} \tag{7.8}$$

(3)

$$\mathcal{L}_F^\alpha[S_F^\alpha(x)^n f(S_F^\alpha(x))] = (-1)^n (D_F^\alpha)^n \mathbf{B}(\omega) \tag{7.9}$$

(4)

$$\mathcal{L}^\alpha[(D_F^\alpha)^n f(S_F^\alpha(x))] = (S_F^\alpha(\omega))^{n\alpha} \mathbf{B}(\omega) - (S_F^\alpha(\omega))^{n\alpha-1} f(S_F^\alpha(0))$$
$$- (S_F^\alpha(\omega))^{n\alpha-2} D_F^\alpha f(x)|_{x=S_F^\alpha(0)} - \cdots - (D_F^\alpha)^{n-1} f(x)|_{x=S_F^\alpha(0)}. \tag{7.10}$$
where $\mathbf{B}(\omega) = \mathcal{L}^\alpha[f(x)]$.

Definition 7.4. The left-sided Riemann-Liouville fractal integral of order β is defined by

$$_a\mathcal{I}_x^\beta f(x) = \frac{1}{\Gamma^\alpha(\beta)} \int_a^x \frac{f(t)}{(S_F^\alpha(x) - S_F^\alpha(t))^{1-\beta}} d_F^\alpha t, \ x > a, \tag{7.11}$$

where $\alpha \neq \beta > 0$.

Definition 7.5. The right-sided Riemann-Liouville fractal integral of order β is defined by

$$_x\mathcal{I}_b^\beta f(x) = \frac{1}{\Gamma^\alpha(\beta)} \int_x^b \frac{f(t)}{(S_F^\alpha(x) - S_F^\alpha(t))^{1-\beta}} d_F^\alpha t \quad x < b. \tag{7.12}$$

where $\alpha \neq \beta > 0$.

Definition 7.6. The left-sided Riemann-Liouville fractal derivative is defined by

$$_a\mathcal{D}_x^\beta f(x) = \frac{1}{\Gamma^\alpha(n-\beta)} (D_F^\alpha)^n \int_a^x \frac{f(t)}{(S_F^\alpha(x) - S_F^\alpha(t))^{-n+\beta+1}} d_F^\alpha t. \tag{7.13}$$

Definition 7.7. The right-sided Riemann-Liouville fractal derivative is defined by

$$_x\mathcal{D}_b^\beta f(x) = \frac{1}{\Gamma^\alpha(n-\beta)} (-D_F^\alpha)^n \int_x^b \frac{f(t)}{(S_F^\alpha(t) - S_F^\alpha(x))^{-n+\beta+1}} d_F^\alpha t. \tag{7.14}$$

Definition 7.8. For $f(t) \in C^n(F)$ the left-sided Caputo fractal derivative is defined by

$$
{}^C_a\mathcal{D}^\beta_x f(x) = \frac{1}{\Gamma^\alpha(n-\beta)} \int_a^x \frac{1}{(S_F^\alpha(x) - S_F^\alpha(t))^{-n+\beta+1}} (D_F^\alpha)^n f(t) d_F^\alpha t
$$

(7.15)

where $n\alpha - \alpha \leq \beta < n\alpha$ and $n = \max(0, -[-\beta])$.

Definition 7.9. The right-sided Caputo fractal derivative is defined by

$$
{}^C_x\mathcal{D}^\beta_b f(x) = \frac{1}{\Gamma^\alpha(n-\beta)} \int_{S_F^\alpha(x)}^{S_F^\alpha(b)} \frac{1}{(S_F^\alpha(x) - S_F^\alpha(t))^{-n+\beta+1}} (-D_F^\alpha)^n f(t) d_F^\alpha t.
$$

(7.16)

Remark 7.2. If we choose $\beta = \alpha$ then we arrive at to the local fractal derivative whose order is equal the dimension of the fractal.

Example 7.1. Consider a fractal function as

$$
f(x) = (S_F^\alpha(x))^2, \quad x \in F,
$$

(7.17)

where $\alpha = 0.63$ for the triadic Cantor set . Then, its left-sided Riemann-Liouville fractal derivative is

$$
{}_0\mathcal{D}^{0.5}_x f(x) = \frac{\Gamma^\alpha(3)}{\Gamma^\alpha(2.5)} S_F^\alpha(x)^{1.5},
$$

(7.18)

and its left-sided Riemann-Liouville fractal integral is

$$
{}_0\mathcal{I}^{0.5}_x f(x) = \frac{\Gamma^\alpha(3)}{\Gamma^\alpha(3.5)} S_F^\alpha(x)^{2.5}
$$

(7.19)

Definition 7.10. The fractal Mittag-Leffler function of one parameter is given by

$$
E_\gamma^\alpha(x) = \sum_{k=0}^\infty \frac{S_F^\alpha(x)^j}{\Gamma^\alpha(\gamma k + 1)}, \quad \gamma > 0.
$$

(7.20)

Definition 7.11. The generalized two parameter η, ν Mittag-Liffler function on fractal F with α-dimension is defined

$$
E_{\eta,\nu}^\alpha(x) = \sum_{k=0}^\infty \frac{S_F^\alpha(x)^k}{\Gamma^\alpha(\eta k + \nu)}, \quad \eta > 0, \quad \nu \in R.
$$

(7.21)

In the special case we have the following results

$$E_{1,1}^{\alpha}(x) = e^{S_F^{\alpha}(x)}, \tag{7.22}$$

$$E_{1,2}^{\alpha}(x) = \frac{e^{S_F^{\alpha}(x)-1}}{S_F^{\alpha}(x)}, \tag{7.23}$$

$$E_{2,1}^{\alpha}(x) = \cosh(S_F^{\alpha}(x)), \tag{7.24}$$

$$E_{2,2}^{\alpha}(x) = \frac{\sinh(S_F^{\alpha}(x))}{S_F^{\alpha}(x)}. \tag{7.25}$$

Lemma 7.1. *The fractal Laplace transform of the Mittag-Leffler function is defined by*

$$\mathcal{L}^{\alpha}\left(S_F^{\alpha}(x)^{\nu-1}E_{\eta,\nu}^{\alpha}(\lambda x^{\eta})\right) = \frac{S_F^{\alpha}(\omega)^{\eta-\nu}}{S_F^{\alpha}(\omega)^{\eta}-\lambda}, \quad S_F^{\alpha}(\omega) > 0, \quad \left|\lambda S_F^{\alpha}(\omega)^{-\eta}\right| < 1. \tag{7.26}$$

Definition 7.12. The fractal Wright function is defined by

$$W_{\eta,\nu}^{\alpha}(x) = \sum_{k=0}^{\infty} \frac{S_F^{\alpha}(x)^k}{k!\,\Gamma^{\alpha}(\eta k+\nu)}, \quad \eta > 0, \ \nu \in R. \tag{7.27}$$

Definition 7.13. The more general Wright function is given by

$${}_1\Psi_1\left[\begin{array}{c}(n+1,1)\\(\sigma n+1,\sigma-\upsilon)\end{array}\middle| (-\mu)\,S_F^{\alpha}(x)^{\sigma-\upsilon}\right] = \sum_{j=0}^{\infty} \frac{\Gamma^{\alpha}(n+j+1)}{\Gamma^{\alpha}(\sigma n+\upsilon+\sigma j)}\frac{S_F^{\alpha}(x)^j}{j!}. \tag{7.28}$$

Definition 7.14. The generalized three-parameter Mittag-Leffler function is defined by:

$$E_{\sigma,\upsilon}^{\alpha,\rho}(x) = \sum_{j=0}^{\infty} \frac{(\rho)_j}{\Gamma^{\alpha}(\sigma j+\upsilon)}\frac{S_F^{\alpha}(x)^j}{j!} = \frac{1}{\Gamma^{\alpha}(\rho)}{}_1\Psi_1\left[\begin{array}{c}(\rho,1)\\(\upsilon,\sigma)\end{array}\middle|, S_F^{\alpha}(x)\right], \tag{7.29}$$

where $\sigma > 0$, $\upsilon > 0$, σ, υ, $S_F^{\alpha}(x) \in R$. By using the fact that

$${}_1\Psi_1\left[\begin{array}{c}(a,\upsilon)\\(\vartheta,\chi)\end{array}\middle| t\right] = \sum_{n=0}^{\infty} \frac{\Gamma^{\alpha}(n\upsilon+a)}{\Gamma^{\alpha}(\chi n+\vartheta)\,n!}. \tag{7.30}$$

Lemma 7.2. *The Laplace transform of the Wright function is defined by*

$$\mathcal{L}_F^\alpha \left(S_F^\alpha(x)^{\sigma n} {}_1\Psi_1 \left[\begin{array}{c} (n+1,1) \\ (\sigma n+1, \sigma - \upsilon) \end{array} \middle| (-\mu)\, S_F^\alpha(x)^{\sigma - \upsilon} \right] \right)$$

$$= n! \frac{S_F^\alpha(\omega)^{(\sigma-\upsilon)-(n\upsilon+1)}}{\left(S_F^\alpha(\omega)^{\sigma-\upsilon} + \mu \right)^{n+1}}. \qquad (7.31)$$

Lemma 7.3. *The fractal Laplace transform of the non-local fractal Caputo derivative is as*

$$\mathcal{L}_F^\alpha \left[{}_0^C D_x^\beta f(x) \right] = \frac{(S_F^\alpha(\omega))^{m\alpha}\, \mathbf{B}(\omega) - (S_F^\alpha(\omega))^{m\alpha-\alpha} f(S_F^\alpha(0))}{(S_F^\alpha(\omega))^{m\alpha-\beta}} +$$

$$\frac{-(S_F^\alpha(\omega))^{m\alpha-2\alpha} D_{F,x}^\alpha f(x)|_{x=S_F^\alpha(0)} - \cdots - D_{F,x}^{m\alpha-\alpha} f(x)|_{x=S_F^\alpha(0)}}{(S_F^\alpha(\omega))^{m\alpha-\beta}}. \qquad (7.32)$$

where $n = \max(0, -[-\beta])$.

Lemma 7.4. *The Laplace transformation of the non-local fractal Riemann-Liouville integral is given by*

$$\mathcal{L}^\alpha[{}_0\mathcal{I}_x^\beta f(x)] = \frac{\mathbf{B}(\omega)}{S_F^\alpha(\omega)^\beta}. \qquad (7.33)$$

Proof. The Laplace transform of the fractal Riemann-Liouville integral is

$$\mathcal{L}^\alpha[{}_0\mathcal{I}_x^\beta f(x)] = \mathcal{L}^\alpha \left[\frac{1}{\Gamma(\beta)} \int_0^x \frac{f(t)}{(S_F^\alpha(x) - S_F^\alpha(t))^{\alpha-\beta}} d_F^\alpha t \right]. \qquad (7.34)$$

Using the Eqs.(7.6) and (7.7) we arrive at

$$\mathcal{L}^\alpha[{}_0\mathcal{I}_x^\beta f(x)] = \frac{1}{\Gamma(\beta)} \mathbf{B}(\omega) \mathcal{L}_F^\alpha [S_F^\alpha(x)^{\beta-1}]$$

$$= \frac{1}{\Gamma(\beta)} \mathbf{B}(\omega) \frac{\Gamma(\beta)}{S_F^\alpha(\omega)^\beta}$$

$$= \frac{\mathbf{B}(\omega)}{S_F^\alpha(\omega)^\beta}. \qquad (7.35)$$

$\square$

Lemma 7.5. *The fractal Laplace transform of the non-local fractal Riemann-Liouville derivative of order $\beta \in [0,1)$ is given by*

$$\mathcal{L}_F^\alpha\{{}_0\mathcal{D}_x^\beta f(x), x, s\} = S_F^\alpha(s)^\beta \mathcal{F}_F^\alpha(s) - \sum_{k=1}^n S_F^\alpha(s)^{n-k}\, {}_0\mathcal{D}_x^{\beta-n+k-1} f(x)|_{S_F^\alpha(0)}, \qquad (7.36)$$

where $n = [\beta] + 1$.

Lemma 7.6. *For given η and ν then the inverse fractal Laplace is*

$$(\mathcal{L}_F^\alpha)^{-1}\left[\frac{S_F^\alpha(\omega)^{\eta-\nu}}{S_F^\alpha(\omega)^\eta + S_F^\alpha(a)}\right] = S_F^\alpha(x)^{\nu-1}E_{\eta,\nu}^\alpha(-S_F^\alpha(a)S_F^\alpha(x)^\eta). \qquad (7.37)$$

Proof. By using the series expansion we have

$$\frac{S_F^\alpha(\omega)^{\eta-\nu}}{S_F^\alpha(\omega)^\eta + S_F^\alpha(a)} = \frac{1}{S_F^\alpha(\omega)^\nu}\frac{1}{1 + \frac{S_F^\alpha(a)}{S_F^\alpha(\omega)^\eta}}$$

$$= \frac{1}{S_F^\alpha(\omega)^\nu}\sum_{n=0}^\infty \left(\frac{-S_F^\alpha(a)}{S_F^\alpha(\omega)^\eta}\right)^n$$

$$= \sum_{n=0}^\infty \frac{-S_F^\alpha(a)^n}{-S_F^\alpha(\omega)^{n\eta+\nu}}. \qquad (7.38)$$

The inverse fractal local Laplace transform of Eq.(7.38) it gives

$$\sum_{n=0}^\infty \frac{(-S_F^\alpha(a))^n S_F^\alpha(x)^{n\eta+\nu-1}}{\Gamma^\alpha(n\eta+\nu)} = S_F^\alpha(x)^{\nu-1}\sum_{n=0}^\infty \frac{(-S_F^\alpha(a)S_F^\alpha(x)^\eta)^n}{\Gamma^\alpha(n\eta+\nu)}$$

$$= S_F^\alpha(x)^{\nu-1}E_{\eta,\nu}^\alpha(-S_F^\alpha(a)S_F^\alpha(x)^\eta). \qquad (7.39)$$

$\square$

Lemma 7.7. *Let $\eta \geq \nu > 0$, and $S_F^\alpha(\omega)^{\alpha-\beta} > |S_F^\alpha(a)|$ then we have*

$$(\mathcal{L}_F^\alpha)^{-1}\left[\frac{1}{(S_F^\alpha(\omega)^\eta + S_F^\alpha(a)S_F^\alpha(\omega)^\nu)^{n+1}}\right]$$

$$= S_F^\alpha(x)^{\eta(n+1)-1}\sum_{k=0}^\infty \frac{-S_F^\alpha(a)^k}{\Gamma^\alpha(k(\eta-\nu)+(n+1)\eta)}\binom{n+k}{k}S_F^\alpha(x)^{k(\eta-\nu)}$$

$$(7.40)$$

Lemma 7.8. *For $\eta \geq \nu$, $\eta > \xi$, $S_F^\alpha(\omega)^{\eta-\nu} > |S_F^\alpha(a)|$, we have*

$$(\mathcal{L}_F^\alpha)^{-1}\left[\frac{S_F^\alpha(\omega)^\xi}{(S_F^\alpha(\omega)^\eta + S_F^\alpha(a)S_F^\alpha(\omega)^\nu + S_F^\alpha(b)}\right]$$

$$= S_F^\alpha(x)^{\eta-\xi-1}\sum_{n=0}^\infty\sum_{k=0}^\infty \frac{(-S_F^\alpha(a))^k(-S_F^\alpha(b))^n}{\Gamma^\alpha(k(\eta-\nu)+(n+1)\eta-\xi)}\binom{n+k}{k}S_F^\alpha(x)^{k(\eta-\nu)+n\eta}.$$

$$(7.41)$$

Now, we write some important composition relations as

$$_a\mathcal{I}_x^\beta \, _a\mathcal{D}_x^\beta f(x) = f(x) - \sum_{j=1}^n \frac{(_a\mathcal{D}_x^{\beta-j}f(x))|_{(S_F^\alpha(a))}}{\Gamma^\alpha(\beta+1-j)}(S_F^\alpha(x) - S_F^\alpha(a))^{\beta-j}, \qquad (7.42)$$

$$
{}_x\mathcal{I}_b^\beta \, {}_x\mathcal{D}_b^\beta f(x) = f(x) - \sum_{j=1}^n \frac{({}_x\mathcal{D}_b^{\beta-j} f(x))|_{(S_F^\alpha(b))}}{\Gamma^\alpha(\beta+1-j)} (S_F^\alpha(b) - S_F^\alpha(x))^{\beta-j}, \quad (7.43)
$$

$$
{}_a\mathcal{I}_x^\beta \, {}_a^C\mathcal{D}_x^\beta f(x) = f(x) - \sum_{j=1}^n \frac{((D_F^\alpha)^j f(x))|_{(S_F^\alpha(a))}}{\Gamma^\alpha(j+1)} (S_F^\alpha(x) - S_F^\alpha(a))^j, \quad (7.44)
$$

$$
{}_x\mathcal{I}_b^\beta \, {}_x^C\mathcal{D}_b^\beta f(x) = f(x) - \sum_{j=1}^n \frac{((D_F^\alpha)^j f(x))|_{(S_F^\alpha(b))}}{\Gamma^\alpha(j+1)} (S_F^\alpha(b) - S_F^\alpha(x))^j. \quad (7.45)
$$

$$
{}_a\mathcal{D}_x^\beta (S_F^\alpha(x) - S_F^\alpha(a))^\eta = \frac{\Gamma^\alpha(\eta+1)}{\Gamma(\eta+1-\beta)} (S_F^\alpha(x) - S_F^\alpha(a))^{\eta-\beta}, \quad (7.46)
$$

$$
{}_a\mathcal{I}_x^\beta (S_F^\alpha(x) - S_F^\alpha(a))^\eta = \frac{\Gamma^\alpha(\eta+1)}{\Gamma^\alpha(\eta+1+\beta)} (S_F^\alpha(x) - S_F^\alpha(a))^{\eta+\beta}, \quad (7.47)
$$

$$
{}_0\mathcal{I}_x^\beta (S_F^\alpha(x))^\eta = \frac{\Gamma^\alpha(\eta+1)}{\Gamma^\alpha(\eta+1+\beta)} (S_F^\alpha(x))^{\eta+\beta}, \quad (7.48)
$$

$$
{}_0\mathcal{D}_x^\beta (S_F^\alpha(x))^\eta = \frac{\Gamma^\alpha(\eta+1)}{\Gamma^\alpha(\eta+1-\beta)} (S_F^\alpha(x))^{\eta-\beta}. \quad (7.49)
$$

Example 7.2. Consider the fractal differential equation as

$$
D_{F,t}^\alpha x(t) = \chi_F(t,\alpha), \quad x(t)|_{t=0} = 0 \quad (7.50)
$$

Then its solution is

$$
x(t) = S_F^\alpha(t) \quad (7.51)
$$

Example 7.3. Consider the fractal differential equation as

$$
D_{F,t}^\alpha x(t) = \chi_F(t,\alpha) A x(t) \quad (7.52)
$$

where A is $n \times n$ matrix, $x \in R$. Then solution of Eq.(7.52) is

$$
x(t) = \exp(S_F^\alpha(t) A) x_0, \quad x(t)|_{t=0} = x_0. \quad (7.53)
$$

This equation suitable as model non-linear dynamics.

Example 7.4. Consider a linear differential equation

$$
{}_0\mathcal{D}_x^{\frac{1}{2}} y(x) = y(x), \quad (7.54)
$$

with the following initial condition, namely

$$
{}_0\mathcal{D}_x^{-\frac{1}{2}} y(x)|_{S_F^\alpha(0)} = 1. \quad (7.55)
$$

By inspection, the solution for the Eq.(7.54) becomes

$$
y(x) = S_F^\alpha(x)^{-\frac{1}{2}} E_{1/2,1/2}^\alpha(-\sqrt{S_F^\alpha(x)}). \quad (7.56)
$$

In Figure 7.2, we have plotted Eq.(7.56).

Example 7.5. Consider linear non-local fractal differential equation as

$$
{}^{C}_{0}D^{\beta}_{x}y(x) + y(x) = 0, \tag{7.57}
$$

with initial condition

$$
y(x)|_{x=S^{\alpha}_{F}(0)} = 1, \quad D^{\alpha}_{F}y(x)|_{x=S^{\alpha}_{F}(0)} = 0. \tag{7.58}
$$

For solving Eq.(7.57), we take Laplace transform of Eq.(7.57) and using Eq.(7.32) we have

$$
\mathbf{B}(\omega) = \mathcal{L}^{\alpha}_{F}[\,{}^{C}_{0}D^{\beta}_{x}f(x)] = \frac{(S^{\alpha}_{F}(\omega))^{\alpha}\mathbf{B}(\omega) - 1}{S^{\alpha}_{F}(\omega))^{\alpha-\beta}} \tag{7.59}
$$

It follows that

$$
\mathbf{B}(\omega) = \frac{S^{\alpha}_{F}(\omega))^{\alpha-\beta}}{1 + S^{\alpha}_{F}(\omega))^{\beta}}. \tag{7.60}
$$

Using the fractal inverse Laplace transform Eq.(7.37) we obtain

$$
y(x) = S^{\alpha}_{F}(x)^{\alpha-1}E^{\alpha}_{\beta,\alpha}(-S^{\alpha}_{F}(x)^{\beta}). \tag{7.61}
$$

Example 7.6. Consider the following non-local differential equation with the following initial condition

$$
{}_{0}D^{\frac{4}{3}}_{F}y(x) - \lambda y(x) = (S^{\alpha}_{F}(x))^{2}, \tag{7.62}
$$

$$
{}_{0}D^{\frac{1}{3}}_{F}y(x)|_{S^{\alpha}_{F}(0)} = 1, \qquad {}_{0}D^{\frac{-1}{6}}_{F}y(x)|_{S^{\alpha}_{F}(0)} = 2. \tag{7.63}
$$

For solving Eq.(7.62) we apply the fractal Laplace transformation on both side of it and we get

$$
S^{\alpha}_{F}(\omega)^{\frac{4}{3}}\mathbf{B}(\omega) - 1 - 2(S^{\alpha}_{F}(\omega))^{\frac{1}{2}} - \lambda\mathbf{B}(\omega) = \frac{2}{S^{\alpha}_{F}(\omega)^{3}}. \tag{7.64}
$$

After some calculation we obtained

$$
\mathbf{B}(\omega) = \frac{1}{S^{\alpha}_{F}(\omega)^{\frac{4}{3}} - \lambda} + \frac{2S^{\alpha}_{F}(\omega)^{\frac{1}{2}}}{S^{\alpha}_{F}(\omega)^{\frac{4}{3}} - \lambda} + \frac{2S^{\alpha}_{F}(\omega)^{-3}}{S^{\alpha}_{F}(\omega)^{\frac{4}{3}} - \lambda}. \tag{7.65}
$$

By computing the inverse fractal Laplace transform we obtain

$$
y(x) = S^{\alpha}_{F}(x)^{\frac{4}{3}}E^{\alpha}_{F,4/3,4/3}(\lambda S^{\alpha}_{F}(x)^{\frac{4}{3}}) + 2S^{\alpha}_{F}(x)^{\frac{-1}{6}}E^{\alpha}_{F,4/3,5/6}(\lambda S^{\alpha}_{F}(x)^{\frac{4}{3}}
$$
$$
+ 2S^{\alpha}_{F}(x)^{\frac{10}{3}}E^{\alpha}_{F,4/3,13/3}(\lambda S^{\alpha}_{F}(x)^{\frac{4}{3}}. \tag{7.66}
$$

7.2 Scale change on the local and non-local fractal derivatives

In this section we preset the scale change of local and non-local fractal derivatives [Golmankhaneh and Baleanu (2016d)].

Definition 7.15. A function $f(S_F^\alpha(x))$ is called fractal homogenous of degree-$m\alpha$ or invariant under fractal re-scaling if we have

$$f(S_F^\alpha(\lambda x)) = \lambda^{m\alpha} f(S_F^\alpha(x)), \tag{7.67}$$

where for some m and for all λ. For the triadic Cantor set we have

$$f\left(S_F^\alpha\left(\frac{1}{3^n}x\right)\right) = \left(\frac{1}{3^n}\right)^\alpha f(S_F^\alpha(x)), \quad \lambda = 1/3^n, n = 1, 2, \ldots \tag{7.68}$$

where $\alpha = 0.6$ which is the dimension of triadic Cantor set.

Corollary 7.1. *The fractal derivative of the fractal homogenous function $f(S_F^\alpha(x))$ re-scaling as follows*

$$D_F^\alpha f(S_F^\alpha(\lambda x)) = \lambda^{m\alpha-\alpha} f(S_F^\alpha(x)), \tag{7.69}$$

Proof. By $x \to \lambda x$, and $S_F^\alpha(\lambda x) = \lambda^\alpha S_F^\alpha(x)$ one can easily obtain the results. $\qquad\square$

Corollary 7.2. *If we re-scale $x \to \lambda x$, then we have*

$$_0D_x^\beta[f(S_F^\alpha(\lambda x))] = \lambda^{\beta\alpha} \ _0D_{\lambda x}^\beta[f(S_F^\alpha(x))] \tag{7.70}$$

7.3 Gauge integral on fractal calculus

In this section we present analogue of gauge integral/Henstock-Kurzweil integral (See Definition 2.61) on fractal calculus [Golmankhaneh and Baleanu (2016b)]. Suppose $\dot{P}$ is a δ-fine partition.

Definition 7.16. We define $\sigma_*^\alpha[F, I]$ as

$$\sigma_*^\alpha[F, I] = \sum_{i=1}^{n} \Gamma(\alpha+1)(x_i - x_{i-1})^\alpha \ \theta(F, [x_{i-1}, x_i]). \tag{7.71}$$

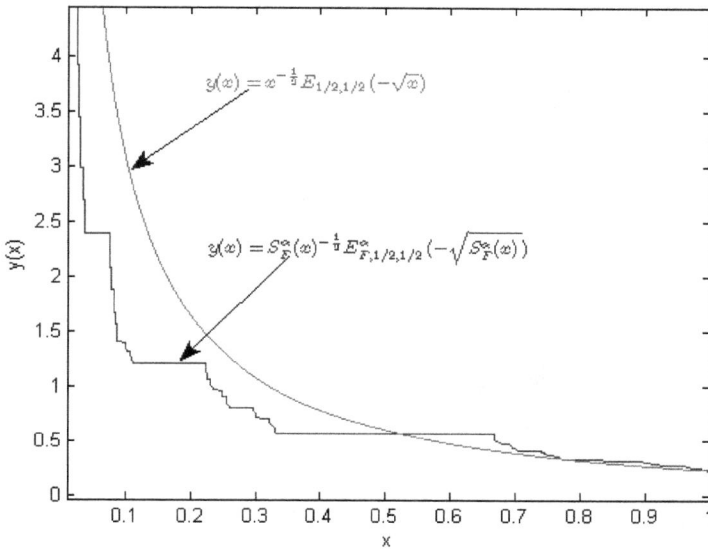

Fig. 7.2: Graph corresponding to Example 7.4

Definition 7.17. The generalized coarse grained mass using the gauge function is defined by

$$
{}^{*}\gamma^{\alpha}_{\delta(t)}(F, a, b) = \inf_{|\dot{P}| < \sup\{\delta(t_i);\ t_i \in [x_{i-1}, x_i]\}} \sigma^{\alpha}_{*}[F, I], \tag{7.72}
$$

where $|\dot{P}| = \max_{1 \leq i \leq n}(x_i - x_{i-1})$.

Definition 7.18. The generalized mass function ${}^{*}\gamma^{\alpha}(F, a, b)$ is defined by

$$
{}^{*}\gamma^{\alpha}(F, a, b) = \lim_{\sup\{\delta(t_i):\ t_i \in [x_{i-1}, x_i]\} \to 0} {}^{*}\gamma^{\alpha}_{\delta(t)}(F, a, b). \tag{7.73}
$$

Definition 7.19. The generalized integral staircase function is defined by

$$
{}^{*}S^{\alpha}_{F}(x) = \begin{cases} {}^{*}\gamma^{\alpha}(F, a_0, x) & \text{if } x \geq a_0 \\ -{}^{*}\gamma^{\alpha}(F, a_0, x) & \text{otherwise.} \end{cases} \tag{7.74}
$$

Definition 7.20. The ${}^{*}\gamma$-dimension is defined by

$$
\begin{aligned}
\dim_{*\gamma}(F \cap [a, b]) &= \inf\{\alpha :\ {}^{*}\gamma^{\alpha}(F, a, b) = 0\} \\
&= \sup\{\alpha :\ {}^{*}\gamma^{\alpha}(F, a, b) = \infty\}.
\end{aligned} \tag{7.75}
$$

Lemma 7.9. *Comparing the $^*\gamma$-dimension with the γ-dimension shows that $^*\gamma$-dimension finer than γ-dimension.*

Proof. Suppose $\dim_\gamma(F \cap [a,b]) = \alpha$, then $\gamma^\epsilon(F,a,b)$ diverges for any $\epsilon < \alpha$. Consequently, for any $k > 0$, there exists $\delta_0 > 0$ such that $\delta < \delta_0 \Rightarrow \gamma_\delta^\epsilon(F,a,b) > k$. For the $\dot{P}$ we choose the $\delta(t)$ such that $\epsilon < \beta$ but $^*\gamma^\epsilon(F,a,b)$ does not diverge meanwhile $\beta < \epsilon < \alpha$. As a result, $^*\gamma^\beta(F,a,b)$ does not diverge. That is $^*\gamma$-dimension is less than γ-dimension.   □

Definition 7.21. A function $f : F \rightarrow R$ is called to be the generalized Riemann integrable on $F \cap I = [a,b]$, if there exists a number $C \in R$ so that for every $\epsilon > 0$ there exists a gauge δ_ϵ on I such that if $\dot{P}$ is δ_ϵ-fine on I then

$$|S(f,\dot{P}) - C| \leq \epsilon. \tag{7.76}$$

We now show that A is the integral of f. For a given $\epsilon > 0$ let $K \in \mathbb{N}$ with $K > 2/\epsilon$. If $\dot{Q}$ is an arbitrary δ_ϵ-fine partition then

$$\begin{aligned}
|S(f,\dot{Q}) - A)| &\leq |S(f,\dot{Q}) - S(f,\dot{P}_K))| \\
&+ S(f,\dot{P}_K) - A \leq 1/K + 1/K < \epsilon.
\end{aligned} \tag{7.77}$$

In view of $\epsilon > 0$ is arbitrary then f is integrable to A.

Definition 7.22. A function $s : F \rightarrow R$ with is called to be a step function on F if there exist a partition $\{[x_{i-1}, x_i]\}_{i=1}^n$ of I as $\theta(F, [x_{i-1}, x_i]) \neq 0$ and real numbers $\{\alpha_i\}_{i=1}^n$ such that

$$s(x) = \alpha_i \quad \text{for} \quad x \in (x_{i-1}, x_i), \ i = 1, \ldots, n. \tag{7.78}$$

Definition 7.23. A function $f : F \rightarrow R$, is regulated on F if for every $\epsilon > 0$ there exists a step function $s_\epsilon : I \rightarrow R$ which is as follows

$$|f(x) - s_\epsilon(x)| < \epsilon \qquad \text{for all} \qquad x \in F. \tag{7.79}$$

Theorem 7.2. *Consider $f : F \rightarrow R$. If f is a regulated function on F then it is integrable and denote as $f \in \Re^*(F)$.*

Proof. For a given $\epsilon > 0$ let $s_\epsilon : F \rightarrow R$ be a step function such that Eq.(7.79) holds. Therefore we have

$$s_\epsilon(x) - \epsilon \leq f(x) \leq s_\epsilon(x) + \epsilon \quad \text{for} \quad x \in I. \tag{7.80}$$

Let $\varphi_\epsilon(x) = s_\epsilon(x) - \epsilon$ and $\psi_\epsilon(x) = s_\epsilon(x) + \epsilon$ for $x \in F$, thus the step functions φ_ϵ and ψ_ϵ are integrable and $\varphi_\epsilon(x) \leq f(x) \leq \psi_\epsilon(x)$ for $x \in F$. Since

$$\int_a^b (\psi_\epsilon - \varphi_\epsilon) \, d_F^\alpha x = \int_a^b 2\epsilon \chi_F \, d_F^\alpha x = 2({}^*S_F^\alpha(b) - {}^*S_F^\alpha(a))\epsilon. \tag{7.81}$$

which completes the proof. □

Theorem 7.3. *A function $f : F \to R$ is called a regulated function if and only if it has all of its one-side limits at every point of support.*

Proof. $\Rightarrow$ First we know that every step function has one side limits at each point so that any regulated function has the same property, let $c \in [a, b)$ we prove that f has a right hand limit at c. For proving this, let $\epsilon > 0$ be given and suppose $s_\epsilon : F \to R$ be a step function such that Eq.(7.79) holds. Since s_ϵ is a step function and $\lim_{x \to c+} s_\epsilon(x)$ exists, there exists $\delta_\epsilon(c) > 0$ such that if $x, y \in (c, c + \delta_\epsilon(c))$ then $s_\epsilon(x) = s_\epsilon(y)$. Therefore if $x, y \in (c, c + \delta_\epsilon(c))$ then

$$|f(x) - f(y)| \leq |f(x) - s_\epsilon(x)| + |s_\epsilon(x) - s_\epsilon(y)| + s_\epsilon(y) - f(y)| \leq 2\epsilon. \tag{7.82}$$

Since $\epsilon > 0$ is arbitrary in view of the Cauchy Criterion which implies that the right hand limit $\lim_{x \to c+} f(x)$ exists. The existence of the left hand limits at $c \in [(a, b]$ can be prove in the same way.

($\Leftarrow$) Suppose f has one side limits at every point of I. The Cauchy criterion for the existence of the one side limits guarantees that for a given $\epsilon > 0$, there is a gauge δ_ϵ on I such that if $t \in I$ and y_1, y_2 are both in $[t - \delta_\epsilon(t), t)$, or are both in $(t, t + \delta_\epsilon(t)]$ then $|f(y_1) - f(y_2)| \leq \epsilon$. Now let $\dot{P} = \{([x_{i-1}, x_i], t_i)\}_{i=1}^n$ be a δ_ϵ-fine partition of I we define $s_\delta(z) = f(z)$ if z is one of the numbers

$$a = x_0 \leq t_1 \leq \cdots \leq x_{i-1} \leq t_i \leq x_i \leq \cdots \leq t_n \leq x_n = b. \tag{7.83}$$

On the interval $(x_{i-1}, t_i) \subseteq [t_i - \delta_\epsilon(t_i), t_i)$ we define $s_\epsilon(x) = f(\frac{1}{2}(x_{i-1}, t_i))$ so that

$$|f(x) - s_\epsilon(x)| = |f(x) - f\left(\frac{1}{2}(x_{i-1}, t_i)\right)| \leq \epsilon. \tag{7.84}$$

Similarly on the interval $(t_i, x_i) \subseteq (t_i, t_i + \delta_\epsilon(t_i)]$ we define $s_\epsilon(x) = f(\frac{1}{2}(t_i + x_i))$ so that

$$|f(x) - s_\epsilon(x)| = |f(x) - f\left(\frac{1}{2}(t_i + x_i)\right)| \leq \epsilon. \tag{7.85}$$

Hence the step function s_ϵ satisfies $|f(x) - s_\epsilon(x)| \leq \epsilon$ for all $x \in F$. Since $\epsilon > 0$ it implies that f is a regulated function. □

Example 7.7. Consider function on fractal sets as follows:

$$f(x) = \begin{cases} 1 & \text{if} \quad x \in \{[a,b] \cap Q \cap F\} \\ 0 & \text{otherwise} \end{cases} \tag{7.86}$$

where Q is set of rational number and $f(x)$ is discontinues at every point of $[a,b]$. If $\{r_k = k \in N\}$ is an enumeration of the rational number in $[a,b]$ and $\epsilon > 0$ we define the gauge as follows

$$\delta_\epsilon(t) = \begin{cases} \epsilon/2^{k+1} & \text{if} \quad t = r_k \\ 1 & \text{otherwise.} \end{cases} \tag{7.87}$$

Then by this gauge we can control contribution of $^*S_F^\alpha(x_i) - {}^*S_F^\alpha(x_{i-1})$ in the Riemann sum $S(f, \dot{P})$. Then this function is $^*F^\alpha$-integrable but it is not F^α-integrable.

7.3.1   *$^*F^\alpha$-differentiable functions*

Here we generalized local fractal derivatives using the gauge function [Golmankhaneh and Baleanu (2016b)].

Definition 7.24. Let $f : F \to R$. For given $\epsilon > 0$ there exists $\delta_\epsilon(t) > 0$, the right $^*F^\alpha$-derivative of f at $x \in F$ is define by

$$y \in F \quad 0 < y - x < \delta_\epsilon(x) \quad \Rightarrow \quad \left| \frac{f(y) - f(x)}{^*S_F^\alpha(y) - {}^*S_F^\alpha(x)} - D_{+F}^\alpha f(x) \right| < \epsilon, \tag{7.88}$$

Definition 7.25. Let $f : F \to R$. For given $\epsilon > 0$ there exists $\delta_\epsilon(t) > 0$, the left $^*F^\alpha$-derivative of f at $x \in F$ is defined by

$$y \in F \quad 0 < x - y < \delta_\epsilon(x) \quad \Rightarrow \quad \left| \frac{f(y) - f(x)}{^*S_F^\alpha(y) - {}^*S_F^\alpha(x)} - D_{-F}^\alpha f(x) \right| < \epsilon. \tag{7.89}$$

Definition 7.26. If $D_{-F}^\alpha f(x) = D_{+F}^\alpha f(x)$ then $f(x)$ is differentiable and denoted by

$$D_{-F}^\alpha f(x) = D_{+F}^\alpha f(x) = \begin{cases} F^* - \lim_{y \to x} \frac{f(y) - f(x)}{^*S_F^\alpha(y) - {}^*S_F^\alpha(x)} & \text{if} \quad x \in F \\ 0 & \text{otherwise.} \end{cases}$$
$$\tag{7.90}$$

7.4   Random variables and processes on fractal

In this section, we preset the basic definition of random variables and random process on fractal sets [Golmankhaneh and Fernandez (2019); Golmankhaneh and Tunç (2020)].

7.4.1 *Random variables on fractal*

Definition 7.27. A random variable on fractal sets (RVF) is defined by

$$X(\zeta) : \mathcal{S} \to F \tag{7.91}$$

where $\mathcal{S}$ is the sample space.

Definition 7.28. The distribution function of a random variable $X(\zeta)$ is defined by

$$F_X(x) = P(X(\zeta) \le x), \ x \in F. \tag{7.92}$$

Definition 7.29. The probability density function of $X(\zeta)$ is defined as

$$f_X(x) = D_F^\alpha F_X(x). \tag{7.93}$$

Definition 7.30. The mean of an RVF is defined as

$$E[X] = \int_{-\infty}^{\infty} x f_X(x) d_F^\alpha x \tag{7.94}$$

Definition 7.31. The variance of an RVF is defined by

$$Var[X] = \int_{-\infty}^{\infty} (x - E[X])^2 f_X(x) d_F^\alpha x \tag{7.95}$$

Definition 7.32. The m-*th* moment of an RVF is defined by

$$E[X^m] = \int_{-\infty}^{\infty} x^m f_X(x) d_F^\alpha x, \ m \in \mathbb{N}. \tag{7.96}$$

Definition 7.33. The moment generating function of a RVC X is defined by

$$M_X(t) = E(e^{tX}) = \int_{-\infty}^{\infty} e^{tx} f_X(x) d_F^\alpha x \tag{7.97}$$

Definition 7.34. The characteristic function of a RVC X is defined by

$$\Psi_X(\omega) = \int_{-\infty}^{\infty} e^{i\omega x} f_X(x) d_F^\alpha x \tag{7.98}$$

where ω is a real number and $i = \sqrt{-1}$.

Definition 7.35. The fractal Shannon entropy of a RVC X is defined by

$$H_X(x) = -\int_X f_X(x) \log(f_X(x)) d_F^\alpha x \qquad (7.99)$$

Definition 7.36. A RVF over $F \in [0,1]$ is called uniform, if its probability mass function (PMF) is given by

$$f_X(x) = \begin{cases} \frac{1}{\Gamma(\alpha+1)}, & x \in F, \\ 0, & \text{otherwise.} \end{cases} \qquad (7.100)$$

The corresponding cumulative distribution function (CDF) of $X(\zeta)$ is

$$F_X(x) = \int_{-\infty}^x f_X(x) d_F^\alpha x = \begin{cases} 0, & x < 0; \\ S_F^\alpha(x), & 0 \le x \le 1; \\ \Gamma(\alpha+1), & x > 1. \end{cases} \qquad (7.101)$$

The mean of the uniform RVF is

$$E[X] = \int_0^1 S_F^\alpha(x) f_X(x) d_F^\alpha x = \int_0^1 \frac{S_F^\alpha(x)}{\Gamma(\alpha+1)} d_F^\alpha x$$
$$= \frac{S_F^\alpha(x)^2}{2\Gamma(\alpha+1)} \bigg|_0^1 = \frac{S_F^\alpha(1)^2}{2\Gamma(\alpha+1)} = \frac{\Gamma(\alpha+1)}{2},$$

and its variance is

$$Var[X] = E[X^2] - E[X]^2$$
$$= \int_0^1 S_F^\alpha(x)^2 f_X(x) d_F^\alpha x - \left(\int_0^1 S_F^\alpha(x) f_X(x) d_F^\alpha x \right)^2$$
$$= \frac{\Gamma(\alpha+1)^2}{3} - \frac{\Gamma(\alpha+1)^2}{4}$$
$$= \frac{\Gamma(\alpha+1)^2}{12}. \qquad (7.102)$$

Definition 7.37. A RVF is called the Gaussian distribution on the fractal sets if its probability mass function is defined by

$$f_X(x) = \frac{1}{\sqrt{2\pi}\sigma} e^{-(x-\mu)^2/(2\sigma^2)}, \qquad (7.103)$$

and its cumulative distribution function is

$$\varphi(x) = \int_{-\infty}^{(x-\mu)/\sigma} e^{-t^2/2} d_F^\alpha t. \qquad (7.104)$$

Definition 7.38. A RVF is called the Rayleigh if its probability mass function is given by

$$f_X(x) = \begin{cases} \frac{x}{\sigma^2}e^{-x^2/(2\sigma^2)}, & 0 \le x; \\ 0, & x < 0, \end{cases} \tag{7.105}$$

and its cumulative distribution function is defined by

$$F_X(x) = \int_0^x \frac{S_F^\alpha(t)}{\sigma^2}e^{-S_F^\alpha(t)^2/(2\sigma^2)}d_F^\alpha t = 1 - e^{-S_F^\alpha(x)^2/(2\sigma^2)}, 0 \le x, \tag{7.106}$$

and it follows

$$F_X(x) = \begin{cases} 1 - e^{-S_F^\alpha(x)^2/(2\sigma^2)}, & 0 \le x; \\ 0, & 0 \le x. \end{cases} \tag{7.107}$$

Definition 7.39. A RVF is said memoryless if its probability mass function is given by

$$f_X(x) = \begin{cases} \lambda e^{-\lambda x}, & 0 < x; \\ 0, & x < 0. \end{cases} \tag{7.108}$$

The corresponding cumulative distribution function is given by

$$F_X(x) = \begin{cases} 1 - e^{-\lambda S_F^\alpha(x)}, & 0 \le x; \\ 0, & x < 0. \end{cases} \tag{7.109}$$

where $\lambda > 0$.

Definition 7.40. A random RVF and its probability mass function with parameter c are defined by

$$f_X(x) = \begin{cases} c^2 x e^{-cx}, & 0 \le x; \\ 0, & x < 0, \end{cases} \tag{7.110}$$

where $c > 0$, and corresponding cumulative distribution function is given by

$$F_X(x) = \begin{cases} e^{-cS_F^\alpha(x)}\left(e^{cS_F^\alpha(x)} - cS_F^\alpha(x) - 1\right), & 0 \le x; \\ 0, & x < 0. \end{cases} \tag{7.111}$$

7.4.2 Random processes on fractals

In this section we define the random processes on fractal [Golmankhaneh and Fernandez (2019); Golmankhaneh and Tunç (2020)].

Definition 7.41. A random process is a family of the random variables. Random processes are defined as functions of two arguments such as follows

$$X(t, \zeta), \quad t \in F, \quad \zeta \in \mathcal{S}, \tag{7.112}$$

where C^η is called the parameter fractal set. For convenience, we suppose $X(t, \zeta_i) = X_i(t)$, which is called sample function.

Definition 7.42. If $X(t, \zeta) = X(t)$ is a random process, then the correlation function is defined by

$$R(t_1, t_2) = E[X(t_1)X(t_2)]. \tag{7.113}$$

Example 7.8. Suppose a particle performing a random walk on the fractal sets points, and in each step moves to one of its right or left neighboring points with different probability. This process is called a simple random walk on fractal sets and is mathematically defined as

$$X(S_F^\alpha(t)) = X(n), \quad n \le S_F^\alpha(t) < n + 1 \quad t \in F, \tag{7.114}$$

where n is the nth step. The mean and the variance function with fractal support are $E[X(t)] = 0, Var[X(t)] = S_F^\alpha(t) \approx t^\alpha$ (See Figures 7.3 and 7.4).

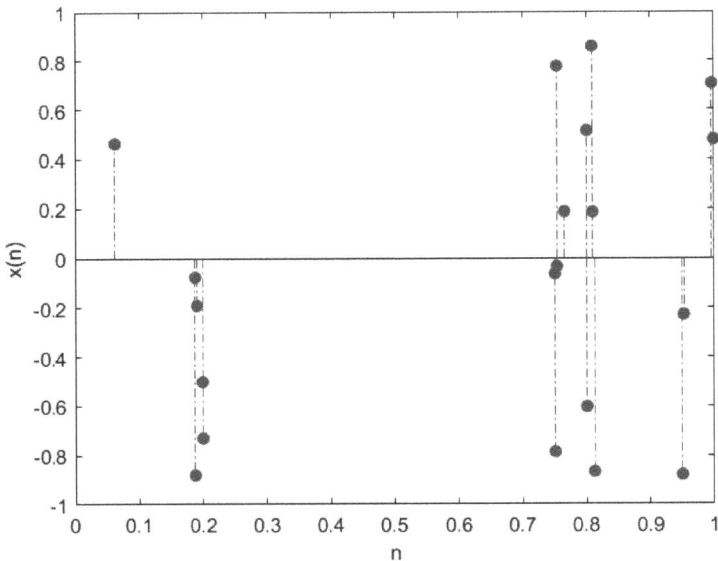

Fig. 7.3: Graph of a random walk on the middle-1/3 Cantor set for $\alpha = 0.5, c = 2$

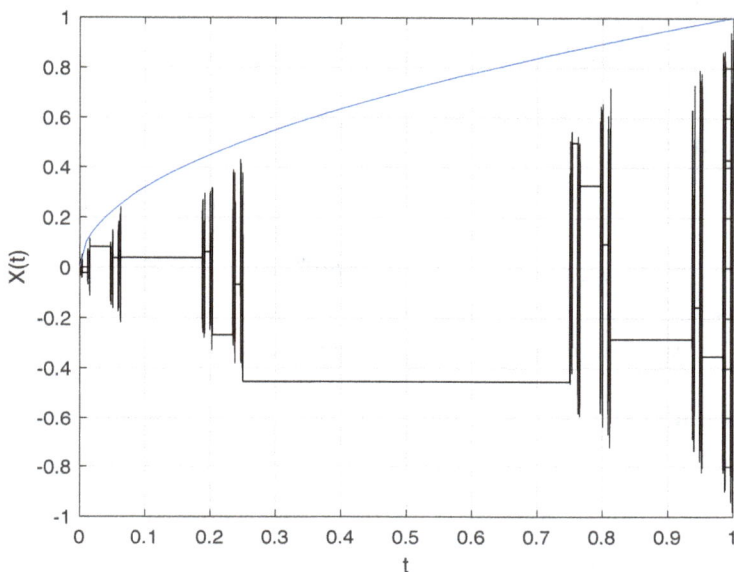

Fig. 7.4: Graph of sample function of the random walk on the middle-1/3 Cantor set and its variance for $\alpha = 0.5, c = 2$ (blue line)

7.5 Mean square fractal calculus

In this section, we present basic concepts of mean square F^α-calculus which are a framework of stochastic F^α-calculus [Golmankhaneh and Tunç (2020)].

Definition 7.43. A class of random variables on fractal sets $X_1, X_2, \ldots, X_n$ is called second order random variables (SRVF) if

$$E[X_1^2], E[X_2^2], \ldots, E[X_1^n] < \infty, \tag{7.115}$$

namely, they are finite.

Definition 7.44. A random process on fractal sets (RPF) $X(t), t \in F$ is called second order if

$$\|X(t)\| = E[X(t)^2] < \infty, \quad t \in F. \tag{7.116}$$

Definition 7.45. A sequence of random variables on fractal sets $X_1, X_2, \ldots, X_n$ is called mean square convergence to X if we have

$$F\!-\!\lim_{n \to \infty} ||X_n - X|| = 0. \tag{7.117}$$

Definition 7.46. Let $X(t)$ be a random process on time fractal sets. If

$$F\!-\!\lim_{\epsilon \to 0} E[X(t + \epsilon) - X(t)]^2 = 0, \quad \epsilon \in F, \tag{7.118}$$

then $X(t)$ is called mean square F-continuous at t.

Definition 7.47. A random process $X(t)$ is said to have mean square F^α-derivative if

$$F\!-\!\lim_{\epsilon \to 0} E\left[\left(\frac{X(t + \epsilon) - X(t)}{S_F^\alpha(\epsilon)} - D_{F,t}^\alpha X(t) \right)^2 \right] = 0. \tag{7.119}$$

Definition 7.48. A random variable is called mean square analytic on fractal sets, if it can be expanded as the convergent series, namely

$$X(t) = \sum_{n=0}^{\infty} \frac{(D_{F,t}^\alpha)^n X(t)|_{t=t_0}}{n!} (S_F^\alpha(t) - S_F^\alpha(t_0)^n. \tag{7.120}$$

Definition 7.49. A mean square F^α-integral of random process $X(t)$ is defined by

$$Z(t) = \int_{t_0}^{t} X(u) d_F^\alpha u$$

$$= \underset{\Delta S_F^\alpha(t_i), \Delta S_F^\alpha(t_k) \to 0}{F\!-\!\lim} E\left[\left(\sum_i X(t_i) \Delta S_F^\alpha(t_i) - \sum_k X(t_k) \Delta S_F^\alpha(t_k) \right)^2 \right]$$

$$= 0. \tag{7.121}$$

Theorem 7.4. *The random process is mean square fractal integrable if following fractal integral exists*

$$\int_{t_0}^{t} \int_{t_0}^{t} R_X(t_1, t_2) d_F^\alpha t_1 d_F^\alpha t_2. \tag{7.122}$$

Proof. Since we have

$$E\left[\left(\sum_i X(t_i)\Delta S_F^\alpha(t_i) - \sum_k X(t_k)\Delta S_F^\alpha(t_k)\right)^2\right]$$

$$= E\left[\sum_i \sum_k X(t_i)X(t_k)\Delta S_F^\alpha(t_i)\Delta S_F^\alpha(t_k)\right.$$

$$+ \sum_i \sum_k X(t_i)X(t_k)\Delta S_F^\alpha(t_i)\Delta S_F^\alpha(t_k)$$

$$\left. - 2\sum_i \sum_k X(t_i)X(t_k)\Delta S_F^\alpha(t_i)\Delta S_F^\alpha(t_k)\right]$$

$$= \left[\sum_i \sum_k R_X(t_i,t_k)\Delta S_F^\alpha(t_i)\Delta S_F^\alpha(t_k) + \sum_i \sum_k R_X(t_i,t_k)\Delta S_F^\alpha(t_i)\Delta S_F^\alpha(t_k)\right.$$

$$\left. - 2\sum_i \sum_k R_X(t_i,t_k)\Delta S_F^\alpha(t_i)\Delta S_F^\alpha(t_k)\right]. \tag{7.123}$$

By taking F^α-limit of both sides of Eq.(7.123) we have

$$\underset{\Delta S_F^\alpha(t_i),\Delta S_F^\alpha(t_k)\to 0}{\text{F-lim}} \sum_i \sum_k R_X(t_i,t_k)\Delta S_F^\alpha(t_i)\Delta S_F^\alpha(t_k), \tag{7.124}$$

or the following integral

$$\int_{t_0}^t \int_{t_0}^t R_X(t_1,t_2)d_F^\alpha t_1 d_F^\alpha t_2 \tag{7.125}$$

which completes the proof. $\square$

Example 7.9. Let us consider the Wiener process and find its mean square fractal integrals as follows:

$$Y(t) = \int_0^t W(t_1)d_F^\alpha t_1. \tag{7.126}$$

Using Theorem 7.4, we lead to

$$\int_0^t \int_0^t R(t_1,t_2)d_F^\alpha t_1 d_F^\alpha t_2$$

$$= \int_0^t \int_0^t D\min(S_F^\alpha(t_1), S_F^\alpha(t_2))d_F^\alpha t_1 d_F^\alpha t_2$$

$$= D\int_0^t \left[\int_0^{t_2} S_F^\alpha(t_1)d_F^\alpha t_1 + \int_{t_2}^t S_F^\alpha(t_2)d_F^\alpha t_1\right]d_F^\alpha t_2$$

$$= D\frac{S_F^\alpha(t)^3}{3}. \tag{7.127}$$

Therefore, Wiener process is a mean square fractal integrable.

Theorem 7.5. *If random process $X(t)$ is mean square F-continuous , then it implies that the mean of $X(t)$ is also F-continuous.*

Proof. Since we have

$$E[(X(t+\epsilon)-X(t))^2] \leq (E[X(t+\epsilon)-X(t)])^2, \qquad (7.128)$$

then Eq.(7.118) follows that

$$F_- \lim_{\epsilon \to 0} E[X(t+\epsilon)-X(t)] = 0, \qquad (7.129)$$

which implies the mean of $X(t)$ is also F-continuous.   □

Example 7.10. Consider Wiener process $W(t), t \in F$ with fractal support. Then, its correlation function is given by

$$R(t_1, t_2) = 2D \min(S_F^\alpha(t_1), S_F^\alpha(t_2)), \quad t_1, t_2 > 0, \qquad (7.130)$$

where D is constant. We use the following inequality

$$|R(t_1, t_2) - R(t, t)| = 2D |\min(S_F^\alpha(t_1), S_F^\alpha(t_2)) - S_F^\alpha(t)| \qquad (7.131)$$
$$\leq 2D \max\left(|S_F^\alpha(t_1) - S_F^\alpha(t)|, |S_F^\alpha(t_2)) - S_F^\alpha(t)|\right).$$

We can write

$$F\_\lim_{t_1,t_2 \to t} \max\left(|S_F^\alpha(t_1) - S_F^\alpha(t)|, |S_F^\alpha(t_2)) - S_F^\alpha(t)|\right) \to t, \qquad (7.132)$$

then we arrive at

$$F\_\lim_{t_1,t_2 \to t} |R(t_1, t_2) - R(t, t)| = 0. \qquad (7.133)$$

Finally, we conclude that $R(t_1, t_2)$ is F-continuous at t.

7.5.1 *Stochastic differential equation on fractal sets*

In this section, we give the series solutions of the stochastic second order mean square fractal derivative differential equation utilizing Frobenius method [Golmankhaneh and Tunç (2020)].

Example 7.11. Let us consider stochastic second order mean square fractal derivative on fractal set

$$(D_{F,t}^\alpha)^2 X(t) + A^2 X(t) = 0, \quad X(0) = Z_0, \quad D_{F,t}^\alpha X(t) = Z_1, \qquad (7.134)$$

where A^2 is Beta random variable and it is independent of Z_0 and Z_1. Our aim is to find a solution of Eq.(7.134) as in the following form

$$X(t) = \sum_{m=0} X_m S_F^\alpha(t)^m, \qquad (7.135)$$

where X_m are RVFs.

By taking second order mean square fractal derivative of both sides of Eq.(7.135), we obtain

$$(D_{F,t}^{\alpha})^2 X(t) = \sum_{m=0} (m+2)(m+1) X_{m+2} S_F^{\alpha}(t)^m. \tag{7.136}$$

Then, by substituting Eq.(7.135) and Eq.(7.136) in to Eq.(7.134), it follows that

$$(D_{F,t}^{\alpha})^2 X(t) + A^2 X(t) =$$
$$\sum_{m=0} \left((m+2)(m+1) X_{m+2} + A^2 X_m \right) S_F^{\alpha}(t)^m = 0. \tag{7.137}$$

So we can write

$$X_{m+2} = -\frac{A^2 X_m}{(m+2)(m+1)}, \quad m > 0, \tag{7.138}$$

and we obtain

$$X(t) = \sum_{i=0} \frac{(-1)^i (A^2)^i X_0}{(2i)!} S_F^{\alpha}(t)^{2i} + \sum_{i=0} \frac{(-1)^i (A^2)^i X_1}{(2i+1)!} S_F^{\alpha}(t)^{2i+1}, \tag{7.139}$$

where $X_0 = Z_0$, and $X_1 = Z_1$. In view of the following

$$\cos(A S_F^{\alpha}(t)) = \sum_{i=0} \frac{(-1)^i (A^2)^i X_0}{(2i)!} S_F^{\alpha}(t)^{2i} \tag{7.140}$$

$$\sin(A S_F^{\alpha}(t)) = \sum_{i=0} \frac{(-1)^i (A^2)^i X_1}{(2i+1)!} S_F^{\alpha}(t)^{2i+1}, \tag{7.141}$$

we have

$$X(t) = Z_0 \cos(A S_F^{\alpha}(t)) + \frac{Z_1}{A} \sin(A S_F^{\alpha}(t)). \tag{7.142}$$

For finding the mean, we have

$$X_M(t) = Z_0 \sum_{i=0}^{M} \frac{(-1)^i (A^2)^i}{(2i)!} S_F^{\alpha}(t)^{2i} + Z_1 \sum_{i=0} \frac{(-1)^i (A^2)^i}{(2i+1)!} S_F^{\alpha}(t)^{2i+1}. \tag{7.143}$$

Then, the mean of Eq.(7.143) is

$$E[X_M(t)] = E[Z_0] \sum_{i=0}^{M} \frac{(-1)^i E[(A^2)^i]}{(2i)!} S_F^{\alpha}(t)^{2i} \tag{7.144}$$
$$+ E[Z_1] \sum_{i=0} \frac{(-1)^i E[(A^2)^i]}{(2i+1)!} S_F^{\alpha}(t)^{2i+1}.$$

In terms of finite terms, we can write

$$E[X_M(t)^2] = \sum_{i=0}^{M} \left(\frac{E[Z_0^2]S_F^\alpha(t)^{4i}}{((2i)!)^2} + \frac{E[Z_1^2]S_F^\alpha(t)^{4i+2}}{((2i+1)!)^2} \right) E[(A^2)^{2i}]$$

$$+ \sum_{i=0}^{M} \sum_{j=0}^{i-1} (-1)^{i+j} \left(\frac{E[Z_0^2]S_F^\alpha(t)^{2(i+j)}}{((2i)!(2j)!)} + \frac{E[Z_1^2]S_F^\alpha(t)^{2(i+j+1)}}{((2i+1)!(2j+1)!)} \right) E[(A^2)^{i+j}]$$

$$+ 2E[Z_0Z_1] \sum_{i=0}^{M} \sum_{j=0}^{M} \frac{(-1)^{i+j}S_F^\alpha(t)^{2(i+j)+1}}{(2i)!(2j+1)!} E[(A^2)^{i+j}]. \qquad (7.145)$$

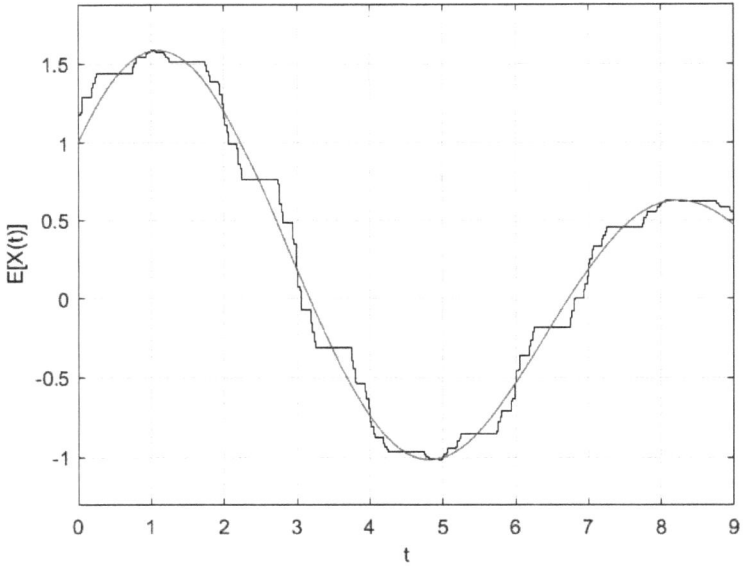

Fig. 7.5: Mean function of random process on the middle-ξ Cantor set in the case of $\alpha = 0.5$, $E[Z_0] = 1$, $E[Z_1] = 1$, and $E[Z_0Z_1] = 1$

In Figure 7.5, we have plotted Eq.(7.146). By choosing some terms of Eq.(7.144) and Eq.(7.145), we get approximation of the mean and variance functions as follows

$$E[X(t)] = 1 + S_F^\alpha(t) - \frac{1}{3}S_F^\alpha(t)^2 - \frac{1}{9}S_F^\alpha(t)^3$$

$$+ \frac{1}{48}S_F^\alpha(t)^4 + \frac{1}{240}S_F^\alpha(t)^5 + \cdots, \qquad (7.146)$$

and

$$Var[X(t)] = 1 + 2S_F^\alpha(t) - \frac{8}{9}S_F^\alpha(t)^3 - \frac{55}{72}S_F^\alpha(t)^4$$
$$+ \frac{77}{540}S_F^\alpha(t)^5 - \frac{253}{8100}S_F^\alpha(t)^6 + \cdots. \qquad (7.147)$$

7.6 Hilbert spaces on fractals

In this section we introduce analogue of Hilbert spaces on fractals [Golmankhaneh (2010)].

Definition 7.50. A functions, $f : F \to \Re$, such that

$$\int_a^b |f(x)|^2 d_F^\alpha x < \infty, \qquad (7.148)$$

which is called α-square F^α-integrable function and denoted by $L_2(F)$.

Definition 7.51. A dot product/inner product in $L_2(F)$ is defined by:

$$< f, g > = \int_a^b f(x)g(x)d_F^\alpha x, \qquad (7.149)$$

where $f, g \in L_2(F)$.

Definition 7.52. A norm in $L_2(F)$ is defined by

$$||f|| = [< f, f >]^{1/2}, \qquad (7.150)$$

where $||f|| = 0$ if only if $f = 0$.

Definition 7.53. A distance between two fractal functions is defined by

$$d(f, g) = ||f - g||. \qquad (7.151)$$

The $L_2(F)$ is called a complete fractal Hilbert space respect to given norm in Eq.(7.150).

Definition 7.54. A set of functions $\{f_n(x)\}$ in $L_2(F)$, are said to be an orthogonal sets if

$$\int_a^b f_n(x)f_m(x)d_F^\alpha x = \delta_{n,m}. \qquad (7.152)$$

where $\delta_{n,m}$ is the Kronecker delta .

Definition 7.55. Let $\{f_n(x)\}$ be a sequence of functions which belong to $L_2(F)$. If there exists a function $f(x) \in L_2(F)$ such that

$$\lim_{n \to \infty} \int_a^b |f_n(x) - f(x)|^2 d_F^\alpha x = 0. \tag{7.153}$$

or for given $\epsilon > 0$ there exists a number $n_0 > 0$ such that

$$||f_n(x) - f(x)|| < \epsilon \quad \text{whenever} \quad n > n_0. \tag{7.154}$$

Then we say that $f_n(x)$ converge uniformly on F to f.

Definition 7.56. The sequence of functions $\{f_n(x)\} \in L_2(F)$ is said to be a Cauchy sequence if

$$\lim_{\substack{m \to \infty \\ n \to \infty}} \int_a^b |f_m(x) - f_n(x)|^2 d_F^\alpha x = 0. \tag{7.155}$$

or, for given $\epsilon > 0$ there exists a number $n_0 > 0$ such that

$$||f_m - f_n|| = \int_a^b |f_m(x) - f_n|^2 d_F^\alpha x < \epsilon \quad \text{whenever} \quad m > n_0, n > n_0. \tag{7.156}$$

7.7   Self-adjoint fractal differential operator

In this section we define self-adjoint fractal differential operator [Golmankhaneh (2010)].

Definition 7.57. Let $\mathbf{L}$ be a fractal local differential operator in the following form:

$$\mathbf{L}f(x) = p_0(x)(D_F^\alpha)^2 f(x) + p_1(x)D_F^\alpha f(x) + p_2(x)f(x). \tag{7.157}$$

Now let calculate

$$< f, \mathbf{L}f > = \int_a^b f(x)\mathbf{L}f(x)d_F^\alpha x \tag{7.158}$$

$$= \int_a^b f(x)\{p_0(x)(D_F^\alpha)^2 f(x) + p_1(x)D_F^\alpha f(x) + p_2(x)f(x)\}d_F^\alpha x.$$

Integrating by parts one and twice, gives

$$< f, \mathbf{L}f > = [f(x)(p_1(x) - D_F^\alpha p_0(x))f(x)]|_{x=a}^b$$

$$+ \int_a^b \{(D_F^\alpha)^2[p_0(x)(f(x)] - D_F^\alpha[p_1(x)f(x)] + p_2(x)f(x)\}f(x)d_F^\alpha x. \tag{7.159}$$

By comprising Eqs.(7.159) and (7.158) we obtain

$$f(x)((D_F^\alpha)^2 p_0(x) - D_F^\alpha p_1(x))f(x) + 2f(x)(D_F^\alpha p_0(x) - p_1(x))D_F^\alpha f(x) = 0,$$
(7.160)

and

$$D_F^\alpha p_0(x) = p_1(x).$$
(7.161)

Thus we can write

$$\bar{\mathbf{L}}[f(x)] = (D_F^\alpha)^2[p_0(x)(f(x)] - D_F^\alpha[p_1(x)f(x)] + p_2(x)f(x) \quad (7.162)$$
$$= p_0(x)(D_F^\alpha)^2 f(x) + (2D_F^\alpha p_0(x) - p_1(x))D_F^\alpha f(x)$$
$$+ ((D_F^\alpha)^2 p_0(x) - D_F^\alpha p_1(x) + p_2(x))f(x),$$

If $\bar{\mathbf{L}}[f(x)] = \mathbf{L}[f(x)]$ the operator $\bar{\mathbf{L}}$ is called fractal self-adjoint.

7.8 Strum-Liouville equation on fractals

In this section we define Strum-Liouville equation on fractals [Golmankhaneh (2010)].
Consider linear fractal differential operator

$$\mathbf{L}[f(x)] = -D_F^\alpha[p(x)D_F^\alpha f(x)] - q(x)f(x).$$
(7.163)

Rewrite Eq.(7.163) in the Strum-Liouville equation form, we have

$$D_F^\alpha[p(x)D_F^\alpha f(x)] + [q(x) + \lambda r(x)]f(x) = 0,$$
(7.164)

or

$$\mathbf{L}[f(x)] = \lambda r(x)f(x),$$
(7.165)

where $p(x) > 0$, $D_F^\alpha p(x)$, $q(x), r(x) > 0$ are F-continuous on $F \subset [a, b] \ \forall \ x \in F$.
The solution of Eq.(7.164) which is indicate by $\{v, u\}$ on given interval $a \le x \le b$ satisfy boundary conditions at the two end points a and b as

$$a_1 v(a) + a_2 D_F^\alpha v(a) = 0 \qquad (a) \qquad (7.166)$$
$$a_1 u(a) + a_2 D_F^\alpha u(a) = 0 \qquad (a)$$
$$b_1 v(b) + b_2 D_F^\alpha v(b) = 0 \qquad (b)$$
$$b_1 u(b) + b_2 D_F^\alpha u(b) = 0 \qquad (b)$$

where a_1, a_2, in Eq.(7.166) are given constants but not both zero and but also these two b_1, b_2.
Eq.(7.164) with boundary condition Eq.(7.166) is called fractal Sturm-Liouville problems .

Theorem 7.6. *Let $f_m(x)$ and $f_n(x)$ be eigenfunctions of the fractal Strum-Liouville problem that correspond to different eigenvalues λ_m and λ_n, respectively. Then $f_m(x)$, $f_n(x)$ are orthogonal on that interval with respect to the weight function $r(x)$.*
In the theorem we may have the following cases:

(1) If $p(a) = 0$, then Eq.(7.166a) can be dropped from the problem.
(2) If $p(b) = 0$, then Eq.(7.166b) can be dropped from the problem.
(3) If $p(a) = p(b)$ then Eq.(7.166) can be replaced by the periodic boundary conditions

$$f(a) = f(b) \qquad D_F^\alpha f(a) = D_F^\alpha f(b). \tag{7.167}$$

The boundary value problem consisting of the Sturm-Liouville Eq.(7.164) and the periodic boundary condition Eq.(7.167) is called a periodic Sturm-Liouville problem.

Proof. Let, $f_m(x), f_n(x)$ satisfy

$$D_F^\alpha[p(x)D_F^\alpha f_m(x)] + [q(x) + \lambda_m r(x)]f_m(x) = 0, \tag{7.168}$$

and

$$D_F^\alpha[p(x)D_F^\alpha f_n(x)] + [q(x) + \lambda_n r(x)]f_n(x) = 0. \tag{7.169}$$

Then, multiplying Eq.(7.168) by $f_n(x)$ and Eq.(7.169) by $-f_m(x)$ and adding them, we get

$$\begin{aligned}
(\lambda_m - \lambda_n)r(x)f_m(x)f_n(x) &= f_m(x)D_F^\alpha[p(x)D_F^\alpha f_n(x)] \\
&\quad - f_n(x)D_F^\alpha[p(x)D_F^\alpha f_m(x)] \\
&= D_F^\alpha[(p(x)D_F^\alpha f_n(x))f_m(x) \\
&\quad - (p(x)D_F^\alpha f_m(x))f_n(x)], \tag{7.170}
\end{aligned}$$

Fractal integrating, we thus obtain

$$(\lambda_m - \lambda_n) \int_a^b r(x)f_m f_n d_F^\alpha x = [p(x)(D_F^\alpha f_n f_m - D_F^\alpha f_m f_n)]_a^b. \tag{7.171}$$

The expression on the right equals

$$p(b)[D_F^\alpha f_n(b)f_m(b) - D_F^\alpha f_m(b)f_n(b)] - p(a)[D_F^\alpha f_n(a)f_m - D_F^\alpha f_m(a)f_n(a)]. \tag{7.172}$$

We now have to consider several case depending on whether p vanishes or does not vanish at a or b. Now we consider the following cases:

(1) If $p(a) = 0$ and $p(b) = 0$ then the expression in Eq.(7.172) is zero. As λ_m and λ_n are distinct. Thus obtain the desired orthogonality

$$\int_a^b r(x) f_m(x) f_n(x) d_F^\alpha x = 0 \qquad (m \neq n), \qquad (7.173)$$

without the use of the boundary conditions Eq.(7.166).

(2) Let $p(b) = 0$, but $p(a) \neq 0$. Then the first line in Eq.(7.172) is zero. We consider the remaining expression in Eq.(7.172) From Eq.(7.166a) we have

$$a_1 f_n(a) + a_2 D_F^\alpha f_n = 0 \qquad (7.174)$$
$$a_1 f_m(a) + a_2 D_F^\alpha f_m = 0.$$

Let $a_2 \neq 0$. We multiply the first equation by $f_m(a)$, the last by $-f_n(a)$ and add,

$$a_2 [D_F^\alpha f_n(a) f_m(a) - D_F^\alpha f_m(a) f_n(a)] = 0. \qquad (7.175)$$

Since $a_2 \neq 0$, the expression in brackets must be zero. This expression is identical with that in the last line of Eq.(7.172). Hence Eq.(7.172) is zero and from Eq.(7.171), we obtain Eq.(7.173) as before. If $a_2 = 0$ then by assumption $a_1 \neq 0$ and the argument of proof is similar.

(3) If $p(a) = 0$ but $p(b) \neq 0$ the proof is similar to Case 2, but instead of Eq.(7.166a) we use Eq.(7.166b).

(4) If $p(a) \neq 0$ and $p(b) \neq 0$ we apply both boundary conditions Eq.(7.166) and proceed as in Case 2 and 3.

(5) Let $p(a) = p(b)$. Then Eq.(7.172) takes the form

$$p(b)[D_F^\alpha f_n(b) f_m(b) - D_F^\alpha f_m(b) f_n(b) - D_F^\alpha f_n(a) f_m(a) + D_F^\alpha f_m(a) f_n(a)]. \qquad (7.176)$$

We may use Eq.(7.166) as before and conclude that the expression in brackets is zero. However we immediately see that this also follows from Eq.(7.167). Hence we replace Eq.(7.166) by Eq.(7.167). Hence Eq.(7.171) yields Eq.(7.173) as before. This completes the proof. $\square$

Example 7.12. Consider the following a fractal differential Sturm-Liouville as [Cetinkaya and Golmankhaneh (2021)]

$$-D_F^{2\alpha} y(x) = \lambda y, \qquad (7.177)$$

with boundary conditions

$$y(0) + D_F^\alpha y(x)|_0 = 0, \quad y(1) - D_F^\alpha y(x)|_1 = 0 \qquad (7.178)$$

To find the solution of Eq.(7.177) we consider three cases:

(1) If $\lambda = 0$, then the general solution of Eq.(7.177) is

$$y(x) = AS_F^\alpha(x) + B \tag{7.179}$$

where $A, B \in R$ are constants.

(2) If $\lambda < 0$ then the general solution of Eq.(7.177) is

$$y(x) = A \exp(\sqrt{-\lambda}S_F^\alpha(x)) + B \exp(-\sqrt{-\lambda}S_F^\alpha(x)) \tag{7.180}$$

(3) If $\lambda > 0$ then the general solution of Eq.(7.177)

$$y(x) = A \cos(\sqrt{\lambda}S_F^\alpha(x)) + B \sin(\sqrt{\lambda}S_F^\alpha(x)) \tag{7.181}$$

Example 7.13. Consider the following fractal differential Sturm-Liouville equation

$$D_F^\alpha y(x) + 3D_F^\alpha y(x) + 2y(x) + \lambda y = 0 \tag{7.182}$$

with boundary conditions

$$y(0) = 0, \quad y(1) = 0 \tag{7.183}$$

The characteristic equation of Eq.(7.182) is

$$r^2 + 3r + 2 + \lambda = 0 \tag{7.184}$$

with zeros

$$r_1 = \frac{-3 + \sqrt{1 - 4\lambda}}{2}, r_2 = \frac{-3 - \sqrt{1 - 4\lambda}}{2} \tag{7.185}$$

To obtain the solution of Eq.(7.182) we consider three cases:

(1) If $\lambda < \frac{1}{4}$ then r_1 and r_2 are real and distinct, so the general solution of equation Eq.(7.182) is

$$y(x) = c_1 \exp(r_1 S_F^\alpha(x)) + c_2 \exp(r_2 S_F^\alpha(x)), \quad A, B \in R \tag{7.186}$$

(2) If $\lambda = \frac{1}{4}$, then $r_1 = r_2 = -3/2$, so the general solution of equation Eq.(7.182) is

$$y(x) = \exp(-3S_F^\alpha(x)/2) + B \exp(c_1 + c_2 S_F^\alpha(x)) \tag{7.187}$$

(3) If $\lambda > \frac{1}{4}$, then $r_1 = -3/2 + i\omega$ $r_2 = -3/2 - i\omega$, where $\omega = \frac{\sqrt{4\lambda-1}}{2}$, so the general solution of equation Eq.(7.182) is

$$y(x) = \exp(-3S_F^\alpha(x)/2)(\sin(n\omega S_F^\alpha(x)), n = 1, 2, 3, \ldots \tag{7.188}$$

Using Eqs.(7.182) and (7.183), the eigenvalues of is

$$\lambda_n = (1 + 4n^2\pi^2)/4. \tag{7.189}$$

7.8.1 *Fractal Lengendre's equation*

Consider fractal Lengendre differential equation such as [Golmankhaneh (2010)]

$$(1 - S_F^\alpha(x)^2)(D_F^\alpha)^2 \, f(x) - 2S_F^\alpha(x)D_F^\alpha f(x) - n(n+1)f(x) = 0, \quad (7.190)$$

where $\lambda = n(n+1)$ and $r(x) = 1$. Since $p(-1) = p(1) \neq 0$ with the both boundary conditions of Eq.(7.166), we have Strum-Liouville problem on the interval $-1 \leq x \leq 1$. Here, apply the power series method to find its solutions. To do this, we substitute

$$f(x) = \sum_{m=0}^{\infty} a_m [S_F^\alpha(x)]^m, \quad (7.191)$$

and its derivatives into Eq.(7.190). Let $n(n+1) = k$. Then we obtain

$$(1 - S_F^\alpha(x)^2) \sum_{m=2}^{\infty} m(m-1)a_m S_F^\alpha(x)^{m-2}$$

$$- 2S_F^\alpha(x) \sum_{m=1}^{\infty} m a_m S_F^\alpha(x)^{m-1} + k \sum_{m=0}^{\infty} a_m S_F^\alpha(x)^m = 0. \quad (7.192)$$

By writing out each series and arranging each power in a column we obtain

$$\begin{aligned} 2a_2 + n(n+1)a_0 &= 0 && \text{coefficients of } S_F^\alpha(x)^0 \\ 6a_3 + [-2 + n(n+1)]a_1 &= 0 && \text{coefficients of } S_F^\alpha(x)^1, \end{aligned} \quad (7.193)$$

and in general when $m = 2, 3, \ldots,$. Thus, we obtain recursion formula as

$$a_{m+2} = -\frac{(n-m)(n+m+1)}{(m+2)(m+1)} a_m \quad m = 0, 1, \ldots. \quad (7.194)$$

Substituting Eq.(7.194) into Eq.(7.191) we obtain

$$f(x) = a_0 f_1(x) + a_1 f_2(x), \quad (7.195)$$

where

$$f_1(x) = 1 - \frac{n(n+1)}{2!} S_F^\alpha(x)^2$$

$$+ \frac{(n-2)n(n+1)(n+3)}{4!} S_F^\alpha(x)^4 - \cdots, \quad (7.196)$$

and

$$f_2(x) = S_F^\alpha(x) - \frac{(n-1)(n+2)}{3!} S_F^\alpha(x)^3$$

$$+ \frac{(n-3)(n-1)(n+2)(n+4)}{5!} S_F^\alpha(x)^5 - \cdots. \quad (7.197)$$

These independent series solutions converge for $|S_F^\alpha(x)| < 1$.

7.8.2 *Fractal Legendre polynomials*

If n in fractal Legendre's equation be a nonnegative integer [Golmankhaneh (2010)]. Then the right side of Eq.(7.194) is zero if $m = n$, namely, $a_{n+2} = 0, a_{n+4} = 0, a_{n+6} = 0, \dots$. If n is even number, then $f_1(x)$ becomes polynomials of degree n. If n is odd number then the same is true for f_2. These polynomials multiplied by some constants, are called fractal Legendre polynomials. The first few of these functions are as

$$P_0^\alpha(x) = 1 \tag{7.198}$$
$$P_1^\alpha(x) = S_F^\alpha(x)$$
$$P_2^\alpha(x) = \frac{1}{2}(3(S_F^\alpha(x))^2 - 1), \ vdots \tag{7.199}$$

$$P_n^\alpha(x) = \sum_{m=0}^{[n/2]} (-1)^m \frac{(2n-2m)!}{2^n m!(n-m)!(n-2m)!} S_F^\alpha(x)^{n-2m}, \tag{7.200}$$

where $[n/2] = n/2$ for n even, $(n-1)/2$ for n odd.

7.8.3 *Fractal Rodrigues' formula*

The Legendre polynomials can be written as

$$P_n^\alpha(x) = \frac{1}{2^n n!}(D_F^\alpha)^n[((S_F^\alpha(x))^2 - 1)^n]. \tag{7.201}$$

which is called fractal Rodrigues' formula . Fractal Legendre polynomials are orthogonal on the interval $-1 \le x \le 1$, namely,

$$\int_{-1}^{1} P_m^\alpha(x)P_n^\alpha(x)d_F^\alpha x = \frac{2}{2n+1}\delta_{nm}, \tag{7.202}$$

where δ_{nm} denotes the Kronecker delta.

7.8.4 *Fractal Hermite's equation*

Consider fractal Hermite's differential equation as [Golmankhaneh (2010)]:

$$(D_F^\alpha)^2 f(x) - 2S_F^\alpha(x)D_F^\alpha f(x) + 2nf(x) = 0, \ -\infty < x < +\infty, \ n \in R \tag{7.203}$$

with boundary conditions of Eq.(7.166). Multiplying Eq.(7.203) by $\exp((-S_F^\alpha(x))^2/2)$ we get self-adjoint form of Eq.(7.203) as

$$D_F^\alpha[e^{(-S_F^\alpha(x))^2} D_F^\alpha f(x)] + 2ne^{(-S_F^\alpha(x))^2} f(x) = 0. \tag{7.204}$$

By comparing Eq.(7.164), Eq.(7.166) and Eq.(7.204) we have

$$a = -\infty, \quad b = \infty, \quad p(x) = e^{(-S_F^\alpha(x))^2}, \quad q(x) = 0,$$
$$r(x) = e^{(-S_F^\alpha(x))^2}, \quad \lambda = 2n. \tag{7.205}$$

To find solution of Eq.(7.204) using power series method we suppose the solution such as

$$f(x) = \sum_{m=0}^{\infty} a_m [S_F^\alpha(x)]^m. \tag{7.206}$$

If we substitute Eq.(7.206) in Eq.(7.204) we get

$$\sum_{m=2}^{\infty} m(m-1)a_m S_F^\alpha(x)^{m-2} - 2S_F^\alpha(x) \sum_{m=1}^{\infty} m a_m S_F^\alpha(x)^{m-1}$$
$$+ 2n \sum_{m=0}^{\infty} a_m S_F^\alpha(x)^m = 0. \tag{7.207}$$

Therefore our recurrence relations become:

$$(m+2)(m+1)a_{m+2} - 2m a_m + 2n a_m = 0 \quad \text{for} \quad m = 0,1,2,3,\ldots. \tag{7.208}$$

After simplification this becomes

$$a_{m+2} = \frac{2(m-n)}{(m+2)(m+1)} a_m \quad \text{for} \quad \text{all} \quad m = 0,1,2,3,\ldots. \tag{7.209}$$

By inserting these values for the coefficients into Eq.(7.206) we obtain

$$f(x) = a_0 f_1(x) + a_1 f_2(x) \tag{7.210}$$

where

$$f_1(x) = 1 + \frac{2n}{2!} S_F^\alpha(x)^2 + \frac{2n(2n-4)}{4!} S_F^\alpha(x)^4 + \cdots. \tag{7.211}$$
$$= 1 + \sum_{k=1}^{\infty} 2n(2n-4)\cdots(2n-4k+4)\frac{S_F^\alpha(x)^{2k}}{(2k)!},$$

and

$$f_2(x) = S_F^\alpha(x) + \frac{2n-2}{2!} S_F^\alpha(x)^3 + \frac{(2n-2)(2n-6)}{5!} S_F^\alpha(x)^5 + \cdots.$$
$$= S_F^\alpha(x) + \sum_{k=1}^{\infty} (2n-2)(2n-6)\cdots(2n-4k+2)\frac{S_F^\alpha(x)^{2k+1}}{(2k+1)!}. \tag{7.212}$$

These linearly independent series solutions converge for $-\infty < S_F^\alpha(x) < \infty$.

7.8.5   *Fractal Hermite's polynomials*

Let n in fractal Hermite's equation be a nonnegative integer [Golmankhaneh (2010)]. For the case of even number $f_1(x)$ reduces to a polynomials of degree and for the case of odd number the same is true for f_2. These polynomials are called fractal Hermite's polynomials. The first few of these functions are

$$H_0^\alpha(x) = 1 \tag{7.213}$$
$$H_1^\alpha(x) = 2S_F^\alpha(x)$$
$$H_2^\alpha(x) = 4S_F^\alpha(x)^2 - 2$$
$$H_3^\alpha(x) = 8S_F^\alpha(x)^3 - 12S_F^\alpha(x), \tag{7.214}$$
$$\vdots$$

The fractal Hermite polynomials is written explicitly as

$$H_n^\alpha(x) = \sum_{m=0}^{\lfloor n/2 \rfloor} \frac{(-1)^m n!}{m!(n-2m)!} (2S_F^\alpha(x))^{n-2m}, \tag{7.215}$$

where $\lfloor n/2 \rfloor$ is floor function that is defined as

$$\lfloor x \rfloor = \max\{n \in Z \mid n \leq x\}. \tag{7.216}$$

7.8.6   *Fractal Rodrigues' formula*

The fractal Hermite polynomials $H_n^\alpha(x)$ can be expressed by Rodrigues'formula

$$H_n^\alpha(x) = (-1)^n e^{(-S_F^\alpha(x))^2} (D_F^\alpha)^n (e^{(-S_F^\alpha(x))^2}) \quad \text{where} \quad n = 0, 1, 2, \ldots. \tag{7.217}$$

Fractal Hermite's polynomials are orthogonal on the interval $-\infty \leq x \leq \infty$, namely,

$$\int_{-\infty}^{\infty} H_m^\alpha(x) H_n^\alpha(x) e^{(-S_F^\alpha(x))^2} d_F^\alpha x = n! 2^n \delta_{nm}, \tag{7.218}$$

where δ_{nm} denotes the Kronecker delta.

Remark 7.3. Hermite and Legendre Polynomials form a complete orthogonal set.

7.9 Scale transform of the fractal Hermite and the Legendre polynomials

Here we present scale transform of Hermite and Legendre polynomials [Golmankhaneh (2010)]. Let scale x with λx. Then the fractal Legendre will change under that as

$$P_n^\alpha(\lambda x) = \sum_{m=0}^{[n/2]} (-1)^m \frac{(2n-2m)!}{2^n m!(n-m)!(n-2m)!} S_F^\alpha(\lambda x)^{n-2m}. \qquad (7.219)$$

As $S_F^\alpha(\lambda x) = \lambda^\alpha S_F^\alpha(x)$, then we get

$$P_n^\alpha(\lambda x) = \sum_{m=0}^{[n/2]} \lambda^{\alpha(n-2m)} (-1)^m \frac{(2n-2m)!}{2^n m!(n-m)!(n-2m)!} S_F^\alpha(x)^{n-2m}. \qquad (7.220)$$

Likewise for the fractal Hermite's polynomials

$$H_n^\alpha(\lambda x) = \sum_{m=0}^{\lfloor n/2 \rfloor} \frac{(-1)^m n!}{m!(n-2m)!} (2S_F^\alpha(\lambda x))^{n-2m} \qquad (7.221)$$

$$= \sum_{m=0}^{\lfloor n/2 \rfloor} \lambda^{\alpha(n-2m)} \frac{(-1)^m n!}{m!(n-2m)!} (2S_F^\alpha(x))^{n-2m}.$$

7.10 Fractal finite difference and fractal derivative

In this section, we define fractal shift operator and fractal difference operator. Furthermore, the fractal differential equation and its corresponding difference equation is given [Golmankhaneh and Cattani (2019)]. Let us define a α-difference equation on F as follows:

$$z_{F,n+1} = \chi_F B \, z_{F,n},$$

where B is an operator. In view of fractal Taylor expansion , we have

$$z_{F,n+1} = \sum_{k=0}^{\infty} \frac{S_F^\alpha(h)^k}{k!} (D_F^\alpha)^k z(t)|_{t=t_n}$$

$$= e^{S_F^\alpha(h) D_F^\alpha} z_n, \qquad (7.222)$$

where

$$z_{n+1} = z(t_{n+1}),$$
$$t_{n+1} = h + t_n, \quad h \in R, \quad z_n \in F,$$
$$D_F^\alpha z_{F,n} = D_F^\alpha z_F(t)|_{t=t_n}, \qquad (7.223)$$

consequently, we have

$$B = e^{S_F^\alpha(h)D_F^\alpha}. \tag{7.224}$$

In view of Eq.(7.222), B is called fractal shift operator . Fractal difference operator is denoted by Δ_F and defined by

$$\Delta_F = B - 1, \tag{7.225}$$

it follows that

$$\Delta_F z_n = z_{n+1} - z_n. \tag{7.226}$$

Using Eq.(7.224) the fractal α-order derivative in terms of the fractal α-difference equation is

$$\begin{aligned} D_F^\alpha &= \frac{1}{S_F^\alpha(h_n)} \ln B \\ &= \frac{1}{S_F^\alpha(h)} \ln(1 + \Delta_F) \\ &= \frac{1}{S_F^\alpha(h)} \sum_{k=1}^{\infty} (-1)^{k+1} \frac{\Delta_F^k}{k!}. \end{aligned} \tag{7.227}$$

Fractal higher α-order derivatives is approximated by

$$(D_F^\alpha)^m \approx \left(\frac{\Delta_F}{S_F^\alpha(h)} \right)^m. \tag{7.228}$$

Example 7.14. Consider $z(t) = S_F^\alpha(t)$, then using Eq.(7.222) we can write

$$\begin{aligned} S_{F,n+1}^\alpha &= \chi_F B S_{F,n}^\alpha \\ &= \sum_{k=0}^{\infty} \frac{S_F^\alpha(h)^k}{k!} (D_F^\alpha)^k S_F^\alpha(t)|_{t=t_n} \\ &= S_F^\alpha(t_n) + S_F^\alpha(h)\chi_F(t_n), \end{aligned} \tag{7.229}$$

where

$$D_F^\alpha S_{F,n}^\alpha = D_F^\alpha S_F^\alpha(t)|_{t=t_n} = \chi_F(t_n). \tag{7.230}$$

By Eq.(7.226) we have

$$\Delta_F S_{F,n}^\alpha = S_{F,n+1}^\alpha - S_{F,n}^\alpha. \tag{7.231}$$

Using Eq.(7.228) we get

$$\begin{aligned} (D_F^\alpha)^2 S_F^\alpha(t) &= \frac{1}{S_F^\alpha(h)^2} (\Delta_F)^2 S_{F,n}^\alpha \\ &= \frac{1}{S_F^\alpha(h)^2} \Delta_F(S_{F,n+1}^\alpha - S_{F,n}^\alpha) \\ &= \frac{1}{S_F^\alpha(h)^2} (S_{F,n+2}^\alpha - 2S_{F,n+1}^\alpha + S_{F,n}^\alpha). \end{aligned} \tag{7.232}$$

We note that Eq.(7.232) is used to obtain numerical solution of the second α-order fractal differential equations.

7.11 Fractal difference and its relation with fractal differential equations

In this section, we show the relation between fractal difference equations and fractal differential equations which should later be used to find numerical solution of the fractal differential equations [Golmankhaneh and Cattani (2019)].
Consider the following fractal α-difference equation

$$z_{n+1} = \chi_F \mathbf{B} z_n, \quad z_n \in F, \quad \mathbf{B} \in R, \tag{7.233}$$

then the solution of Eq.(7.233) is

$$z_n = \chi_F \mathbf{B}^n z_0. \tag{7.234}$$

If we set $z_n = z(t_n)$, then

$$S_F^\alpha(t_{n+1}) = S_F^\alpha(t_n) + S_F^\alpha(h), \quad t_n, \quad t_{n+1} \in F, \quad h \in (0,1), \tag{7.235}$$

and

$$S_F^\alpha(t_n) = n S_F^\alpha(h). \tag{7.236}$$

It follows that Eq.(7.233) becomes

$$\frac{z(t_n + h) - z(t_n)}{S_F^\alpha(h)} = \frac{\mathbf{B} - 1}{S_F^\alpha(h)} z(t_n), \tag{7.237}$$

then the corresponding fractal differential equation is

$$D_F^\alpha z(t) = \frac{\mathbf{B} - 1}{S_F^\alpha(h)} z(t). \tag{7.238}$$

The solution of Eq.(7.238) is

$$z(t) = z(0) \exp\left(\frac{\mathbf{B} - 1}{S_F^\alpha(h) S_F^\alpha(t)}\right). \tag{7.239}$$

7.12 Numerical method for solving fractal differential equation

In this section, we present an analogue of Euler method in the fractal calculus by using fractal difference operator [Golmankhaneh and Cattani (2019)].
Consider α-order fractal differential equation as

$$D_F^\alpha z(t) = f(S_F^\alpha(t), z(t)), \quad z(t_0) = z_0. \tag{7.240}$$

In view of Eq.(7.228) we obtain

$$z_{n+1}(t) = z_n(t) + S_F^\alpha(h)f(S_F^\alpha(t), z(t)). \tag{7.241}$$

By substituting fractal Taylor expansion of $f(S_F^\alpha(t), z(t))$, in Eq.(7.241) we get approximate analytical solution of Eq.(7.240) as follows:

$$z(t+h) = z(t_0) + S_F^\alpha(h)D_F^\alpha z(t) + \frac{1}{2}S_F^\alpha(h)^2(D_F^\alpha)^2 z(t) + O(S_F^\alpha(h)^3). \tag{7.242}$$

The numerical solution of the fractal differential equation is obtained by utilizing Equation (7.241), which might be called fractal Euler method. Considering conjugacy of ordinary calculus and fractal calculus, the fractal local truncation error (FLTE) is given by

$$FLTE = z(t_0+h) - z(t_1) = \frac{1}{2}S_F^\alpha(h)^2(D_F^\alpha)^2 z(t)|_{t=t_0} + O(S_F^\alpha(h)^3), \tag{7.243}$$

Eq.(7.243) is valid, if we have

$$\forall\, t \in F, \quad \exists\, M > 0, \quad |(D_F^\alpha)^3 z(t)| < M. \tag{7.244}$$

Recalling conjugacy of ordinary calculus and fractal calculus, the fractal global truncation error is given by

$$\forall\, t \in F, \exists\, N > 0, \; |(D_F^\alpha)^2 z(t)| < N, \tag{7.245}$$

and f is the fractal Lipschitz continuous , namely

$$|f(S_F^\alpha(t), z_1) - f(S_F^\alpha(t), z_2)| < Q|z_1 - z_2|. \tag{7.246}$$

In the same manner, the fractal bounded global truncation error (FGTE) is given by

$$|FGTE| \leq \frac{S_F^\alpha(h)N}{2Q}\left(\exp(Q(S_F^\alpha(t) - S_F^\alpha(t_0)) - 1)\right), \tag{7.247}$$

since $S_F^\alpha(h) \leq h^\alpha$ then we obtain

$$|FGTE| \approx h^\alpha. \tag{7.248}$$

For this reason, the fractal Euler method is also called α-order Euler method.

Example 7.15. Consider the following fractal Cauchy problem

$$D_F^\alpha z(t) = z(t), \quad z(0) = 1. \tag{7.249}$$

The exact solution Eq.(7.249) is

$$z(t) = \exp(S_F^\alpha(t)). \tag{7.250}$$

Applying Eq.(7.242), we proceed with corresponding difference equation as follows

$$z_{n+1}(t) = z_n(t) + S_F^\alpha(h)z(t), \tag{7.251}$$

so that by using the fractal Euler method to approximate solution of Eq.(7.249) we get

$$z(t+h) = 1 + S_F^\alpha(h)z(t) + \frac{1}{2}S_F^\alpha(h)^2(D_F^\alpha)^2 z(t) + O(S_F^\alpha(h)^3). \tag{7.252}$$

In Figure 7.6 as well as in the following figures, the smooth colored lines represent the standard results with $\alpha = 1$.

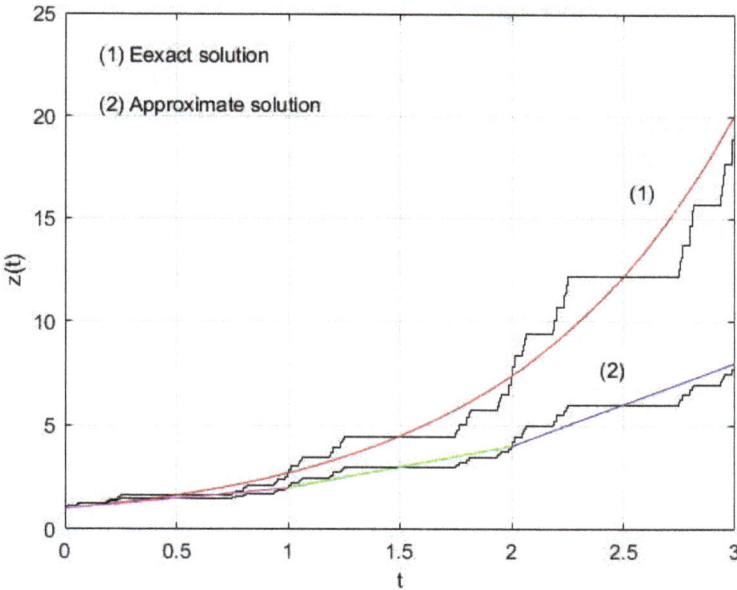

Fig. 7.6: Graph of exact solution and approximate solution using numerical fractal Euler method where step sizes $h = 1$ and $\alpha = 0.63$

Example 7.16. Consider the fractal time varying harmonic oscillator as follows

$$(D_F^\alpha)^2 z(t) + z(t) = 0, \quad z(0) = 1, \quad D_F^\alpha z(t)|_{t=0}. \tag{7.253}$$

In view of conjugacy of fractal and standard calculus, the solution of
Eq.(7.253) is $z(t) = \cos(S_F^\alpha(t))$. The corresponding fractal difference equation is

$$\frac{1}{S_F^\alpha(h)^2}(z_{n+2} - 2z_{n+1} + z_n) + z_n = 0, \qquad (7.254)$$

with initial conditions

$$z_0 = 0, \ z_1 = 0. \qquad (7.255)$$

Using the standard techniques one can get the solution of recurrence relations Eq.(7.254) as follows

$$z_n = \frac{1}{2}((1 + iS_F^\alpha(h))^n + (1 - iS_F^\alpha(h))^n), \qquad (7.256)$$

where z_n is real number and $i = \sqrt{-1}$. By Eq.(7.236) we have

$$z_n = \cos(S_F^\alpha(t_n)) + \frac{1}{2}S_F^\alpha(h)S_F^\alpha(t_n)\cos(S_F^\alpha(t_n)) + O(S_F^\alpha(h)^2). \qquad (7.257)$$

The approximate solution of Eq.(7.254) for the case $S_F^\alpha(t_n)S_F^\alpha(h) \ll 1$ is

$$z_n = \cos(S_F^\alpha(t_n)), \qquad (7.258)$$

and

$$S_F^\alpha(h) \to 0 \Rightarrow z_n = \cos(S_F^\alpha(t_n)), \qquad (7.259)$$

which is expected.

7.13   Laplace equations on fractal cubes

In this section, we give basic tools of calculus for fractal Cantor cubes [Golmankhaneh and Nia (2021)]. Let us start by definition of a Cantor cubes viz $\mathbf{F} = C^\iota \times C^\iota \in R^2$ where C^ι is the thin Cantor-like set and $\times$ indicted Cartesian product. The thin Cantor-like set is generated by deleting fraction of length $(=\iota)$ from the middle of a line and iterating this operation from the previous steps (For more details about thin Cantor-like set (See ref. [DiMartino and Urbina (2014a)]). In Figure 7.7 we present $\mathbf{F} = C^{1/3} \times C^{1/3}$ which is called pre-fractal because of finite iteration of the thin Cantor-like set $(=C^{1/3})$ was used in Cartesian product. The dimension of the thin Cantor-like set is given by:

$$Dim_{C^\iota} = \frac{\log 2}{\log 2 - \log(1 - \iota)}. \qquad (7.260)$$

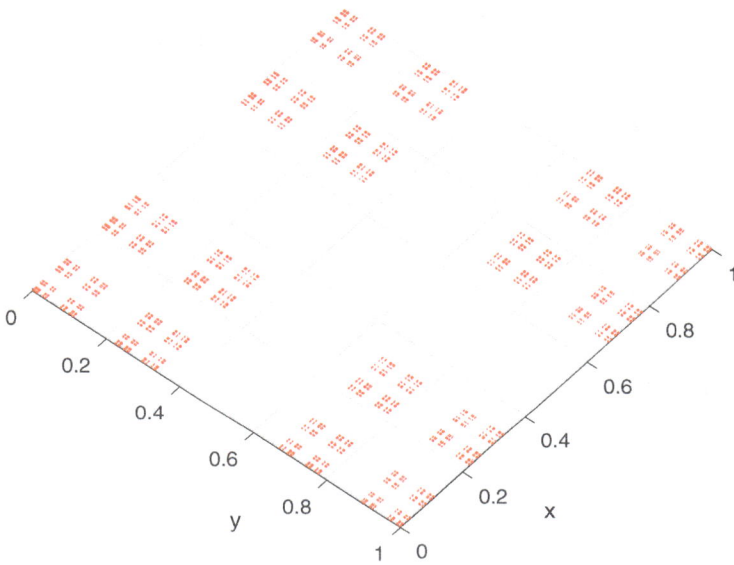

Fig. 7.7: Graph of Cantor cubes which create by Cartesian product two thin Cantor-like with $Dim_{C^\iota} = \alpha = \beta = 0.62$

The dimension of the Cantor cube is defined by [Falconer (2004)]:

$$Dim_{\mathbf{F}} = Dim_{C^\iota} + Dim_{C^\iota} = 2Dim_{C^\iota}. \qquad (7.261)$$

Definition 7.58. The flag function for the Cantor cubes $\mathbf{F}$ is defined as follows:

$$\Omega(\mathbf{F}, J) = \begin{cases} 1, & \text{if } \mathbf{F} \cap J \neq \emptyset; \\ 0, & \text{otherwise}, \end{cases} \qquad (7.262)$$

where $J = [b_1, b_2] \times [c_1, c_2]$ and $b_1, b_2, c_1, c_2 \in R$.

Definition 7.59. A subdivision of the box $J = [b_1, b_2] \times [c_1, c_2]$ is a set of the following form:

$$\mathbf{P}_{[b_1,b_2] \times [c_1,c_2]} = \{b_1 = x_0, x_1, x_2, \ldots, x_n = b_2\} \times \{c_1 = y_0, y_1, y_2, \ldots, y_m = c_2\}, \qquad (7.263)$$

where $\times$ denotes the Cartesian product.

Definition 7.60. The mass function for a Cantor cubes $\mathbf{F}$ and a subdivision $\mathbf{P}_{[b_1,b_2] \times [c_1,c_2]}$ for any $0 < \eta < 1, 0 < \epsilon < 1$ is defined by:

$$\gamma^{\eta,\epsilon}(\mathbf{F}, b_1, b_2, c_1, c_2)$$

$$= \lim_{\delta \to 0} \inf_{\mathbf{P}_{[b_1,b_2] \times [c_1,c_2]}:|\mathbf{P}| \le \delta_1, \delta_2} \sum_{j=1}^{m} \sum_{i=1}^{n} \Gamma(\eta+1)\Gamma(\epsilon+1)(x_i - x_{i-1})^\eta (y_j - y_{j-1})^\epsilon$$

$$\times \Omega(\mathbf{F}, [x_{i-1}, x_i] \times [y_{j-1}, y_j])$$

$$= \lim_{\delta \to 0} \inf_{\mathbf{P}_{[b_1,b_2] \times [c_1,c_2]}:|\mathbf{P}| \le \delta_1, \delta_2} K_{\delta_1, \delta_2}, \tag{7.264}$$

where $|\mathbf{P}|$ is defined by:

$$|\mathbf{P}| = \max_{1 \le i \le n,\, 1 \le j \le m} (x_i - x_{i-1}) \times (y_j - y_{j-1}) \tag{7.265}$$

and $\Gamma(*)$ is standard Gammma function.

Definition 7.61. The integral staircase function $S_{\mathbf{F}}^{\eta,\epsilon}(x,y)$ for the Cantor cubes $\mathbf{F}$ is defined by:

$$S_{\mathbf{F}}^{\eta,\epsilon}(x,y) = \begin{cases} \gamma^{\eta,\epsilon}(\mathbf{F}, b_0, c_0, x, y), & \text{if} \quad x \ge b_0, \; y \ge c_0; \\ -\gamma^{\eta,\epsilon}(\mathbf{F}, b_0, c_0, x, y), & \text{otherwise,} \end{cases} \tag{7.266}$$

where b_0, c_0 are real numbers and fixed. The integral staircase functions for a Cantor cube is plotted in Figure 7.8.

Definition 7.62. The γ_2-dimension of $\mathbf{F} \cap ([b_1, b_2] \times [c_1, c_2])$ is given by:

$$\dim_{\gamma_2}(\mathbf{F} \cap ([b_1, b_2] \times [c_1, c_2])) = \inf\{\max\{\eta, \epsilon\} : \gamma^{\eta,\epsilon}(\mathbf{F}, b_1, b_2, c_1, c_2) = 0\}$$
$$= \sup\{\max\{\eta, \epsilon\} : \gamma^{\eta,\epsilon}(\mathbf{F}, b_1, b_2, c_1, c_2) = \infty\}.$$

In Figure 7.9, we have presented that the intersection point of blue line with surface of $K_{\delta_2', \delta_1'}/K_{\delta_2, \delta_1}$ that leads to dimension of $\mathbf{F}$, namely

$$\dim_{\gamma_2}(\mathbf{F} \cap ([b_1, b_2] \times [c_1, c_2])) = 2Dim_{C^i} \tag{7.267}$$

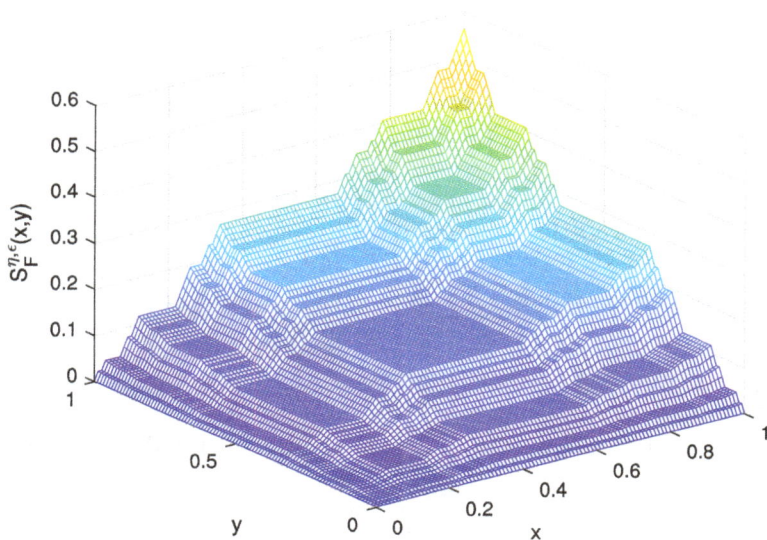

Fig. 7.8: Graph of staircase function of Cantor cubes for the case $Dim_{C^\iota} = \alpha = \beta = 0.62$, shows that fractal measure is not zero while its area is zero

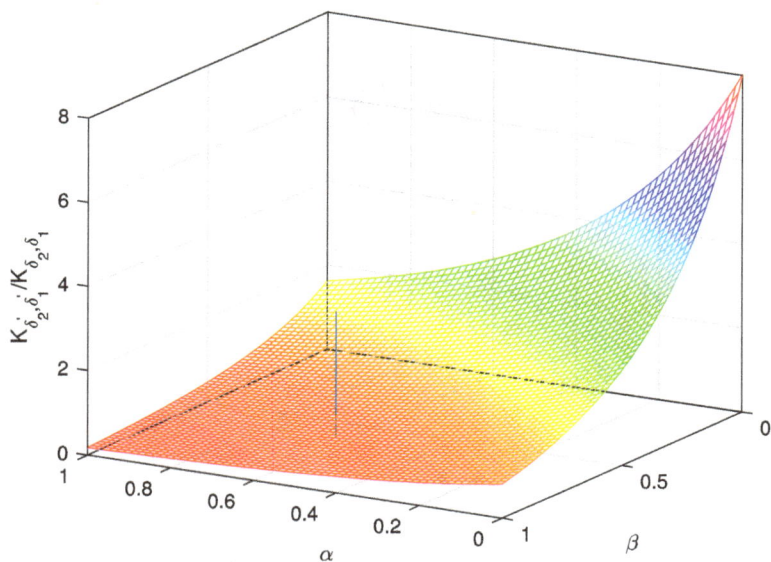

Fig. 7.9: The point of intersection of the blue line and the surface indicates the Cantor cubes viz $Dim_{\mathbf{F}} = 1.2$

Definition 7.63. The characteristic function of Cantor cubes $\mathbf{F}$ is defined by (See Figure 7.10):

$$\chi_F(x,y) = \begin{cases} \Gamma(\eta+1)\Gamma(\epsilon+1), & (x,y) \in \mathbf{F} \\ 0, & \text{othewise.} \end{cases} \tag{7.268}$$

Definition 7.64. The fractal integral (**F**-integral) on Cantor cube $\mathbf{F} \subset R^2$ is defined by:

$$\int_{(b_1,b_2)}^{(c_1,c_2)} h(x,y)d_{\mathbf{F}}^{\eta}x d_{\mathbf{F}}^{\epsilon}y \approx \sum_{j=1}^{m}\sum_{i=1}^{n} h(x,y)\left(S_{\mathbf{F}}^{\eta,\epsilon}(x_i,y_j) - S_{\mathbf{F}}^{\eta,\epsilon}(x_{i-1},y_{j-1})\right). \tag{7.269}$$

If $\mathbf{F}$ is a ζ-perfect set.

Definition 7.65. The **F**-partial derivative of $h(x,y)$ with respect to x is defined as:

$$D_{\mathbf{F},\mathbf{x}}^{\eta}h(x,y) = \begin{cases} \mathbf{F}\text{-}\lim_{(x',y)\to(x,y)} \frac{h(x',y)-h(x,y)}{S_{\mathbf{F}}^{\eta,\epsilon}(x',y)-S_{\mathbf{F}}^{\eta,\epsilon}(x,y)}, & \text{if} \quad (x,y) \in \mathbf{F}, \\ 0, & \text{otherwise,} \end{cases} \tag{7.270}$$

if the limit exists.

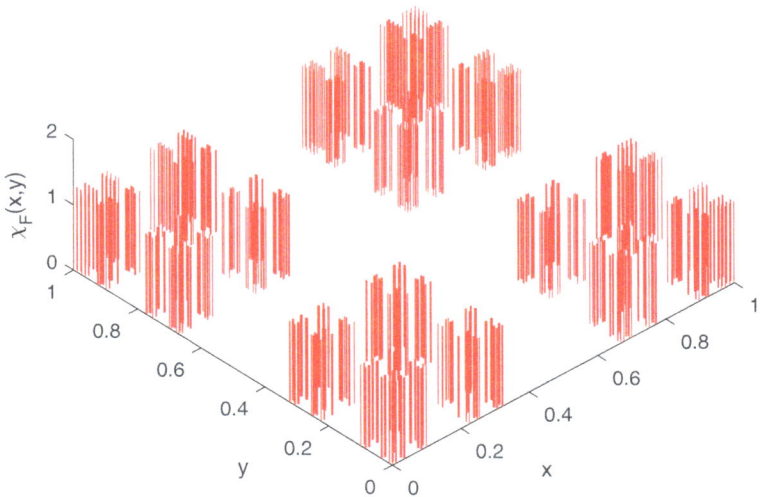

Fig. 7.10: Graph of characteristic function on Cantor cubes for the case of $Dim_{C^{\iota}} = \alpha = \beta = 0.62$

7.14 Laplace equations on fractals

In this section, we suggest the fractal homogeneous and non-homogeneous Laplace equation and their solutions [Golmankhaneh and Nia (2021)].

Example 7.17. Consider the fractal homogeneous Laplace equation with boundary conditions as follows:

$$\nabla^2_{F,K}\Omega = D^{2\alpha}_{F,x}\Omega + D^{2\beta}_{K,y}\Omega = 0, \tag{7.271}$$

with the boundary conditions:

$$\Omega(0,y) = h(y) = S^\alpha_K(y), \quad \Omega(1,y) = 0, \quad \Omega(x,0) = 0, \quad \Omega(x,1) = 0, \tag{7.272}$$

where F, K are fractal sets, namely, $\mathbf{F} = F \times K$. Now, we want to find solution of Eq.(7.271). Let the solution be as follows:

$$\Omega(x,y) = \psi(x)\varphi(y). \tag{7.273}$$

By substituting Eq.(7.273) into Eq.(7.271), we obtain

$$D^{2\alpha}_{F,x}\psi(x) - \kappa\psi(x) = 0, \quad \psi(1) = 0, \tag{7.274}$$

and

$$D^{2\beta}_{K,y}\varphi(y) + \kappa\varphi(y) = 0, \quad \varphi(0) = 0, \quad \varphi(1) = 0. \tag{7.275}$$

It follows that:

$$\varphi_n(y) = \sin\left(\frac{n\pi S^\beta_C(y)}{\Gamma(\beta+1)}\right), \quad \kappa_n = \left(\frac{n\pi}{\Gamma(\beta+1)}\right)^2, \quad n = 1,2,3,\ldots, \tag{7.276}$$

where $S^\beta_C(1) = \Gamma(\beta+1)$. By considering Eq.(7.276) and Eq.(7.274) we have:

$$D^{2\alpha}_{F,x}\psi(x) - \left(\frac{n\pi}{\Gamma(\beta+1)}\right)^2\psi(x) = 0, \quad \psi(1) = 0. \tag{7.277}$$

The solution of Eq.(7.277) by using conjugacy of fractal calculus with the ordinary calculus, is

$$\psi(x) = b_1 \cosh\left(\frac{n\pi S^\alpha_F(x)}{\Gamma(\beta+1)}\right) + b_2 \sinh\left(\frac{n\pi S^\alpha_F(x)}{\Gamma(\beta+1)}\right), \quad \psi(1) = 0. \tag{7.278}$$

or,

$$\psi(x) = b_1 \cosh\left(\frac{n\pi(S^\alpha_F(x) - \Gamma(\alpha+1))}{\Gamma(\beta+1)}\right)$$
$$+ b_2 \sinh\left(\frac{n\pi(S^\alpha_F(x) - \Gamma(\alpha+1))}{\Gamma(\beta+1)}\right), \quad \psi(1) = 0. \tag{7.279}$$

Then applying $\psi(1) = 0$, we obtain:

$$\psi(x) = b_2 \sinh\left(\frac{n\pi(S_F^\alpha(x) - \Gamma(\alpha+1))}{\Gamma(\beta+1)}\right). \tag{7.280}$$

Hence, we get:

$$\Omega_n(x,y) = G_n \sinh\left(\frac{n\pi(S_F^\alpha(x) - \Gamma(\alpha+1))}{\Gamma(\beta+1)}\right)\sin\left(\frac{n\pi S_K^\beta(y)}{\Gamma(\beta+1)}\right), \quad n = 1,2,3,\ldots. \tag{7.281}$$

Then we can write:

$$\Omega(x,y) = \sum_{n=1}^\infty G_n \sinh\left(\frac{n\pi(S_F^\alpha(x) - \Gamma(\alpha+1))}{\Gamma(\beta+1)}\right)\sin\left(\frac{n\pi S_K^\beta(y)}{\Gamma(\beta+1)}\right). \tag{7.282}$$

Utilizing

$$\Omega(0,y) = h(y) = \sum_{n=1}^\infty G_n \sinh\left(\frac{n\pi(-\Gamma(\alpha+1))}{\Gamma(\beta+1)}\right)\sin\left(\frac{n\pi S_K^\beta(y)}{\Gamma(\beta+1)}\right), \tag{7.283}$$

then we arrive at:

$$G_n = \frac{-2}{\pi\sinh\left(\frac{n\pi\Gamma(\alpha+1)}{\Gamma(\beta+1)}\right)}\int_0^1 h(y)\sin\left(\frac{n\pi S_K^\beta(y)}{\Gamma(\beta+1)}\right)d_F^\beta y. \tag{7.284}$$

Using Eq.(7.272) we get:

$$G_n = \frac{-2\Gamma(\beta+1)}{n\pi^2 \sinh\left(\frac{n\pi\Gamma(\alpha+1)}{\Gamma(\beta+1)}\right)}\left[1 - \cos\left(\frac{n\pi}{\Gamma(\beta+1)}\right)\right]. \tag{7.285}$$

Example 7.18. Let us consider the fractal non-homogeneous Laplace equation as follows:

$$D_{F,x}^{2\alpha}Z(x,y) + D_{K,y}^{2\beta}Z(x,y) = \sin(\pi S_K^\alpha(x))\sin(\pi S_F^\beta(y)), \tag{7.286}$$

with the boundary equation

$$Z(x,y) = 0, \quad 0 \le x,y \le 1, \quad D_{F,x}^\alpha Z(0,y) = -\frac{\sin(\pi S_F^\beta(y))}{2\pi}. \tag{7.287}$$

In view of conjugacy of fractal calculus with the ordinary calculus, solution of Eq.(7.287) will be as:

$$Z(x,y) = \frac{\sin(\pi S_F^\alpha(x))\sin(\pi S_F^\beta(y))}{-2\pi^2}. \tag{7.288}$$

The smooth version of solution can be written using $S_F^\alpha(x) < x^\alpha, S_F^\alpha(y) < y^\beta$ as follows:

$$Z(x,y) \approx \frac{\sin(\pi x^\alpha)\sin(\pi y^\beta)}{-2\pi^2}. \tag{7.289}$$

In Figures 7.11 and 7.12, we have presented the solution of Eq.(7.288) for different values of dimension of support of function.

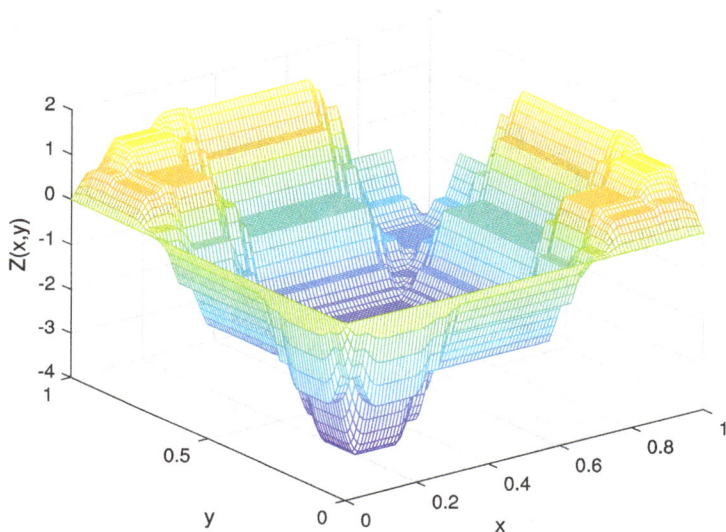

Fig. 7.11: Graph of Eq.(7.288) for the case of $\alpha = \beta = 0.62$

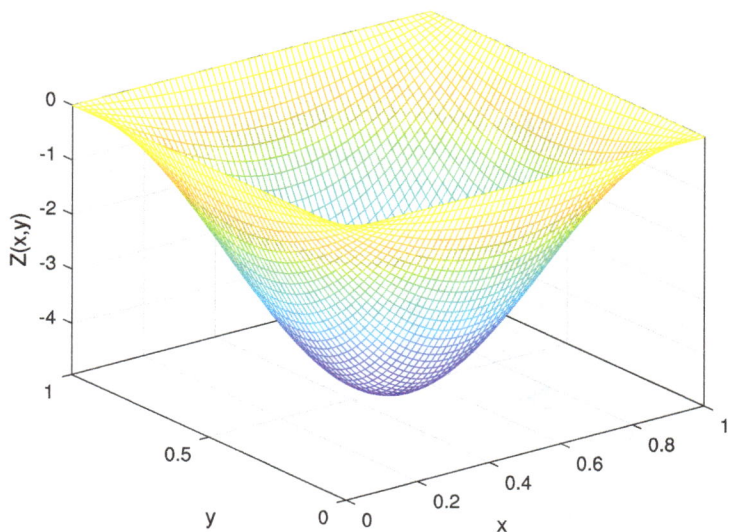

Fig. 7.12: Graph of Eq.(7.288) for the case of $\alpha = \beta = 1$

7.15   Fractal calculus on Cantor tartan spaces

In this section, we define fractal calculus on Cantor tartan spaces [Golmankhaneh and Fernandez (2018)].

Fractals are sets with self-similar properties such that their fractal dimension exceeds their topological dimension. We shall consider calculus on the Cantor tartan, a space $\mathbb{F}$ which is established by first taking the Cartesian product F of a Cantor set and a continuous interval and then taking the union of two orthogonal copies of this F. In other words:

$$\mathbb{F} = F_1 \cup F_2 \subset [0,1]^2, \tag{7.290}$$

$$F_1 = C \times [0,1], \quad F_2 = [0,1] \times C, \tag{7.291}$$

where $C \subset [0,1]$ is a Cantor set of some fractal dimension less than 1 [Falconer (2004)]. We can consider intersections of the Cantor tartan $\mathbb{F}$ with a box $I = [a_1, a_2] \times [b_1, b_2]$, a_1, a_2, b_1, $b_2 \in R$. The sketches in Figure 7.13 show finite iterations (approximations to fractal space) of the Cantor tartan with different dimensions.

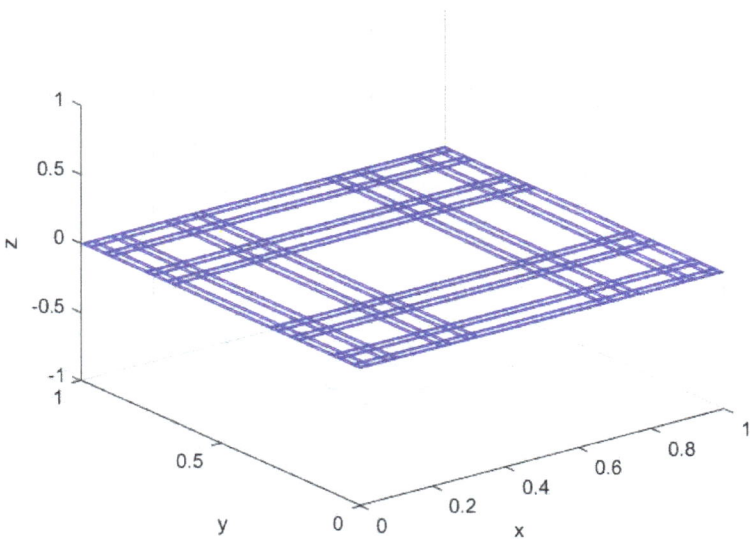

Fig. 7.13: Cantor tartan with dimension 1.63

Definition 7.66. The flag function for the Cantor tartan $\mathbb{F} = F_1 \cup F_2 \subset \Re^2$ and the box $I = [a_1, a_2] \times [b_1, b_2]$, which is denoted by $\Psi(\mathbb{F}, I)$, is defined as follows:

$$\Psi(\mathbb{F}, I) = \begin{cases} 1, & \text{if } \mathbb{F} \cap I \neq \emptyset, \\ 0, & \text{otherwise.} \end{cases} \tag{7.292}$$

A subdivision of the box $I = [a_1, a_2] \times [b_1, b_2]$ is a set of the following form:

$$\mathbb{P}_{[a_1,a_2] \times [b_1,b_2]} = \{a_1 = x_0, x_1, x_2, \ldots, x_n = a_2\} \times \{b_1 = y_0, y_1, y_2, \ldots, y_m = b_2\}, \tag{7.293}$$

where $\times$ denotes the Cartesian product.

Definition 7.67. For a Cantor tartan space $\mathbb{F}$ and a subdivision $\mathbb{P}_{[a_1,a_2] \times [b_1,b_2]}$ as above, and for any η, ϵ between 0 and 1, we define

$$\sigma^{\eta,\epsilon}\left(\mathbb{F}, \mathbb{P}_{[a_1,a_2] \times [b_1,b_2]}\right)$$
$$= \sum_{j=1}^{m} \sum_{i=1}^{n} \Gamma(\eta+1)\Gamma(\epsilon+1)(x_i - x_{i-1})^\eta (y_j - y_{j-1})^\epsilon$$
$$* \Psi(\mathbb{F}, [x_{i-1}, x_i] \times [y_{j-1}, y_j]). \tag{7.294}$$

For a Cantor tartan space $\mathbb{F}$ and box $I = [a_1, a_2] \times [b_1, b_2]$ as above, and for any η, ϵ between 0 and 1, the mass function $\gamma^{\eta,\epsilon}(\mathbb{F}, a_1, a_2, b_1, b_2)$ is defined as

$$\gamma^{\eta,\epsilon}(\mathbb{F}, a_1, a_2, b_1, b_2) = \lim_{\delta \to 0} \left[\inf_{\mathbb{P}_{[a_1,a_2] \times [b_1,b_2]} : |\mathbb{P}| \leq \delta} \sigma^{\eta,\epsilon}\left(\mathbb{F}, \mathbb{P}_{[a_1,a_2] \times [b_1,b_2]}\right) \right], \tag{7.295}$$

where $|\mathbb{P}|$ is defined as

$$|\mathbb{P}| = \max_{1 \leq i \leq n, \ 1 \leq j \leq m} (x_i - x_{i-1}) \times (y_j - y_{j-1}). \tag{7.296}$$

Lemma 7.10. *For a Cantor tartan space $\mathbb{F}$ and fixed $\eta, \epsilon \in (0, 1)$ as above, the mass function $\gamma^{\eta,\epsilon}(\mathbb{F}, a_1, a_2, b_1, b_2)$ is continuous and monotonically increasing in each of the four real variables a_1, a_2, b_1, b_2.*

Lemma 7.11. *For a Cantor tartan space $\mathbb{F}$ and fixed $\eta, \epsilon \in (0, 1)$ as above, the mass function $\gamma^{\eta,\epsilon}(\mathbb{F}, a_1, a_2, b_1, b_2)$ is zero for any a_1, a_2, b_1, b_2 such that $\mathbb{F}$ does not intersect $(a_1, a_2) \times (b_1, b_2)$.*

Proof. If $\mathbb{F}$ does not intersect $I = [a_1, a_2] \times [b_1, b_2]$, then the flag function $\Psi(\mathbb{F}, [x_{i-1}, x_i] \times [y_{j-1}, y_j])$ is zero for all i, j, so we are done.

So we assume that $\mathbb{F}$ intersects the boundary of I. Fix $\nu > 0$; we aim to find a subdivision $\mathbb{P} = \mathbb{P}_{[a_1,a_2]\times[b_1,b_2]}$ such that $\sigma^{\eta,\epsilon}(\mathbb{F},\mathbb{P}) < \nu$. In order to do this, we choose $\mathbb{P}$ such that, in the notation of (7.293), we have

$$(x_1 - x_0)^\eta (y_j - y_{j-1})^\epsilon < \frac{\nu}{m\Gamma(\eta+1)\Gamma(\epsilon+1)}, \tag{7.297}$$

$$(x_n - x_{n-1})^\eta (y_j - y_{j-1})^\epsilon < \frac{\nu}{m\Gamma(\eta+1)\Gamma(\nu+1)}, \tag{7.298}$$

$$(x_i - x_{i-1})^\eta (y_1 - y_0)^\epsilon < \frac{\nu}{n\Gamma(\eta+1)\Gamma(\epsilon+1)}, \tag{7.299}$$

$$(x_i - x_{i-1})^\eta (y_m - y_{m-1})^\epsilon < \frac{\nu}{n\Gamma(\eta+1)\Gamma(\epsilon+1)}, \tag{7.300}$$

for all i and j. Thus, by the definition (7.294), we have $\sigma^{\eta,\epsilon}(\mathbb{F},\mathbb{P}) < \nu$. Since $\epsilon > 0$ was arbitrary, this is enough to prove $\gamma^{\eta,\epsilon}(\mathbb{F},a_1,a_2,b_1,b_2) = 0$ as required. $\square$

Definition 7.68. The integral staircase function $S_{\mathbb{F}}^{\eta,\epsilon}(x,y)$ for the Cantor tartan $\mathbb{F}$ is defined by (See Figure 7.14):

$$S_{\mathbb{F}}^{\eta,\epsilon}(x,y) = \begin{cases} \gamma^{\eta,\epsilon}(\mathbb{F},a_0,b_0,x,y), & \text{if } x \geq a_0, \ y \geq b_0; \\ -\gamma^{\eta,\epsilon}(\mathbb{F},a_0,b_0,x,y), & \text{otherwise}, \end{cases} \tag{7.301}$$

where a_0, c_0 are real numbers chosen according to convenience (e.g. often we might choose $a_0 = c_0 = 0$).

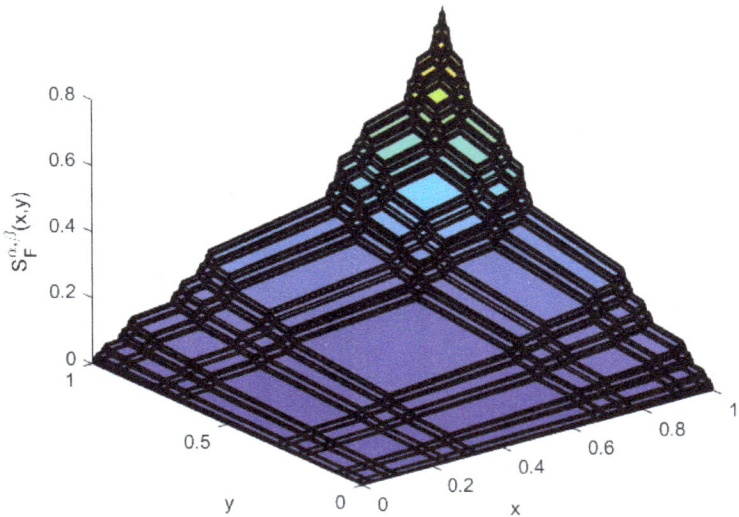

Fig. 7.14: The staircase functions for Cantor tartan with dimension 1.63

Definition 7.69. A point (x, y) is called a point of change of a function $h(x, y)$ if it is not constant over any open set $[a_1, a_2] \times [b_1, b_2]$ containing (x, y). The set of all points of change of $h(x, y)$ is indicated by $Sch(h)$. The set $Sch(S_{\mathbb{F}}^{\eta,\epsilon}(x, y))$ is called ζ-perfect if $S_{\mathbb{F}}^{\eta,\epsilon}(x, y)$ is finite for all $(x, y) \in \mathbb{F}$. If $\mathbb{F}$ is a ζ-perfect set, then we define the F^η-partial derivative of $h(x, y)$ with respect to x as

$$^x D_{\mathbb{F}}^\eta h(x, y) = \begin{cases} \mathbb{F}\text{-}\lim_{(x',y)\to(x,y)} \frac{h(x',y)-h(x,y)}{S_{\mathbb{F}}^{\eta,\epsilon}(x',y)-S_{\mathbb{F}}^{\eta,\epsilon}(x,y)}, & \text{if } (x,y) \in \mathbb{F}, \\ 0, & \text{otherwise,} \end{cases}$$

(7.302)

if the limit exists, where $\mathbb{F}$-lim is defined as the limit taken within the set $\mathbb{F}$. Similarly, we define the F^ϵ-partial derivative of $h(x, y)$ with respect to y as

$$^y D_{\mathbb{F}}^\epsilon h(x, y) = \begin{cases} \mathbb{F}\text{-}\lim_{(x,y')\to(x,y)} \frac{h(x,y')-h(x,y)}{S_{\mathbb{F}}^{\eta,\epsilon}(x,y')-S_{\mathbb{F}}^{\eta,\epsilon}(x,y)}, & \text{if } (x,y) \in \mathbb{F}, \\ 0, & \text{otherwise,} \end{cases}$$

(7.303)

if the limit exists.

Example 7.19. Let us consider the following function, supported on the Cantor tartan with different fractal dimensions:

$$f(x, y) = \sin(x\chi_F^\eta(x)) \sin(y\chi_F^\epsilon(y)), \ (x, y) \in \mathbb{F}, \quad (7.304)$$

where χ_F^η, χ_F^ϵ are characteristic functions for the fractal sets F_1, F_2 whose union is $\mathbb{F}$. (The indices η, ϵ denote the dimension of the respective Cantor sets used to form the sets F_1 and F_2.) The fractal integral of $f(x, y)$ on the Cantor tartan $\mathbb{F} \subset [0, 1] \times [0, 1]$ is as follows:

$$g(x, y)\Big|_{(x=y=1)} = \int_0^x \int_0^y \sin(\chi_F^\eta x) \sin(\chi_F^\epsilon y) d_F^\eta x' d_F^\epsilon y'\Big|_{(x=y=1)}$$

$$= \int_0^1 \Gamma(\epsilon+1) \cos\left(\frac{S_F^\epsilon(y')}{\Gamma(\epsilon+1)}\right) \sin(\chi_F^\eta x)\Big|_0^1 d_F^\eta x'$$

$$= \int_0^1 \left[\Gamma(\epsilon+1) \cos\left(\frac{S_F^\epsilon(1)}{\Gamma(\epsilon+1)}\right) \sin(\chi_F^\eta x)\right.$$

$$\left. - \Gamma(\epsilon+1) \cos\left(\frac{S_F^\epsilon(0)}{\Gamma(\epsilon+1)}\right) \sin(\chi_F^\eta x)\right] d_F^\eta x'$$

$$= \Gamma(\epsilon+1)\Gamma(\eta+1)\left(\cos(1) \cos\left(\frac{S_F^\epsilon(x')}{\Gamma(\eta+1)}\right) - \cos\left(\frac{S_F^\epsilon(x')}{\Gamma(\eta+1)}\right)\right)\Big|_0^1$$

$$= \Gamma(\epsilon+1)\Gamma(\eta+1)(\cos(1)-1)^2, \quad (7.305)$$

since here $S_F^\eta(1) = \Gamma(1+\eta)$, $S_F^\epsilon(1) = \Gamma(1+\epsilon)$, $S_F^\eta(0) = S_F^\epsilon(0) = 0$. We note the following special cases as examples:

$$g(x,y)\bigg|_{(x=y=1)} = \begin{cases} 0.170, \ \eta = \epsilon = 0.63 \ \text{(Cantor tartan of dimension 1.63)} \\ 0.165, \ \eta = \epsilon = 0.43 \ \text{(Cantor tartan of dimension 1.43)} \end{cases} \tag{7.306}$$

Example 7.20. Consider a function with Cantor tartan support as follows:

$$f(x,y) = \sin(x\chi_F^\eta(x) + y\chi_F^\epsilon(y)), \ (x,y) \in \mathbb{F} \tag{7.307}$$

The fractal integral of Eq.(7.307) is as follows:

$$\begin{aligned} g(x,y)\bigg|_{(x=y=1)} &= \int_0^x \int_0^y \sin(\chi_F^\eta x + \chi_F^\epsilon y)d_F^\eta x' \ d_F^\epsilon y'\bigg|_{(x=y=1)} \\ &= \int_0^1 -\Gamma(\epsilon+1)\cos\left(\chi_F^\eta x + \frac{S_F^\epsilon(y)}{\Gamma(\epsilon+1)}\right)\bigg|_0^1 d_F^\eta x' \\ &= \int_0^1 \left[-\Gamma(\epsilon+1)\cos(\chi_F^\eta x + 1) + \Gamma(\epsilon+1)\cos(\chi_F^\eta x)\right] d_F^\eta x' \\ &= -\Gamma(\eta+1)\Gamma(\epsilon+1)\sin\left(\frac{S_F^\eta(x')}{\Gamma(\eta+1)}+1\right) \\ &\quad + \Gamma(\eta+1)\Gamma(\epsilon+1)\sin\left(\frac{S_F^\eta(x')}{\Gamma(\eta+1)}\right)\bigg|_0^1 \\ &= \Gamma(\eta+1)\Gamma(\epsilon+1)[2\sin(1) - \sin(2)] \end{aligned} \tag{7.308}$$

since $S_F^\eta(1) = \Gamma(1+\eta)$, $S_F^\epsilon(1) = \Gamma(1+\epsilon)$, $S_F^\eta(0) = S_F^\epsilon(0) = 0$. We note the following special cases as examples:

$$g(x,y)\bigg|_{(x=y=1)} = \begin{cases} 0.622, \ \eta = \epsilon = 0.63 \ \text{(Cantor tartan of dimension 1.63)} \\ 0.607, \ \eta = \epsilon = 0.43 \ \text{(Cantor tartan of dimension 1.43)} \end{cases} \tag{7.309}$$

Example 7.21. Consider a function on a Cantor tartan space as follows:

$$f(x,y) = S_F^\eta(x)^2 + S_F^\epsilon(y)^2. \tag{7.310}$$

The fractal partial derivatives of $f(x,y)$ with respect to x and y are:

$$D_{F,x}^\eta f(x,y) = 2S_F^\eta(x), \quad D_{F,y}^\eta f(x,y) = 2S_F^\eta(y). \tag{7.311}$$

7.16   New measure based on the staircase function

In this section, we present new measure on fractal sets using staircase function.

Definition 7.70. Let (M, ρ) be a metric space, so we define

$$\mathcal{M}_\delta^\alpha(S) = \inf \left\{ \sum_{i=1}^\infty \Gamma(\alpha+1)(diam\ U_i)^\alpha : \bigcup_{i=1}^\infty U_i \supseteq S,\ \ diam\ U_i < \delta \right\}$$

$$(7.312)$$

where the infimum is over all countable covers of S by sets U_i, and

$$diam\ U = \sup\{\rho(x,y) : x, y \in U\},\ \ diam\ \emptyset = 0. \tag{7.313}$$

Then we can define

$$\mathcal{M}^\alpha(S) = \lim_{\delta \to 0} \mathcal{M}_\delta^\alpha(S) = \sup_{\delta > 0} \mathcal{M}_\delta^\alpha(S), \tag{7.314}$$

which is called the α-dimensional new measure of Borel set S. This leads to definitions of the new dimension as follows:

$$\dim_{\mathcal{M}}(S) = \inf\{\alpha \geq 0 : \mathcal{M}^\alpha(S) = 0\}$$
$$= \sup\left(\{\alpha \geq 0 : \mathcal{M}^\alpha(S) = \infty\} \cup \{0\}\right), \tag{7.315}$$

where $\inf \emptyset = \infty$.

Remark 7.4. The Definition 7.70 can be used to integrate the functions which are not F^α-integrable.

Chapter 8

Applications of Fractal Calculus

In this section, we give applications of fractal calculus in different branches of physics [Satin *et al.* (2013); Satin and Gangal (2019, 2016); Ashrafi and Golmankhaneh (2018); Golmankhaneh and Ashrafi (2017); Golmankhaneh and Balankin (2018)].

8.1 Motion in Fractally distributed medium

If a particle moves in medium with fractal structure to model this motion one can use fractal differential equation. To obtain this we consider one dimensional motion of a particle undergoing friction[Parvate (2009); Parvate and Gangal (2009, 2011)]. If the frictional force is proportional to the velocity the equation of motion nonfractal medium as

$$\frac{dv}{dt} = -k(x)v \tag{8.1}$$

where $k(x)$ is the coefficient of friction. One can rewrite Eq.(8.1) as

$$\frac{dv}{dx}\frac{dx}{dt} = -k(x)v. \tag{8.2}$$

Identifying $dx/dt = v$ and assuming $v \neq 0$, Eq.(8.2) becomes

$$\frac{dv}{dx} = -k(x)v. \tag{8.3}$$

Now, if we fractalize Eq.(8.3) to include fractal underlying medium we have

$$D_F^\alpha v(x) = -k(x) \tag{8.4}$$

where the set F is the support of $k(x)$ which describe the underlying fractal medium , and α is the γ-dimension of F.

Remark 8.1. If F is not α-perfect, then one can choose $\text{Sch}(S_F^\alpha)$ instead.

The function $k(x)$ is called fractional coefficient of friction due to its physical dimensions. The solution of (8.4) is

$$v(x) = v_0 - \int_{x_0}^{x} k(x') d_F^\alpha x' \qquad (8.5)$$

where v_0 and x_0 are the initial velocity and position respectively. For a case $k(x) = \kappa \chi_F(x)$ where κ is a constant Eq.(8.5) reduces to

$$v(x) = v_0 - \kappa(S_F^\alpha(x) - S_F^\alpha(x_0)). \qquad (8.6)$$

The time required for the particle to reach the position x is

$$t(x) = \int_{x_0}^{x} \frac{1}{v(x')} dx' \qquad (8.7)$$

Remark 8.2. We note that the extreme cases of Eq.(8.6) are

(1) If F is empty (frictionless), then we have

$$v(x) = v_0 \qquad (8.8)$$

(2) If $F = R$ (uniform medium) then

$$v(x) = v_0 - \kappa(x - x_0). \qquad (8.9)$$

8.2 Relaxation in glassy materials

Let $\mu(t)$ denote the polarization or magnetic moment in glassy materials which are fractal time processes in relaxation phenomena [Parvate (2009); Parvate and Gangal (2009, 2011)]. Assume that Sch$(\mu) \subset F$ i.e. $\mu(t)$ changes only on a fractal subset F of time, F is α-perfect for some $\alpha \in (0, 1]$,the fractal rate of change of $\mu(t)$ (i.e. the F^α-derivative) is proportional to $\mu(t)$ itself with a negative sign, then we can write

$$D_F^\alpha \mu(t) = -k\mu(t)\chi_F(t) \qquad (8.10)$$

where $k > 0$ is a proportionality constant. To find solution of Eq.(8.10), we apply ϕ both side it so we get

$$\frac{dy}{du} = -ky \qquad (8.11)$$

where $y = \phi[\mu]$, $y(u) = \mu(t)$ for $u = S_F^\alpha(t)$, and $dy/du = D_F^\alpha \mu(t)$. Then the solution of Eq.(8.11) is

$$y = A \exp(-ku). \qquad (8.12)$$

Applying ϕ^{-1} to y, one obtain a solution of Eq.(8.10) as

$$\mu(t) = A \exp(-kS_F^\alpha(t)). \qquad (8.13)$$

This solution evolves on the fractal F. For the case of Cantor set S_F^α is bounded above and below by constant multiples of t^α. We know that the relaxation function is defined by

$$\omega(t) = \frac{\mu(t)\mu(0) >}{< \mu^2(0) >} \tag{8.14}$$

and empirically has the form

$$\omega(t) = \exp[-(t/\tau)^\tau], \quad 0 < \alpha < 1 \tag{8.15}$$

Therefore, Eq.(8.10) is a possible model for such processes.

8.3 A fractal time diffusion equation

Let us consider the fractal diffusion equation as [Parvate (2009); Parvate and Gangal (2009, 2011)]:

$$D_{F,t}^\alpha W(x,t) = \frac{\chi_F(t)}{2} \frac{\partial^2}{\partial x^2} W(x,t), \tag{8.16}$$

where the density $W(x,t), (x,t) \in R \times F$, $D_{F,t}^\alpha$ denotes the partial F^α-derivative with respect to time t. The solution of Eq.(8.16) by using conjugacy is

$$W(x,t) = \frac{1}{(2\pi S_F^\alpha(t))^2} \exp(\frac{-x^2}{2S_F^\alpha(t)}, \quad W(x,0) = \delta(x). \tag{8.17}$$

This is recognized as a sub-diffusive solution. For the Cantor set S_F^α is known to be bounded by kt^α, k constant.

Remark 8.3. We note some researcher has studied anomalous-diffusion using Fractional Brownian motion model and Fractal-time processes . While the Fractional Brownian motion model is local in time but the Fractal-time model is nonlocal. Both are non-Markovian . Meanwhile, the Eq.(8.16) model is Markovian . More, Fractional Brownian motion model is modeled by the following diffusion equation

$$\frac{\partial}{\partial t} P(x,t) = \alpha D t^{\alpha-1} \frac{\partial^2}{\partial x^2} P(x,t), \tag{8.18}$$

The solution of Eq.(8.18) is

$$P(x,t) = \frac{1}{(4\pi D t^\alpha)^{\frac{1}{2}}} \exp\left(\frac{x^2}{4D t^\alpha}\right) \tag{8.19}$$

which is describing Gaussian transport . The $W(x,t)$ using $at^\alpha \le S_F^\alpha(t) \le bt^\alpha$ leads to $P(x,t)$ which is Markovian and causal. We must note that local operator equations are suitable to formulate causal behaviour.

8.4   Fokker-Plank equation on fractal curves

In this section we present the fractal Fokker-Plank equation that was obtained using the Chapmann-Kolmogorov equation. Fractal Fokker-Plank equation is a model for diffusion processes in the disorder media which have self-similar properties and fractal structures. Fractal diffusion equation is solved with local initial condition.

Definition 8.1. If a particle randomly movies on the fractal curves F and its probability to be at $\theta \in F$ at time t is $V(\theta, t)$. Then the probability of the particle to be in the segment $C(a, b)$ is defined by

$$\int_{C(a,b)} V(\theta, t) d_F^\alpha \theta, \tag{8.20}$$

Theorem 8.1. *If the Chapmann-Kolmogorov equation on fractal curves is given by*

$$V(\theta, t + \tau) = \int_{C(a,b)} P(\theta, t + \tau | \theta', t) V(\theta', t) d_F^\alpha \theta', \tag{8.21}$$

where $\theta, \theta' \in F$, and $P(\theta, t + \tau | \theta', t)$ is transition probability of the particle to move from point θ at time t to point θ' at time $t + \tau$. Then, the fractal Fokker-Planck equation is

$$\frac{\partial V(\theta, t)}{\partial t} = -D_F^\alpha \big(A^{(1)}(\theta, t) V(\theta, t) \big) + D_F^{2\alpha} \big(A^{(2)}(\theta, t) V(\theta, t) \big), \tag{8.22}$$

where $A^{(1)}$ and $A^{(2)}$ are called the fractal drift and fractal diffusion coefficient , respectively.

Proof. Let $\Delta \equiv \Delta(\theta, \theta') = J(\theta) - J(\theta')$. Then the intergrand of Eq.(8.21) turns to

$$P(\theta, t + \tau | \theta', t) V(\theta', t) = P(J^{-1}(J(\theta) - \Delta + \Delta), t + \tau | J^{-1}(J(\theta) - \Delta), t) \\ \times V(J^{-1}(J(\theta) - \Delta), t). \tag{8.23}$$

The Taylor expansion of integrand of Eq.(8.21) gives

$$V(\theta, t + \tau) - V(\theta, t) = \int \sum_{n=1}^{\infty} \frac{(-1)^n}{n!} \Delta^n (D_F^\alpha |_\theta)^n \{ P(J^{-1}(J(\theta) \\ + \Delta), t + \tau | \theta, t) V(\theta, t) d_F^\alpha \theta'. \tag{8.24}$$

In view of conjugacy of fractal calculus with ordinary calculus we can write

$$\phi_\theta H(J(\theta), \theta') = H(\theta, \theta'),$$
$$\phi_{\theta'} H(\theta, J(\theta')) = H(\theta, \theta'), \tag{8.25}$$

Let $y = J(\theta)$, $y = J(\theta')$ and suppose $S_F^\alpha(a_0) \leq y, y' \leq S_F^\alpha(a_0)$. By setting

$$\tilde{\Delta}(\theta, y') = \phi_{\theta'} \Delta(\theta, \theta')$$
$$\Delta'(y, y') = \phi_\theta \tilde{\Delta} = \phi_\theta \circ \phi_{\theta'} \Delta(\theta, \theta'),$$
(8.26)

we have

$$\tilde{\Delta}(\theta, y') = J(\theta) - y', \quad \text{and} \quad \Delta'(y, y') = y - y'.$$
(8.27)

Let $V' = \phi_\theta(V)$ and $P' = \phi_{\theta'} \circ \phi_\theta P$. By conjugacy of fractal calculus with ordinary calculus Eq.(8.24) becomes

$$V'(y, t + \tau) - V'(y, t) = \sum_{n=1}^{\infty} \frac{(-1)^n}{n!} \int_{S_F^\alpha(a_0)}^{S_F^\alpha(b_0)} (\Delta')^n$$
$$\left(\frac{\partial}{\partial y} \right)^n \{ P'(y + \Delta', t + \tau | y, t) V'(y, t) dy' \quad (8.28)$$

By setting $dy' = -d\Delta$ it follows that

$$V'(y, t + \tau) - V'(y, t) = - \sum_{n=1}^{\infty} \frac{(-1)^n}{n!} \int_{y - S_F^\alpha(a_0)}^{y - S_F^\alpha(b_0)} (\Delta')^n$$
$$\left(\frac{\partial}{\partial y} \right)^n \{ P'(y + \Delta', t + \tau | y, t) V'(y, t) d\Delta \quad (8.29)$$

The transitional moments are given by

$$\tilde{M}_n(y, t, \tau) = \phi_{\theta'} M_n(\theta', t, \tau)$$
$$= \int_{y - S_F^\alpha(b_0)}^{y - S_F^\alpha(a_0)} (\Delta')^n P'(y + \Delta', t + \tau | y, t) d\Delta' \quad (8.30)$$

Therefore

$$M_n(\theta', t, \tau) = \int_{C(a_0, b_0)} (J(\theta) - J(\theta'))^n P(\theta, t + \tau | \theta', t) d_F^\alpha \theta \quad (8.31)$$

By substituting Eq.(8.30) into Eq.(8.29), and using conjugacy we have

$$V(\theta, t + \tau) - V(\theta, t) = \sum_{n=1}^{\infty} (-D_F^\alpha)^n \left\{ \frac{M_n(\theta, t, \tau)}{n!} V(\theta, t) \right\} \quad (8.32)$$

By expanding of $\tilde{M}_n$ and M_n in Taylor series, we get

$$\frac{\tilde{M}_n(y, t, \tau)}{n!} = (\tilde{A})^n(y, t)\tau + O(\tau^2) \quad (8.33)$$

and

$$\frac{\tilde{M}_n(\theta, t, \tau)}{n!} = A^n(\theta, t)\tau + O(\tau^2) \quad (8.34)$$

where the term of order τ^0 vanishes since for $\tau = 0$ the transition probability is $P'(y, t|y', t) = \delta(y - y')$ that leads to vanishing moments. By taking only the linear terms of τ we have the Kramers-Moyal expansion

$$\frac{\partial}{\partial t}V'(y, t) = \sum_{n=1}^{\infty}\left(-\frac{\partial}{\partial y}\right)^n \{(\tilde{A})^n(y, t)V'(y, t)\} \qquad (8.35)$$

and its conjugacy is

$$\frac{\partial}{\partial t}V'(y, t) = \sum_{n=1}^{\infty}(-D_F^\alpha)^n\{(A)^n(\theta, t)V'(\theta, t)\} \qquad (8.36)$$

Then by considering $n = 2$ we have

$$\frac{\partial}{\partial t}V'(y, t) = (-D_F^\alpha)\{(A)^1(\theta, t)V'(\theta, t)\} + (-D_F^\alpha)^2\{(A)^2(\theta, t)V'(\theta, t) \qquad (8.37)$$

which completes the proof. □

Example 8.1. Consider the fractal Fokker-Planck equation as follows

$$\frac{\partial}{\partial t}V'(y, t) = (-D_F^\alpha)\{(A)^1(\theta, t)V'(\theta, t)\} + (-D_F^\alpha)^2\{(A)^2(\theta, t)V'(\theta, t) \qquad (8.38)$$

with the Gaussian transition probability

$$P(\theta, t + \tau|\theta', t) = \frac{1}{\sqrt{\pi\tau}}\exp\left\{-\frac{(J(\theta) - J(\theta))^2}{\tau}\right\} \qquad (8.39)$$

For solving Eq.(8.38), we can write Eq.(8.31) as

$$M_n(\theta', t, \tau) = \frac{1}{\sqrt{\pi\tau}}\int_{C(a_0, b_0)}(J(\theta) - J(\theta'))^n$$
$$\times \exp\left\{-\frac{(J(\theta) - J(\theta'))^2}{\tau}\right\}d_F^\alpha\theta. \qquad (8.40)$$

The conjugate equation for Eq.(8.40) is

$$\tilde{M}_n(y', t, \tau) = \frac{1}{\sqrt{\pi\tau}}\int_{S_F^\alpha(a_0)}^{S_F^\alpha(b_0)}(y - y')^n\exp\left\{-\frac{(y - y')^2}{\tau}\right\}dy. \qquad (8.41)$$

where $S_F^\alpha(a_0) \ll y'$ and $S_F^\alpha(b_0) \gg y'$. For the cases $S_F^\alpha(a_0) \to -\infty$ and $S_F^\alpha(b_0) \to +\infty$ we have $\tilde{M}_1 = 0$ and $\tilde{M}_2 = 0$. Let $n = 1$ and $n = 2$. Then using Eq.(8.33) we obtain $\tilde{A}^{(1)} = 0$, $\tilde{A}^{(2)} = \frac{1}{4}$ so the first term on the RHS of Eq.(8.37) vanishes and A^2 is a constant. Therefore we arrive at

$$\frac{\partial}{\partial t}V(\theta, t) = A(D_F^\alpha)^2 V(\theta, t) \qquad (8.42)$$

which is called the diffusion equation on the fractal curve F with fractal diffusion coefficient A. The conjugacy of Eq.(8.42) is

$$\frac{\partial V'(y,t)}{\partial t} = A\frac{\partial^2}{\partial y^2}V'(y,t), \tag{8.43}$$

where $V' = \phi[V]$. Given the initial condition $V'(y,0) = \delta(y)$ the solution of Eq.(8.43) will be

$$V'(y,t) = \frac{1}{\sqrt{2\pi At}}\exp\left(-\frac{y^2}{2At}\right) \tag{8.44}$$

By applying ϕ^{-1} we get

$$V(\theta,t) = \frac{1}{2\pi At}\exp\left(-\frac{J(\theta)^2}{2At}\right) \tag{8.45}$$

where $J(\theta) = S_F^\alpha(t)$. The second absolute moment $< L^2 >$ using Eq.(8.45) is given by

$$< L^2 >= \int_{C(c_1,c_2)} L(\theta)^2 V(\theta,t)d_F^\alpha\theta \sim t^\mu. \tag{8.46}$$

By $\mu = 1/\alpha$ we can characterize the anomalous behavior, namely, if

$$\mu < 1 \Rightarrow 1/\alpha < 1, \tag{8.47}$$

then we have subdiffusive behavior e.g. for the von Koch curve $\mu = 0.794$.

8.5 Random walk on fractal curves

In this section, we present discrete random walk on fractal curves and its continuum limit which is called diffusion equation on fractal curve.

8.5.1 *Discrete random walk on fractal curves*

At the Nth step of the simple random walk on a fractal curve is defined by using recursion relation as follows

$$C(N,\theta,\theta') = C(N-1;\theta,J^{-1}(J(\theta')-\Delta)) + C(N-1;\theta,J^{-1}(J(\theta')+\Delta)), \tag{8.48}$$

where $C(N,\theta,\theta')$ is the number of walks start at θ and end in θ'. For the case of large N and small Δ one can write as follows:

$$C(N+1,\theta,\theta') = \{C(N;\theta,J^{-1}(J(\theta')-\Delta)) + C(N;\theta,J^{-1}(J(\theta')+\Delta)) - 2C(N,\theta,\theta')\} + 2C(N,\theta,\theta'). \tag{8.49}$$

Using Taylor expanding , Eq.(8.49) can be rewritten as

$$C(N+1,\theta,\theta') = \left(\Delta^2 D^{2\alpha}_{F_{\theta'}} C(N,\theta,\theta') + \frac{1}{12}\Delta^4 D^{4\alpha}_{F_{\theta'}} C(N,\theta,\theta') + \cdots\right)$$
$$+ 2C(N,\theta,\theta')$$
$$\sim \Delta^2 D^{2\alpha}_{F_{\theta'}} C(N,\theta,\theta') + O(\Delta^4) + 2C(N,\theta,\theta'). \qquad (8.50)$$

Then we have

$$C(N+1,\theta,\theta') - 2C(N,\theta,\theta') \sim \Delta^2 D^{2\alpha}_{F_{\theta'}} C(N,\theta,\theta'). \qquad (8.51)$$

Under the above assumptions, one can write

$$P(N,\theta,\theta') = \frac{1}{2^N} C(N,\theta,\theta'). \qquad (8.52)$$

which is the probability of random walker to start from θ and arrive at θ' after N steps. Then we have

$$C(N,\theta,\theta') = 2^N P(N,\theta,\theta'),$$
$$C(N+1,\theta,\theta') = 2^{N+1} P(N+1,\theta,\theta'). \qquad (8.53)$$

Substituting Eq.(8.53) into (8.51), one can obtain

$$P(N+1,\theta,\theta') - P(N,\theta,\theta') \sim \frac{\Delta^2}{2} D^{2\alpha}_{F_{\theta'}} P(N,\theta,\theta'), \qquad (8.54)$$

By considering limit case of Eq.(8.54) we obtain:

$$\frac{\partial}{\partial N} P(N,\theta,\theta') = \frac{\Delta^2}{2} D^{2\alpha}_{F_{\theta'}} P(N,\theta,\theta'). \qquad (8.55)$$

By conjugacy between fractal calculus and ordinary calculus, it follows immediately that

$$P(N,\theta,\theta') = \frac{1}{\sqrt{2\pi N}} \exp\left(\frac{-(J(\theta)-J(\theta'))^2}{2\Delta^2 N}\right). \qquad (8.56)$$

For a continuum case let $t = N\tau$ where N is number of steps in the discrete walk and τ is the duration between the consecutive steps. By this substituting Eq.(8.56) changes to Eq.(8.45). For calculating first and second moments let $L(\theta) = L(\mathbf{w}(u)) = |\mathbf{w}(u)|$ which is Euclidean distance from the origin on the fractal curve upto a point $\theta = \mathbf{w}(u)$. For the random walk starting from origin ($\theta' = 0$), one can rewrite Eq.(8.56) by setting $N \to \infty$ and $\Delta \to 0$ as follows:

$$P(\mathbf{w}(t) = \theta) = \frac{1}{\sqrt{2\pi At}} \exp\left(-\frac{(J(\theta) = S^\alpha_F(u))^2}{2At}\right), \qquad (8.57)$$

which is similar to Eq.(8.45). Thus the first moment can be obtained by using the following way

$$< L >= \int_{C(-\infty,+\infty)} L(\theta)P(\theta)d_F^{\alpha}\theta. \tag{8.58}$$

Replacing Eq.(8.57) into Eq.(8.58), one can arrive at

$$< L >= 2\frac{1}{\sqrt{2\pi At}} \int_{C(0,+\infty)} L(\theta) \exp\left(-\frac{(J(\theta))^2}{2At}\right) d_F^{\alpha}\theta. \tag{8.59}$$

In view of the following equations

$$c_1[L(\mathbf{w}(u)]^{\alpha} < S_F^{\alpha}(u) < c_2[L(\mathbf{w}(u)]^{\alpha}$$
$$S_F^{\alpha}(u) \sim L(\mathbf{w}(u)]^{\alpha} \sim L^{\alpha}, \tag{8.60}$$

one can rewrite Eq.(8.58), by using Eq.(8.60) in the following form

$$< L >= 2\frac{1}{\sqrt{2\pi At}} \int_{C(0,+\infty)} L \exp\left(-\frac{L^{2\alpha}}{2At}\right) dL^{\alpha}. \tag{8.61}$$

By change of variable $L^{\alpha}/\sqrt{2\pi At} = z$ in Eq.(8.61) we have

$$< L >= Const.t^{1/2\alpha} \int_0^{\infty} z^{1/\alpha} \exp(-z^2)dz \tag{8.62}$$

which leads to

$$< L > \sim t^{1/2\alpha}. \tag{8.63}$$

In the case of von Koch curve $1/\alpha = 0.79$ and $1/2\alpha = 0.396$.

8.5.2 *Random walks with Non-Gaussian distributions on fractals*

The non-Gaussian stable distributions random walks in the Fourier space on the fractal curve is suggested by

$$L_{0,\mu}(\psi) = \exp(-|J(\psi)|^{\mu}), \tag{8.64}$$

where $\psi = \mathbf{w}(k)$ and $S_F^{\alpha}(k) = J(\psi)$. The fractal inverse Fourier transform of Eq.(8.64), gives

$$\tilde{L}_{0,\mu}(\theta) = \frac{1}{2\pi} \int_{C(-\infty,\infty)} \exp(iJ(\psi)J(\theta) - |J(\psi)|^{\mu})d_F^{\alpha}\psi, \tag{8.65}$$

The conjugate of Eq.(8.65) is

$$\bar{L}_{0,\mu}(y = J(\theta)) = \frac{1}{2\pi} \int_{-\infty}^{\infty} \exp(i\tilde{k}y - |\tilde{k}|^{\mu})d\tilde{k} \tag{8.66}$$

For $\mu = 2$ one can obtain the Gaussian distribution as

$$\bar{L}_{0,\mu}(y = J(\theta)) = \frac{1}{\sqrt{2\pi}} \exp\left(-\frac{y^2}{2}\right) \tag{8.67}$$

Applying the fractalizing operator ϕ^{-1} to Eq.(8.67) we get

$$\tilde{L}_{0,\mu}(\theta) = \frac{1}{\sqrt{2\pi}} \exp\left(-\frac{(J(\theta) = S_F^\alpha(u))^2}{2}\right) \tag{8.68}$$

Likewise for the Cauchy case when $\mu = 1$ we have

$$\tilde{L}_{0,\mu}(\theta) = \frac{1}{\pi(1 + (J(\theta) = S_F^\alpha(u))^2)} \tag{8.69}$$

Now by discretizing the integral in Eq.(8.66) by replacing $d\tilde{k}$ in terms of equal steps of size $\Delta\tilde{k}$, we obtain

$$\bar{L}_{0,\mu}(y = J(\theta)) = \sum_{m=-\infty}^{\infty} \exp[i\tilde{k}_m y - |\tilde{k}_m|^\mu]\Delta\tilde{k} \tag{8.70}$$

where $\tilde{k}_m = m\Delta\tilde{k}$. If $\mu \neq 1$ or $\mu \neq 2$ the expansion of Eq.(8.70) for large argument is given by

$$\bar{L}_{0,\mu}(y = J(\theta)) = \frac{1}{\pi} \sum_{m=1}^{\infty} (-1)^{m+1} \frac{y^{-(\mu m+1)}}{m!} \Gamma(1 + m\mu) \sin(\pi\mu m/2) \tag{8.71}$$

The leading term of Eq.(8.71) $(m = 1)$ is

$$\bar{L}_{0,\mu}(y = J(\theta)) = \frac{1}{\pi y^{(\mu+1)}} \Gamma(1 + \mu) \sin(\pi\mu m/2) \tag{8.72}$$

Applying the fractalizing transformation to Eq.(8.72) we get

$$\tilde{L}_{0,\mu}(y = J(\theta)) = \frac{1}{\pi S_F^\alpha(u)^{(\mu+1)}} \Gamma(1 + \mu) \sin(\pi\mu m/2)$$
$$\sim L^{-\alpha(\mu+1)} \tag{8.73}$$

for large values of $J(\theta) = S_F^\alpha(u)$ or L where L is the Euclidean distance . For large values of L, one can see that

$$< L > = \begin{cases} \text{infinite,} & \mu \leq 1/\alpha; \\ \text{finite,} & \mu > 1/\alpha. \end{cases} \tag{8.74}$$

and the second moment is

$$< L^2 > = \begin{cases} \text{diverges,} & \mu \leq 2/\alpha; \\ \text{finite,} & \mu > 2/\alpha. \end{cases} \tag{8.75}$$

We note that a scaling down of the absolute moments for Levy distributions due to the fractal structure of the underlying space. Eq.(8.65) indicates the behaviour of the stable law $L_{0,\mu}$ for values of θ for which $S_F^\alpha(t)$ is large. This behaviour is deferent from ordinary law for large deviation, when the underlying space is not fractal when the stable law decreases as $y^{-(1+\mu)}$. Thus we see that from a Euclidean perspective, the behaviour gets a direct contribution from the exponent of the distribution and the dimension of the curves, which is a striking difference. By taking straight line approximations on fractal curves, the fractal dimension contributes only indirectly. While using fractal calculus, the fractal dimension plays a direct role in changing the nature of heavy tails of the distribution.

8.5.3 *First passage time on fractal curves*

The first passage time , or the time required to reach a certain point $\theta = w(t)$ on the curve F.

Definition 8.2. The first passage time $T(\theta)$ to a point $\theta \in F$ is defined by

$$T(\theta) = \inf\{t : \Theta(t) = \theta\} \tag{8.76}$$

where $\Theta(t)$ is the position of a random walker on F at time t, which started at a point θ_0 at time $t = 0$.

Here we present the relation between the first passage time and the mass covered by the random walker on the fractal curve in this time. Let the walker performs forward and backward steps along the curve. The mass covered in the nth step is given by $\gamma_F^\alpha(F, u_n, u_{n-1})$ corresponding to parameter value lying between u_{n-1} and u_n. Considering on the S_F^α-axis, the resultant displacement of the particle performing a random walk of M steps, some of which is in forward direction and rest in the backward direction on the curve, so as to reach an arbitrary point $\mathbf{w}(u_k)$ on F, is given by sum of γ_F^α over M steps. For $k < M$ such that $M = k + r$, and step size Δ, for every sequence $\{u_1, u_2, \ldots, u_M\}$ such that

$$S_F^\alpha(u_M) = k\Delta. \tag{8.77}$$

Let τ denote the time between successive steps on the fractal curve F. The maximum mass that a random walker covers in a given time t, is $t = M\tau$ where M is the total number of steps taken in time t. By using $k = M - r$ we can rewrite Eq.(8.77) as

$$S_F^\alpha(u_M) = \left(\frac{t}{\tau} - r\right)\Delta \tag{8.78}$$

It is clear that $S_F^\alpha(u_M)$ is maximum when $r = 0$ on the RHS of Eq.(8.78). Hence

$$S_F^\alpha(u_M)|_{max} = \frac{t}{\tau}\Delta \qquad (8.79)$$

Next we want to calculate the minimal time required by a random walker to cover certain mass on the curve. Conversely, let $t = M\tau$ be the time taken for M steps on the curve F, and $M = k + r$, then

$$k = \frac{t - r\tau}{\tau} \qquad (8.80)$$

and one can rewrite Eq.(8.77) as

$$S_F^\alpha(u_M) = \frac{t - r\tau}{\tau}\Delta \qquad (8.81)$$

Thus,

$$t = r\tau + \frac{\tau S_F^\alpha(u_M)}{\Delta} \qquad (8.82)$$

Thus we see that t is minimum when $r = 0$, so we have

$$t_{min} = \frac{\tau}{\Delta}S_F^\alpha(u_M). \qquad (8.83)$$

8.6 Langevin equation on fractal curves

If we consider a particle performing random motion on F, then such a motion in the continuum limit can be described by a Langevin equation . The velocity of a particle moving on the fractal curve is denoted by α-velocity and is defined by

$$\begin{aligned}
v^\alpha(t) &= \lim_{t \to t'} \frac{S_F^\alpha(u(t) - S_F^\alpha(u(t')}{t - t'} \\
&= \frac{d}{dt}S_F^\alpha(u(t)),
\end{aligned} \qquad (8.84)$$

where t is time. If $J(\theta) = S_F^\alpha(u(t))$, then the Langevin equation is defined in the over-damped case as follows:

$$\frac{dJ(\theta(t))}{dt} = \eta(t), \qquad (8.85)$$

where $\eta(t)$ is the noise in the system. The solution of Eq.(8.85) is in the following form

$$J(\theta(t)) = \int_0^t \eta(t')dt'. \qquad (8.86)$$

8.6.1 *Model of noise*

Consider a Lavey distributed noise $p(\eta)$ as the following

$$p(\eta) = \mu \eta_0^\mu \eta^{-1-\mu} \tag{8.87}$$

where η_0 is a lower cutoff, introduced to ensure normalization of the distribution $p(\eta)$. Fourier transform of Eq.(8.87) gives

$$p(k) = < e^{(-ik\eta)} > = \int d\eta \exp(-ik\eta) p(\eta)$$
$$= \exp(-D\eta_0^\mu |k|^\mu), \tag{8.88}$$

where D is a dimensionless geometric factor , $0 < \mu < 2$ is the scaling index . The $\eta(t)$ in Eq.(8.85) is the instantly correlated Levy white noise at a particular instant of time. The microscopic steps η_i with distribution $p(\eta_i)$ are discrete. Then the corresponding difference equation for Langevin equation can be written in terms of discrete time steps $\Delta = t/n$ where n is the number of steps or divisions on the time axis. The Langevin equation is thus discredited as

$$\frac{J(\theta_{n+1}) - J(\theta_n)}{\Delta} = \eta_n, \tag{8.89}$$

where $J(\theta_n) = J(\theta(t_n))$ and $\eta_n = \eta(t_n)$. Let we choose the expression for noise as a stable Levy process of the form given in Fourier space as

$$P(k,t) = \exp[-D\eta_0^\mu \Delta^{\mu-1} |k|^\mu t] \tag{8.90}$$

To keep the coefficient D fixed eliminate time step Δ the cutoff η_0 has to be renormalized by $\eta_0^\mu \Delta^{\mu-1} = 1$. We can appropriate Eq.(8.89) as follows

$$\frac{< J(\theta_{n+1}) - J(\theta_n) >_\theta}{\Delta} = < \eta_n >_\eta . \tag{8.91}$$

Explicitly evaluating the moment $< \eta_n >_\eta$ by using the expression for $p(\eta)$ above gives a finite value for $1 < \mu < 2$, which is

$$\frac{< J(\theta_{n+1}) - J(\theta_n) >_\theta}{\Delta} = const.\eta_0. \tag{8.92}$$

Thus we see that as $\Delta \to 0$, $\eta_0 \to \infty$, i.e. the cutoff moves to infinity, which hold for $\eta_0^\mu \Delta^{\mu-1} = 1$ in this range for μ. For $0 < \mu < 1$, in order for the renormalization $\eta_0^\mu \Delta^{\mu-1} = 1$, to hold $\eta_0 \to 0$ for $\Delta \to 0$. This renormalization is same for ordinary Langevin equation.

8.6.2 *Solution of the Langevin equation*

In this section we preset the solution of Eq.(8.85).

Definition 8.3. The delta function on fractal curve is defined by

$$\delta_F^\alpha(\theta(t) - \theta(t')) = \delta(J(\theta(t)) - J(\theta(t'))). \tag{8.93}$$

The associated probability distribution for a particle located at point θ on F at time t is given by

$$p(\theta_0, t) = \; < \delta_F^\alpha(\theta_0 - \theta(t)) >_\theta. \tag{8.94}$$

In view of Eq.(5.51) one can write

$$\delta_F^\alpha(\theta) = \frac{1}{2\pi} \int_{C(-\infty,\infty)} \exp(iJ(\theta)J(\psi)) d_F^\alpha \psi \tag{8.95}$$

Hence Eq.(8.94) can be written as

$$p(\theta, t) = \; < \frac{1}{2\pi} \int d_F^\alpha \psi \exp(iJ(\psi)[J(\theta) - J(\theta'(t))]) >_{\theta'} \tag{8.96}$$

or

$$p(\theta, t) = \int d_F^\alpha \theta' p(\theta') \left(\frac{1}{2\pi} \int d_F^\alpha \psi \exp(iJ(\psi)[J(\theta) - J(\theta'(t))]) \right) \tag{8.97}$$

It follows that

$$p(\theta, t) = \frac{1}{2\pi} \left[\int d_F^\alpha \psi \int d_F^\alpha \theta' p(\theta') \exp(iJ(\psi)J(\theta)) \exp(-iJ(\psi)J(\theta'(t))) \right] \tag{8.98}$$

or

$$p(\theta, t) = \frac{1}{2\pi} \int d_F^\alpha \psi \exp(iJ(\psi)J(\theta)) < \exp(-iJ(\psi)J(\theta'(t)) >_{\theta'}. \tag{8.99}$$

Therefore we get

$$p(\theta, t) = \frac{1}{2\pi} \int d_F^\alpha \psi \exp(iJ(\psi)J(\theta)) \tilde{p}(\psi, t). \tag{8.100}$$

By using Eq.(5.52) the inverse fractal Fourier transform of Eq.(8.100) we have

$$\tilde{p}(\psi, t) = \int d_F^\alpha \exp(-iJ(\psi)J(\theta)) p(\theta, t), \tag{8.101}$$

or

$$\tilde{p}(\psi, t) = \; < \exp(-iJ(\psi)J(\theta)) >_\theta. \tag{8.102}$$

Substituting the value of $J(\theta)$ from Eq.(8.86) we have

$$\tilde{p}(\psi,t) = < \exp(-iJ(\psi)\left\{\int_0^t \eta(t')dt'\right\}) >_\eta. \qquad (8.103)$$

Discretizing the integral in Eq.(8.103) gives

$$\tilde{p}(\psi,t) = \prod_{n=0}^N < \exp[-iJ(\psi)\eta(t_n)\Delta >, \qquad (8.104)$$

where $t_n = n\Delta$ and the interval $[0,t]$ is divided into N equal parts and $\Delta = (t-0)/N$. In view of the model of noise in Eq.(8.88) and comparing Eq.(8.104) we obtain

$$\tilde{p}(\psi,t) = \prod_{n=0}^N \exp[-D_1|J(\psi)\Delta|^\mu]. \qquad (8.105)$$

where $D_{\eta_0}^\mu = D_1$. Using the renormalization $D_1\Delta^{\mu-1} \to D$ and rewritten Eq.(8.105) becomes

$$\tilde{p}(\psi,t) = \exp\left[-D|J(\psi)|^\mu \int_0^t dt'\right], \qquad (8.106)$$

or

$$\tilde{p}(\psi,t) = \exp[-(D|J(\psi) = S_F^\alpha(k)|^\mu t)]. \qquad (8.107)$$

For $\mu = 2$, we have

$$\tilde{p}(\psi,t) = \exp[-(DS_F^\alpha(k)^2 t)]. \qquad (8.108)$$

The fractal Fourier transform for the above equation gives

$$\tilde{p}(\psi,t) = \frac{1}{\sqrt{2\pi Dt}} \exp(-\frac{S_F^\alpha(u)^2}{2Dt}). \qquad (8.109)$$

We see that solution Eq.(8.109) is the same solution obtained for diffusion equation Eq.(8.45).

8.7 Sub- and super-diffusion on fractal

Let us consider diffusion equation on fractal $F \subset E^n$ (Euclidean space) time-space is given by [Golmankhaneh and Balankin (2018)]:

$$D_{F,t}^\beta p(r,t) = \mathbb{K} \, D_{F,r}^{2\alpha} p(r,t), \qquad (8.110)$$

where $\mathbb{K}$ is the fractal coefficient of diffusion, $r = (\sum_{i=1}^n x_i^2)^{1/2}$, and

$$\frac{\beta}{\alpha} = \gamma = \frac{2}{D_w}, \quad D_w = \frac{2dim_H(F)}{d_s}, \qquad (8.111)$$

where d_s is spectral dimension and D_w is the fractal dimension of random walk. The connectivity dimension (or chemical dimension) d_l is defined by

$$d_l = \frac{dim_H(F)}{d_{min}}, \tag{8.112}$$

where d_{min} is fractal dimension minimum path between two point randomly chosen points on the fractal which has the following form Cantor sets

$$d_{min} = \frac{2dim_H(F)(dim_H(F)+1)}{n+(n+2)dim_H(F)}. \tag{8.113}$$

The solution of Eq.(8.110) is given by

$$p(r,t) = \frac{S_F^\beta(t)^{-1/2}}{\sqrt{4\pi\mathbb{K}}} \exp\left(\frac{-S_F^\alpha(r)^2}{4\mathbb{K}S_F^\beta(t)}\right). \tag{8.114}$$

Table 8.1: Comparing of fractional space approach with fractal approach

Markovian random process on $F \subset E^n$	d_s	d_{min}	γ	Type of diffusion in numerical simulation
Lévy flights	n	$\dfrac{dim_H(F)}{d_s}$	$\dfrac{1}{d_{min}}$	Super-diffusion
Lévy walk	$2dim_H(F) - n$	1	$2 - \dfrac{n}{dim_H(F)}$	Sub-diffusion
Scaled random walk (n=1)	$\dfrac{2dim_H(F)}{dim_H(F)+1}$	Eq.(8.113)	$\dfrac{2}{1+dim_H(F)}$	Super-diffusion

In Table 8.1, Markovian random processes on fractal are defined and characterized by the number of effective dynamical degrees of freedom (d_s), fractal dimension of the minimum path d_{min}, and anomalous diffusion exponent γ.

8.8 Fractal over-damped Langevin equation

In this section we present the Langevin equation on fractals [Golmankhaneh (2019a)]. Let us consider over-damped the fractal Langevin equation as the following form:

$$D^{\alpha}_{F,t}x(t) = \sqrt{2D_0}\,\zeta(t),\tag{8.115}$$

where D_0, is the fractal coefficient of diffusion and $\zeta(t)$ stochastic process with the following

$$< \zeta(t_2)\zeta(t_1) > = \delta^{\gamma}_F(t_2 - t_1),\tag{8.116}$$
$$< \zeta(t) > = 0,\tag{8.117}$$

which is called δ-correlation and its mean square displacement (MSD) is given by

$$< S^{\alpha}_F(x)^2 > = 2D_0 S^{\gamma}_F(t),\tag{8.118}$$

where α is the fractal space dimension and γ is the fractal time dimension. Taking into consideration $S^{\gamma}_F(t) < t^{\gamma}$, we can have

$$< x(t)^2 > \sim 2D_0 t^{\gamma/\alpha}.\tag{8.119}$$

Remark 8.4. Note that

(1) If $\gamma/\alpha = 1$, then we have normal diffusion.
(2) If $\gamma/\alpha < 1$, then we have sub-diffusion.
(3) If $\gamma/\alpha > 1$, then we have supper-diffusion.

8.9 Fractal under-damped Langevin equation

The fractal under-damped Langevin equation with inertial term is considered as [Golmankhaneh (2019a)]

$$(D^{\gamma}_{F,t})^2 x(t) + \gamma_0 D^{\gamma}_{F,t}x(t) = \sqrt{2D_0}\,\gamma_0\zeta(t),\tag{8.120}$$

where γ_0 is the fractal friction coefficient and D_0 is defined by

$$D_0 = \frac{T_0}{m\gamma_0},\tag{8.121}$$

which is called the fractal diffusion constant. The correlation of the velocity $D^{\gamma}_{F,t}x(t) = v(t)$ in Eq.(8.120) is defined by

$$< v(t_1)v(t_2) > = \frac{T_0}{m}\exp\left(-\gamma_0|S^{\gamma}_F(t_2) - S^{\gamma}_F(t_1)|\right).\tag{8.122}$$

Using Eq.(8.122), one can write the mean square displacement (MSD) the following form

$$< S_F^\alpha(x)^2 > = 2D_0 S_F^\gamma(t) + \frac{2D_0}{\gamma_0}(e^{-\gamma_0 S_F^\gamma(t)} - 1). \qquad (8.123)$$

By virtue of the upper bound of $S_F^\gamma(t) < t^\gamma$, one can rewrite Eq.(8.123) in the form

$$< x(t)^2 > \sim \left[2D_0 t^\gamma + \frac{2D_0}{\gamma_0}(e^{-\gamma_0 t^\gamma} - 1)\right]^{1/\alpha}. \qquad (8.124)$$

For the short time, namely, $t \ll \frac{1}{\gamma_0}$, Eq.(8.123) leads to the following form

$$< S_K^\alpha(x)^2 > \sim D_0 \gamma_0 S_K^\gamma(t)^2, \qquad (8.125)$$

and in the case of long time $(t \gg \frac{1}{\gamma_0})$ we arrive at Eq.(8.118).

8.10   Fractal scaled Brownian motion

The scaled Brownian motion is modeled by the fractal stochastic Langevin equation as the following equation [Golmankhaneh (2019a)]

$$D_{F,t}^\gamma x(t) = \sqrt{2D(t)}\, \zeta(t), \quad t \in F, \qquad (8.126)$$

where $< \zeta(t_1), \zeta(t_2) >= \delta_F^\alpha(t_2 - t_1)$, and $< \zeta(t) >= 0$. Assume that the diffusion coefficient as follows depends on time

$$D(t) = D_0 \left(1 + \frac{t}{\tau_0}\right)^{\nu-1}, \quad t \in F, \qquad (8.127)$$

where τ_0 is the characteristic time of the mobility variation. Using Eq.(8.127), MSD is easily obtained as follows:

$$< S_F^\alpha(x)^2 > = 2 \int_0^t D(t')d_F^\gamma t' = \frac{2D_0\tau_0}{\nu}\left(\left[1 + \frac{S_F^\gamma(t)}{\tau_0}\right]^\nu - 1\right),$$

$$\sim \frac{2D_0\tau_0}{\nu}\left(\left[1 + \frac{t^\gamma}{\tau_0}\right]^\nu - 1\right). \qquad (8.128)$$

For short times $(t \ll \tau_0)$ Eq.(8.128) leads

$$< S_F^\alpha(x)^2 > \sim 2D_0\, S_F^\gamma(t), \qquad (8.129)$$

and at long times $(t \gg \tau_0)$, we have

$$< S_F^\alpha(x)^2 > \sim S_F^\gamma(t)^\nu. \qquad (8.130)$$

The results above indicate both sub-and super-diffusive processes on fractals. The ultra-slow fractal scaled Brownian motion is obtained by choosing $\nu = 0$ in Eq.(8.127), it follows that

$$D(t) = D_0 \left(1 + \frac{t}{\tau_0}\right)^{-1}, \quad t \in F. \tag{8.131}$$

Accordingly, the MSD has the logarithmic time dependence, namely,

$$< S_F^\alpha(x)^2 > = 2D_0\tau_0 \log\left(1 + \frac{S_F^\gamma(t)}{S_F^\gamma(\tau_0)}\right) \tag{8.132}$$

$$\sim 2D_0\tau_0 \log\left(1 + \frac{t^\gamma}{\tau_0^\gamma}\right). \tag{8.133}$$

At long times $t \gg \tau_0$, Eq.(8.132) has the following form

$$< S_F^\alpha(x)^2 > \sim \log(S_F^\gamma(t)), \tag{8.134}$$

and smooth version of Eq.(8.134) is given by the form

$$< x(t)^2 > \sim (\log(t^\gamma))^{1/\alpha}. \tag{8.135}$$

8.11   Diffraction fringes from fractal sets

The Fraunhofer diffraction model is applied to explain the diffraction of waves when the diffraction pattern is viewed at a long distance from the diffracting object (See Figure 8.1). The fractal Fourier transform is used for

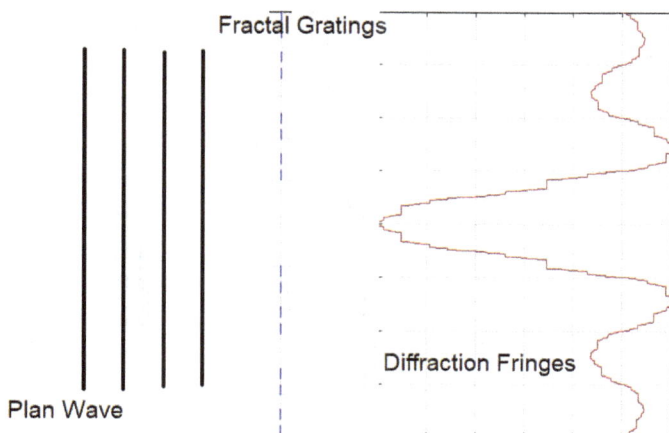

Fig. 8.1: Diffraction outline of the fractal set

finding diffraction patterns from fractal gratings as follows [Golmankhaneh and Baleanu (2016a)]:

The plane wave on a fractal set is defined by

$$f(t) = \begin{cases} \exp i S_F^\alpha(\omega_0) S_F^\alpha(t), & -a/2 < t < a/2; \\ 0, & \text{otherwise}. \end{cases} \tag{8.136}$$

where a is constant, ω_0 the frequency and t is the time. In Figure 8.2, we have plotted real part wave function which given by Eq.(8.136) for the case of triadic Cantor set and $S_F^\alpha(\omega_0) = 6$.

The fractal Fourier transform of Eq.(8.136), gives

$$g(\omega) = \sqrt{\frac{2}{\pi}} \frac{\sin\left(S_F^\alpha(a/2)(S_F^\alpha(\omega) - S_F^\alpha(\omega_0))\right)}{S_F^\alpha(\omega) - S_F^\alpha(\omega_0)} \tag{8.137}$$

In Figure 8.3, we have presented the diffraction fringes corresponding to the fractal set by choosing triadic Cantor set and $a = 2$.

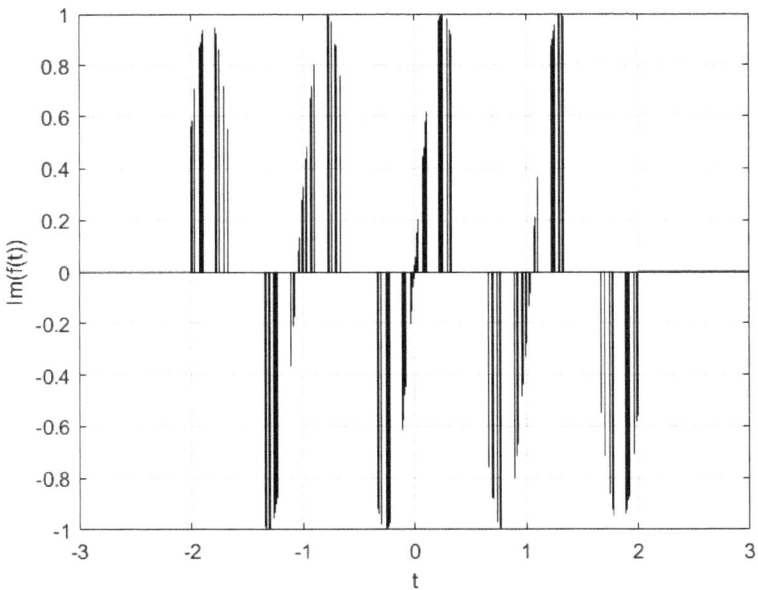

Fig. 8.2: Graph of real part of $f(t)$ on triadic Cantor set

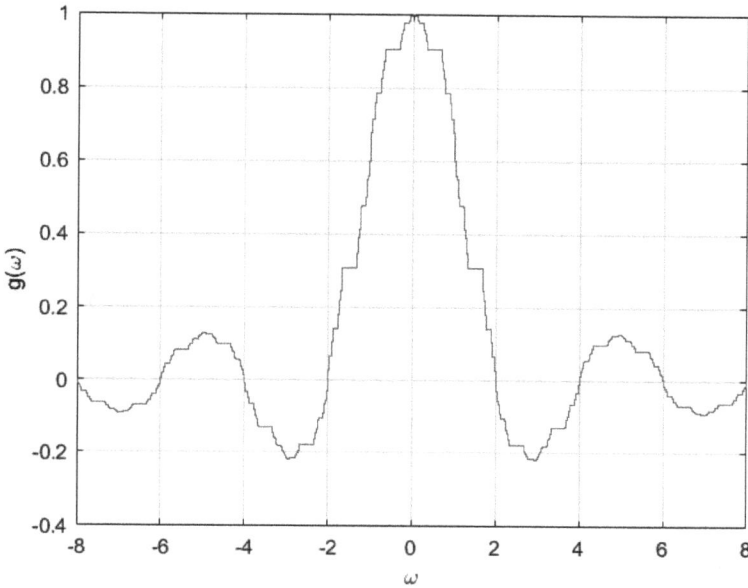

Fig. 8.3: Graph of $g(\omega)$ in the case of triadic Cantor set, $a = 2$, and $\omega_0 = 0$

8.12 Schrödinger equation on fractal

The Schrödinger equation in fractal space with dimension of α is suggested as [Golmankhaneh *et al.* (2015); Golmankhaneh (2019b)]:

$$\frac{-\hbar^2}{2m}(D_{F,x}^\alpha)^2 \psi^\alpha(x,t) + v(x)\psi^\alpha(x,t) = i\hbar\frac{\partial \psi^\alpha(x,t)}{\partial t}, \qquad (8.138)$$

where

$$v(x) = \begin{cases} 0, & 0 \le x \le 1; \\ \infty, & \text{elsewhere,} \end{cases} \qquad (8.139)$$

with boundary conditions

$$\psi^\alpha(0,t) = \psi^\alpha(1,t) = 0. \qquad (8.140)$$

Conjugacy between the fractal calculus and ordinary calculus, one can obtain solution of Eq.(8.138) as the following form

$$\psi^\alpha(x,t) = \sum_{n=1} e^{-iE_n^\alpha t/\hbar}\varphi_n^\alpha(x), \qquad (8.141)$$

where

$$E_n^\alpha = \frac{\pi^2 \hbar^2 n^2}{2m S_F^\alpha(1)}$$

$$= \frac{\pi^2 \hbar^2 n^2}{2m \Gamma(\alpha+1)}, \tag{8.142}$$

which are called eigenvalues of the Schrödinger equation on fractal spaces and it shows that energy levels shifts to higher values. One can find solution of Eq.(8.138) by conjugacy of fractal calculus with ordinary calculus as follows:

$$\varphi_n^\alpha(x) = \sqrt{\frac{2}{S_F^\alpha(1)}} \sin\left(\frac{\pi n S_F^\alpha(x)}{S_F^\alpha(1)}\right)$$

$$= \sqrt{\frac{2}{\Gamma(\alpha+1)}} \sin\left(\frac{\pi n S_F^\alpha(x)}{\Gamma(\alpha+1)}\right). \tag{8.143}$$

In Figure 8.4, we have plotted Eq.(8.143).

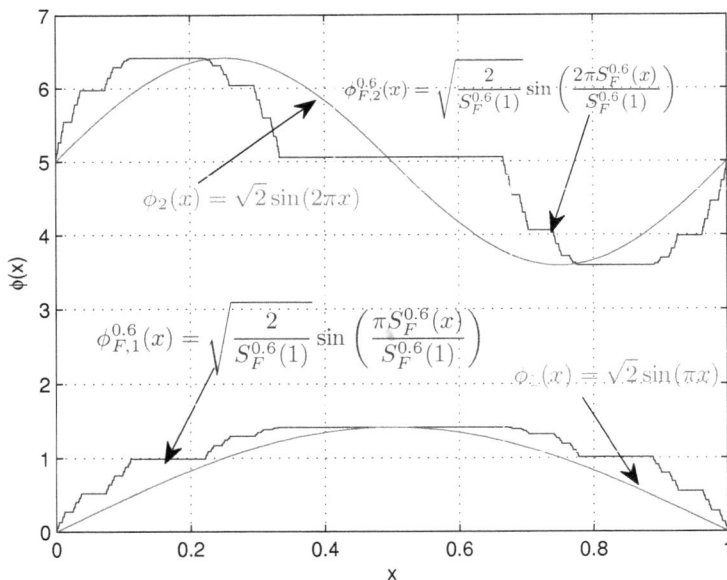

Fig. 8.4: The graph of eigenfunctions of Eq.(8.138)

As $S_F^\alpha(x) \le x^\alpha$, then one can write

$$\varphi_n^\alpha(x) \sim \sqrt{\frac{2}{\Gamma(\alpha+1)}} \sin\left(\frac{\pi n x^\alpha}{\Gamma(\alpha+1)}\right), \qquad (8.144)$$

which are called the eigenfunctions of the Schrödinger equation on fractal spaces. It shows how the solutions are changing with the dimension of the underlying space.

In Figures 8.5 and 8.6, the energy levels of the Schrödinger equations and corresponding wave functions for the different dimensions were plotted.

8.13 Partition function of systems on fractal

In this section, we present a generalized thermodynamics framework based on fractal space and temperature. The analogue of canonical ensemble partition function involving fractal temperature set is given by [Golmankhaneh (2019b)]:

$$z^{\alpha,\mu} = \sum_{n=1} \exp\left(-\frac{E_n^\alpha}{k_B S_F^\mu(T)}\right), \qquad (8.145)$$

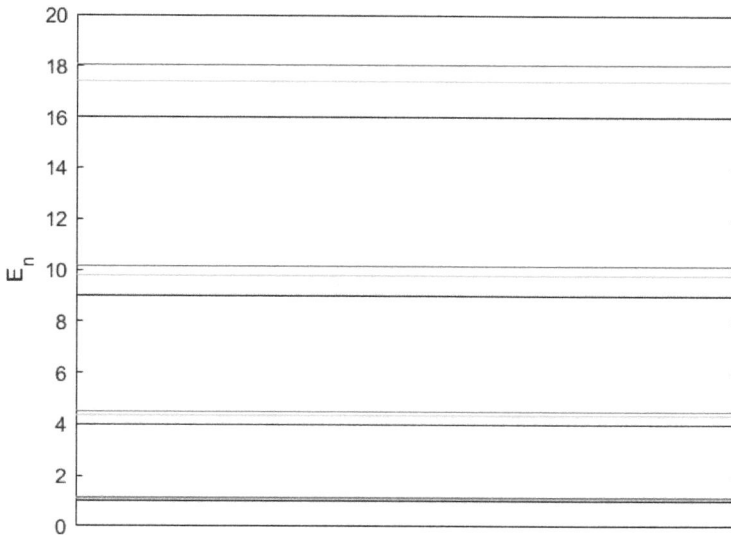

Fig. 8.5: Energy levels versus by space dimension (black), $\alpha = 1$, $\alpha = 0.5$ (blue), $\alpha = 0.75$ (green)

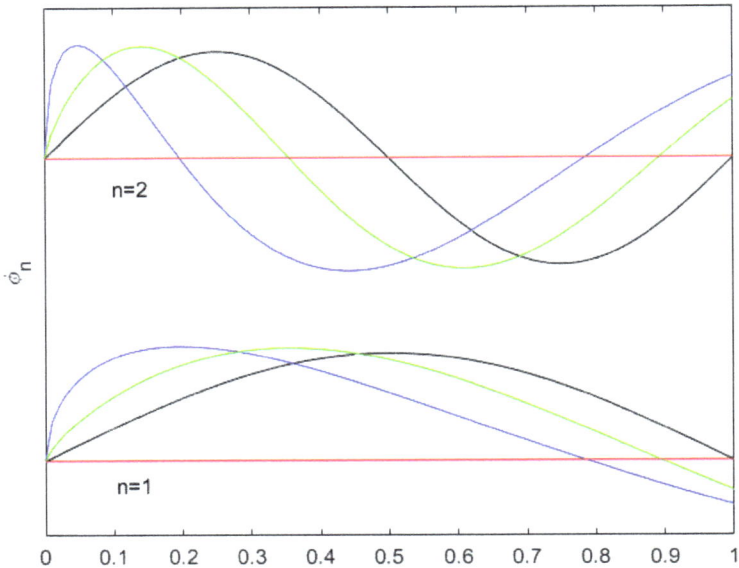

Fig. 8.6: Wave function corresponding to space dimension (black), $\alpha = 1$, $\alpha = 0.5$ (blue), $\alpha = 0.75$ (green)

where k_B is the fractal Boltzmann constant and μ is the fractal dimension of $T \in F$ temperature. The probability of each microstate n of system is suggested by

$$P_n = \frac{1}{z^{\alpha,\mu}} \exp\left(-\frac{E_n^\alpha}{k_B S_F^\mu(T)}\right), \tag{8.146}$$

According to the conjugacy of the fractal calculus with ordinary calculus, the energy of each microstate in the space and the fractal temperature is given by

$$E^{\alpha,\mu} = N k_B S_F^\mu(T)^2 D_T^\mu \ln z$$
$$\sim N k_B T^{2\mu} D_T^\mu \ln z, \tag{8.147}$$

where N is the number of particles in the system. The fractal heat capacity is defined by

$$C^{\alpha,\mu} = D_T^\mu E^{\alpha,\mu}. \tag{8.148}$$

The analogue of the Boltzmann's entropy on fractals is given by

$$\mathcal{M}^{\alpha,\mu} = N k_B \ln z + N k_B S_F^\mu(T) D_T^\mu \ln z. \tag{8.149}$$

The analogue of the Helmholtz's free energy is defined by

$$H^{\alpha,\mu} = E^{\alpha,\mu} - T\mathcal{M}^{\alpha,\mu}, \tag{8.150}$$

or

$$H^{\alpha,\mu} = -Nk_B \ln z^{\alpha,\mu}. \tag{8.151}$$

Let the energy function of the paramagnetic salt, which includes fractal temperature, be as follows

$$E^{\alpha,\mu} = -N B \nu_B \tanh\left(\frac{\nu_B B}{k_B S_F^{\mu}(T)}\right), \tag{8.152}$$

where B is the applied external magnetic field to the paramagnetic salt , and ν_B is the Bohr magneton . Then, by submitting Eq.(8.152) into Eq.(8.148) we arrive at

$$C^{\alpha,\mu} = Nk_B \left(\frac{2\nu_B B}{k_B S_F^{\mu}(T)}\right)^2 \frac{\exp(-\frac{E_n^{\alpha}}{k_B S_F^{\mu}(T)})}{\left(1 + \exp(\frac{2\nu_B B}{k_B S_F^{\mu}(T)})\right)^2}. \tag{8.153}$$

Using the upper bound of $S_F^{\mu}(T) < T^{\mu}$, Eq.(8.153) becomes

$$C^{\alpha,\mu} \sim Nk_B \left(\frac{2\nu_B B}{k_B T^{\mu}}\right)^2 \frac{\exp(\frac{2\nu_B B}{k_B T^{\mu}})}{\left(1 + \exp(\frac{2\nu_B B}{k_B T^{\mu}})\right)^2}. \tag{8.154}$$

In Figure 8.7, we have sketched the heat capacity of the paramagnetic salt.

Suppose a system with two energy levels as

$$E_0 = 0,$$
$$E_1 = \frac{\pi^2 \hbar^2}{2m\Gamma(\alpha+1)}, \tag{8.155}$$

where m is the mass of the particle and $\hbar$ is the reduced Planck constant. The corresponding partition function of this system is

$$z^{\alpha,\mu} = 1 + \exp\left(-\frac{\pi^2 \hbar^2}{2mk_B \Gamma(\alpha+1) S_F^{\mu}(T)}\right). \tag{8.156}$$

Replacing Eq.(8.156) into Eq.(8.147), we can obtain

$$E^{\alpha,\mu} = \frac{\frac{k_B a}{\Gamma(\alpha+1)}}{1 + \exp\left(-\frac{a}{\Gamma(\alpha+1) S_F^{\mu}(T)}\right)} \exp\left(-\frac{a}{\Gamma(\alpha+1) S_F^{\mu}(T)}\right), \tag{8.157}$$

where

$$a = \frac{\pi^2 \hbar^2}{2mk_B}. \tag{8.158}$$

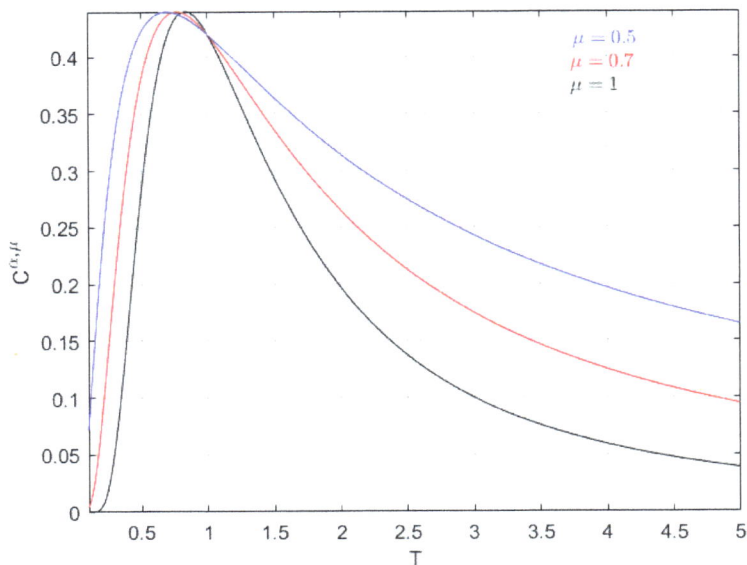

Fig. 8.7: Graph of the heat capacity of paramagnetic salt for different fractal dimensions

Next we want to get the fractal heat capacity by combining (8.148) with (8.157) we get

$$C^{\alpha,\mu} = \frac{k_B a^2}{\Gamma(\alpha+1)^2 S_F^{\mu}(T)^2} \frac{1}{\left(1 + \exp(-\frac{a}{\Gamma(\alpha+1)S_F^{\mu}(T)})\right)^2} \exp\left(-\frac{a}{\Gamma(\alpha+1)S_F^{\mu}(T)}\right).$$

(8.159)

Utilizing the upper bound of staircase function $(S_F^{\mu}(T) < T^{\mu})$ we get

$$C^{\alpha,\mu} \sim \frac{k_B a^2}{\Gamma(\alpha+1)^2 T^{2\mu}} \frac{1}{\left(1 + \exp(-\frac{a}{\Gamma(\alpha+1)T^{\mu}})\right)^2} \exp\left(-\frac{a}{\Gamma(\alpha+1)T^{\mu}}\right).$$

(8.160)

The fractal analogue of the Einstein solid model involving fractal temperature and using the energy function for the photon is suggested by

$$E^{\alpha,\mu} = \frac{3N\mathcal{E}}{\exp\left(\frac{\mathcal{E}}{k_B S_F^{\mu}(T)}\right) - 1},$$

(8.161)

where $\mathcal{E} = h\nu$(photon energy), ν (frequency of light), and h is the Planck constant. Then, the heat capacity of the photon system using Eq.(8.148) is derived and suggested by

$$C^{\alpha,\mu} = \frac{3N\mathcal{E}^2}{k_B S_F^\mu(T)^2} \frac{\exp\left(\frac{\mathcal{E}}{k_B S_F^\mu(T)}\right)}{\left[\exp\left(\frac{\mathcal{E}}{k_B S_F^\mu(T)}\right) - 1\right]^2}. \tag{8.162}$$

Using $S_F^\mu(T) \leq T^\mu$ one can get smooth version as

$$C^\mu = \frac{3N\mathcal{E}^2}{k_B T^{2\mu}} \frac{\exp\left(\frac{\mathcal{E}}{k_B T^\mu}\right)}{\left[\exp\left(\frac{\mathcal{E}}{k_B T^\mu}\right) - 1\right]^2}. \tag{8.163}$$

The fractal energy function in the Dulong-Petit solid model is given by

$$E^{\alpha,\mu} = 3Nk_B S_F^\mu(T). \tag{8.164}$$

The heat capacity of the photon system utilizing Eq.(8.148) is

$$C^\mu = 3Nk_B \frac{1}{\Gamma(1+\mu)}. \tag{8.165}$$

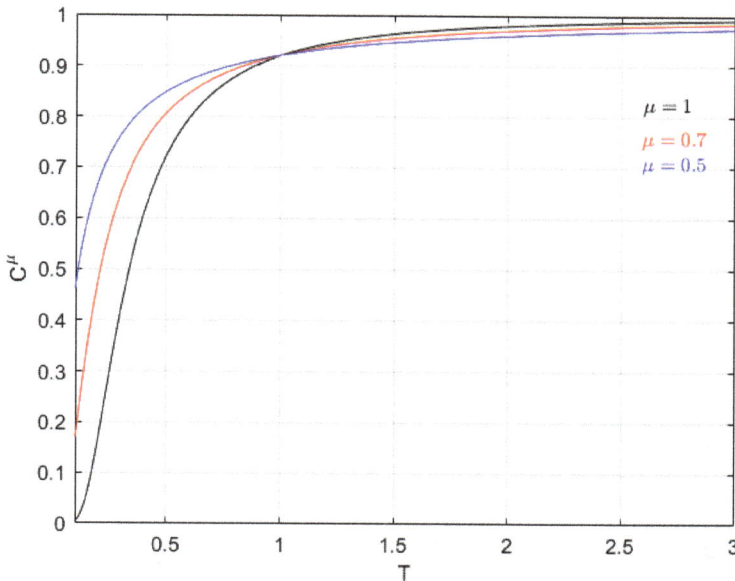

Fig. 8.8: The Graph of heat capacity versus fractal temperature with different dimensions using the Einstein solid model

Fractal Calculus and Its Applications

The fractal energy function in the Debye solid model is given by

$$E^{\alpha,\mu} = \frac{3Nk_B\pi^4 S_F^\mu(T)^4}{5\theta_D^3}. \tag{8.166}$$

Thus, applying Eq.(8.148) one can get the heat capacity for the Debye solid model as

$$C^\mu = \frac{12}{5}\pi^4 Nk_B \left(\frac{S_F^\mu(T)}{\theta_D}\right)^3. \tag{8.167}$$

Then we can write the smooth version of Eq.(8.167) using $S_F^\mu(T) \leq T^\mu$ as

$$C^\mu \sim \frac{12}{5}\pi^4 Nk_B \frac{T^{3\mu}}{\theta_D^3}. \tag{8.168}$$

Finally, the Einstein, Debay, and Dulong-Petit models involving fractal space and fractal temperature can affect the thermal capacity.

In Figures 8.7, 8.8, and 8.9 the heat capacity of paramagnetic salt for different fractal dimensions, the Einstein solid model for the case of fractal temperature with different dimensions, and the Debye solid model for the fractal temperature with different dimensions, respectively.

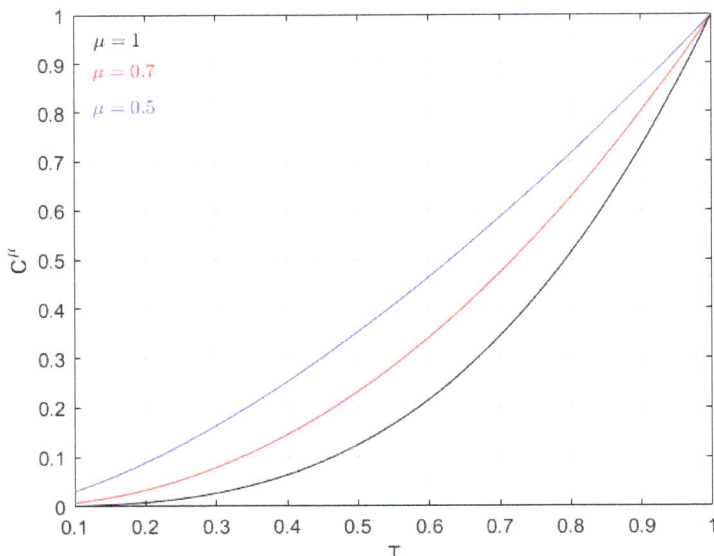

Fig. 8.9: Graph of heat capacity versus the fractal temperature with different dimensions using the Debye solid model

8.14 Density of states on fractal spaces

The density of states (DOS) which is indicated by $g(E)$ in 0-, 1-, 2- and 3-dimensional system were given as [Golmankhaneh (2019b)]

$$g(E) \propto \delta(E), \qquad 0 - dimension, \qquad (8.169)$$
$$g(E) \propto E^{-1/2}, \qquad 1 - dimension, \qquad (8.170)$$
$$g(E) \propto constant, \qquad 2 - dimension, \qquad (8.171)$$
$$g(E) \propto E^{1/2}, \qquad 3 - dimension. \qquad (8.172)$$

Subsequently, the density of states for the fractal system with non-integer dimensions is suggested by

$$g(E) \propto E^{-\alpha_1/2}, \qquad 0 < \alpha_1 < 1, \qquad (8.173)$$
$$g(E) \propto E^{\alpha_2/2}, \qquad 1 < \alpha_2 < 3, \qquad (8.174)$$

where $\alpha_2 \neq 2$. In Figures 8.10, 8.11, 8.12, and 8.13 we have plotted DOS for the systems with different integer and non-integer dimensions.

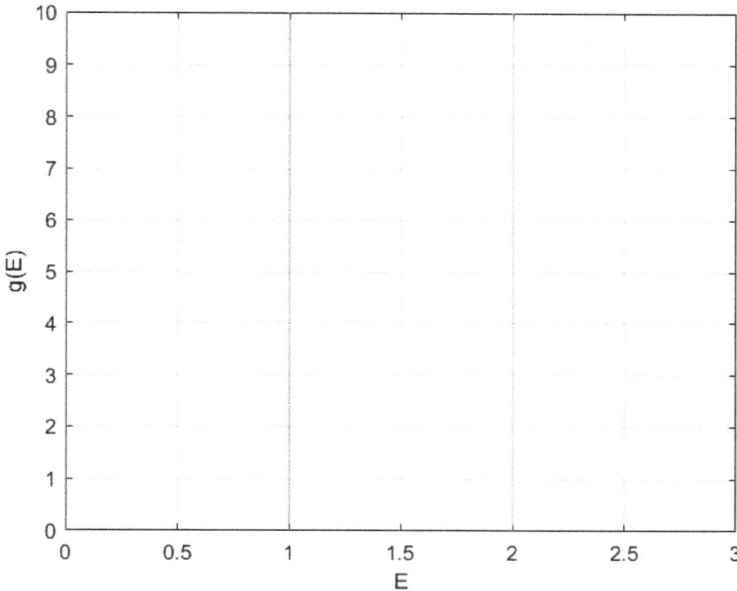

Fig. 8.10: Graph of DOS for the system with dimension 0

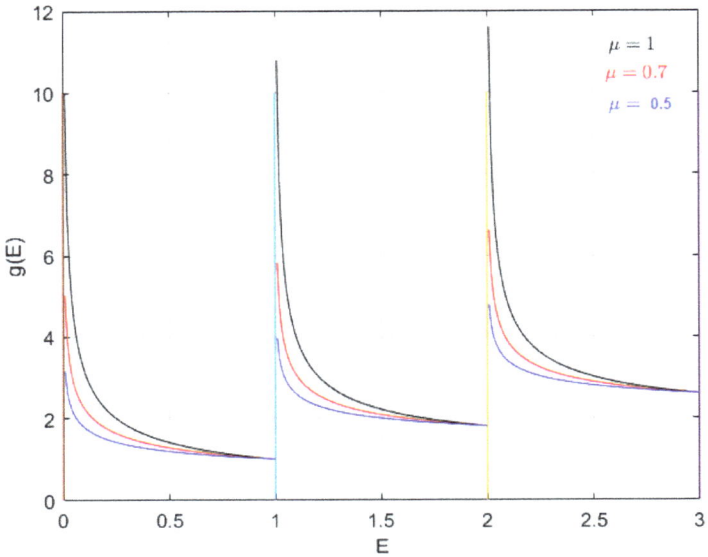

Fig. 8.11: Graph of DOS for the system with dimensions 1, 0.7 and 0.5

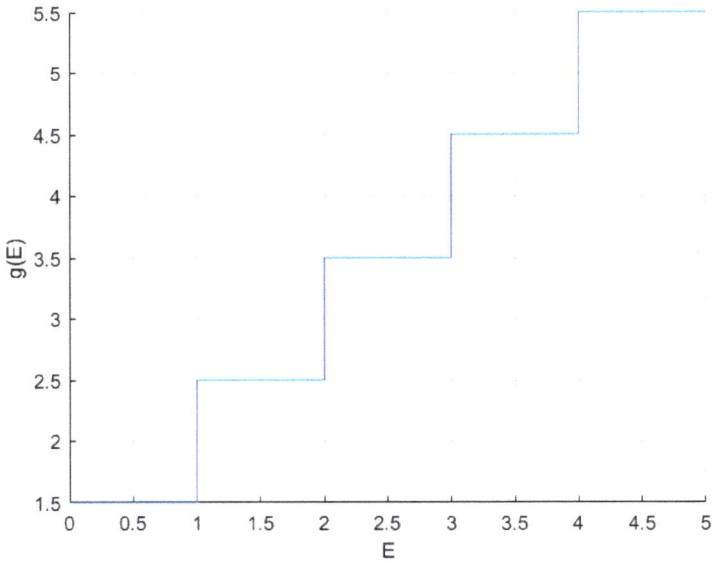

Fig. 8.12: Graph of DOS for the system with dimension 2

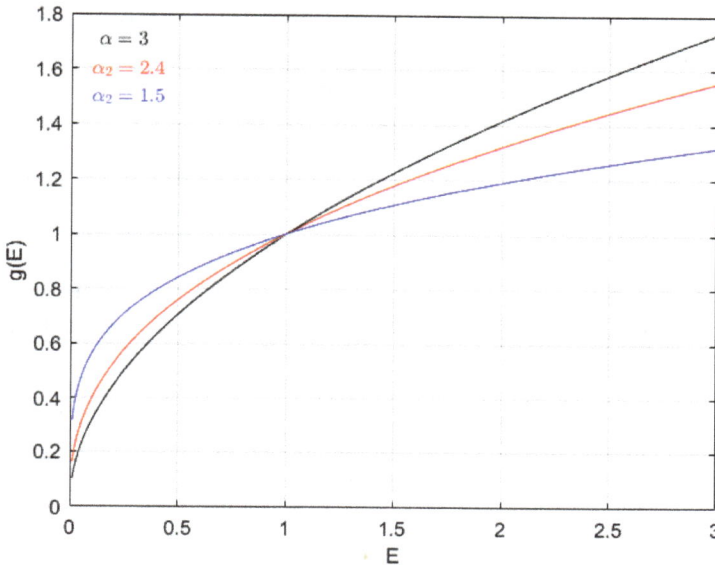

Fig. 8.13: Graph of DOS for the system with dimensions 3, 2.4 and 1.5

8.15 Newton's second law on fractal time

A point particle with mass m that moves in space-fractal time [Golmankhaneh *et al.* (2013a)].
The Newton's second law is suggested as

$$f = m_K (D_{F,t}^{\alpha})^2 x(t), \quad t \in F, \tag{8.175}$$

where $x(t)$ is the position of particle, f is a force applied on particle and the physical dimension of $[m_K] = (Mass)(Time)^{\alpha}$, where $m_K = m$.

Example 8.2. The simple harmonic oscillator motion equation in space-fractal time using Eq.(8.175) is

$$-kx = m_F (D_{F,t}^{\alpha})^2 x(t), \quad x(0) = A, \quad D_{F,t}^{\alpha} x(t)|_0 = 0. \tag{8.176}$$

where k is the spring constant and $f = -kx$ is the force applied to the particle. The solution of Eq.(8.176) is

$$x(t) = A \cos(\omega S_F^{\alpha}(t)), \tag{8.177}$$

where $\omega = \sqrt{k/m_F}$ is a constant. In view of $S_F^{\alpha}(t) < t^{\alpha}$, we can write

$$x(t) \propto A \cos(\omega t^{\alpha}). \tag{8.178}$$

Example 8.3. Consider particles moving along the fractal Cesàro curve, which absorbs the particles on it [Golmankhaneh (2017)]. Then a mathematical model for this phenomenon is suggested by

$$D_F^\alpha \rho(\theta) = -k\rho(\theta), \quad k \text{ is constant.} \tag{8.179}$$

where $\rho(\theta)$ is the density of particles on the fractal Cesàro curve. The solution of Eq.(8.179) is

$$\rho(\theta) = \rho(0)\exp(-kJ(\theta)) \tag{8.180}$$

where $J(\theta) = S_F^\alpha(t)$ is staircase function of the fractal Cesàro curve. In Figure 8.14 we have plotted Eq.(8.180) for the cases of $k = 1$, $\rho(0) = 1$, $\alpha = 1.7$.

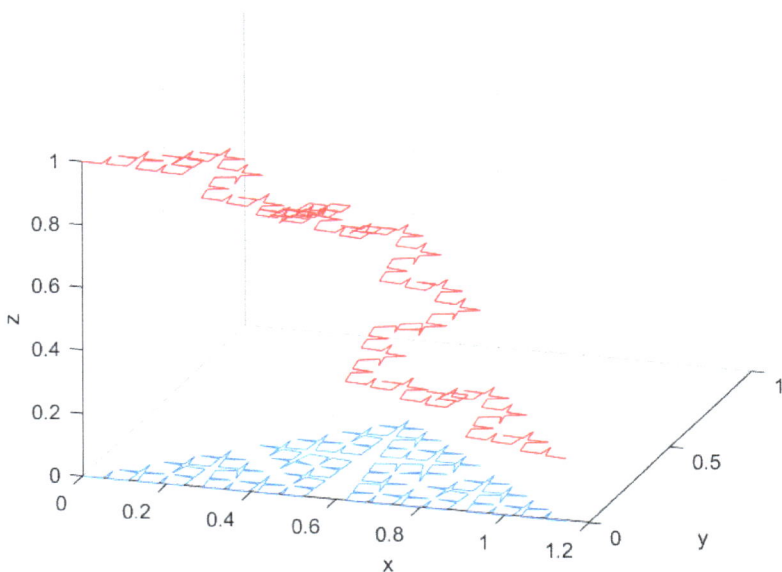

Fig. 8.14: Graph of solution of Eq.(8.180)

Example 8.4. Consider the following differential equation on the fractal Cesàro curve:

$$(D_F^\alpha)^2 f(\theta) + 9f(\theta) = \cos(\theta), \quad \theta \in F. \tag{8.181}$$

with boundary conditions

$$D_F^\alpha f(\theta) = 5, \quad f(\theta) = -5/3. \tag{8.182}$$

Using the conjugacy, solution of Eq.(8.181) is

$$f(\theta) = \lambda \cos(3J(\theta)) + \frac{5}{3}\sin(3J(\theta)) + \frac{1}{8}\cos(J(\theta)), \tag{8.183}$$

where $\lambda \in R$ is constant.

8.16 Series resistor, an inductor, and a capacitor involving fractal time

In this section, we present two examples as applications in which temporal evolution with a fractal structure takes place. In part 8.16.1, we take a series resistor, an inductor, and a capacitor (RLC) circuit, and in part 8.16.2, we take a parallel RLC circuit, which are widely used in the field of physics and engineering [Golmankhaneh *et al.* (2021c); Banchuin (2022)].

8.16.1 *Series RLC circuit involving fractal time*

Consider a series RLC circuit such that its dynamics, which include fractal time, is as follows:

$$D_F^{2\alpha}i(t) + \frac{R}{L}D_F^{\alpha}i(t) + \frac{1}{LC}i(t) = 0, \qquad 0 < \alpha \le 1, \ t \in \kappa, \tag{8.184}$$

using the initial conditions $i(0) = A$, and $D_F^{\alpha}i(t)|_{t=0} = A$. In Eq.(8.184), R, L, and C are the resistor, inductor, and the capacitor, respectively. Applying the fractal Laplace transform to both sides of Eq.(8.184), we get

$$i(\omega) = \frac{AS_F^{\alpha}(\omega)}{S_F^{\alpha}(\omega)^2 + \frac{R}{L}S_F^{\alpha}(\omega) + \frac{1}{L*C}} + \frac{A}{S_F^{\alpha}(\omega)^2 + \frac{R}{L}S_F^{\alpha}(\omega) + \frac{1}{LC}}$$
$$+ \frac{R}{L}\frac{A}{S_F^{\alpha}(\omega)^2 + \frac{R}{L}S_F^{\alpha}(\omega) + \frac{1}{LC}}, \tag{8.185}$$

applying the inverse Laplace to Eq.(8.185), then gives

$$i(t) = \frac{A\sqrt{C}(2L+R)}{2\sqrt{-4L+CR^2}}\left(-1 + \exp\left(\frac{\sqrt{-4L+CR^2}}{\sqrt{C}L}S_F^{\alpha}(t)\right)\right)$$
$$* \exp\left(-\frac{R\sqrt{C} + \sqrt{-4L+CR^2}}{2L}S_F^{\alpha}(t)\right)$$
$$+ \sqrt{-4L+CR^2}\left(1 + \exp\left(\frac{\sqrt{-4L+CR^2}}{\sqrt{C}L}S_F^{\alpha}(t)\right)\right), \tag{8.186}$$

In view of the upper bound of the staircase function $(S_F^\alpha(t) \le t^\alpha)$, we have

$$i(t) \sim \frac{A\sqrt{C}(2L+R)}{2\sqrt{-4L+CR^2}}\left(-1+\exp\left(\frac{\sqrt{-4L+CR^2}}{\sqrt{C}L}t^\alpha\right)\right)$$

$$* \exp\left(-\frac{R\sqrt{C}+\sqrt{-4L+CR^2}}{2L}t^\alpha\right)$$

$$+\sqrt{-4L+CR^2}\left(1+\exp\left(\frac{\sqrt{-4L+CR^2}}{\sqrt{C}L}t^\alpha\right)\right). \quad (8.187)$$

In Figures 8.15 and 8.16, we have plotted Eq.(8.186) and Eq.(8.187), respectively.

8.16.2 *Parallel RLC circuit involving fractal time*

Consider a parallel RLC circuit, if its dynamics includes fractal time, so to model it, we can write the following differential equation as [Golmankhaneh *et al.* (2021c); Banchuin (2022)]

$$D_F^{2\alpha}i(t) + \frac{1}{RC}D_F^\alpha i(t) + \frac{1}{LC}i(t) = 0, \quad 0 < \alpha \le 1, \ t \in F, \quad (8.188)$$

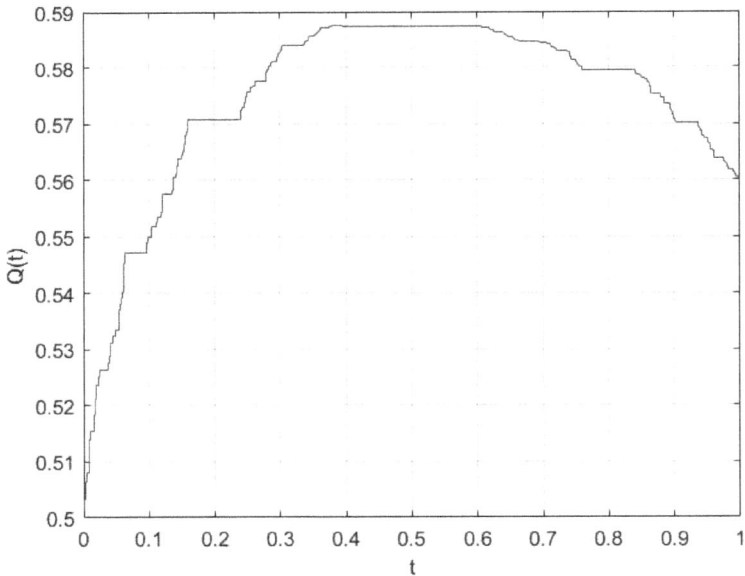

Fig. 8.15: Graph of Eq.(5.67) using $A = 0.5$, $R = 3$, $L = 0.8$, $C = 2$.

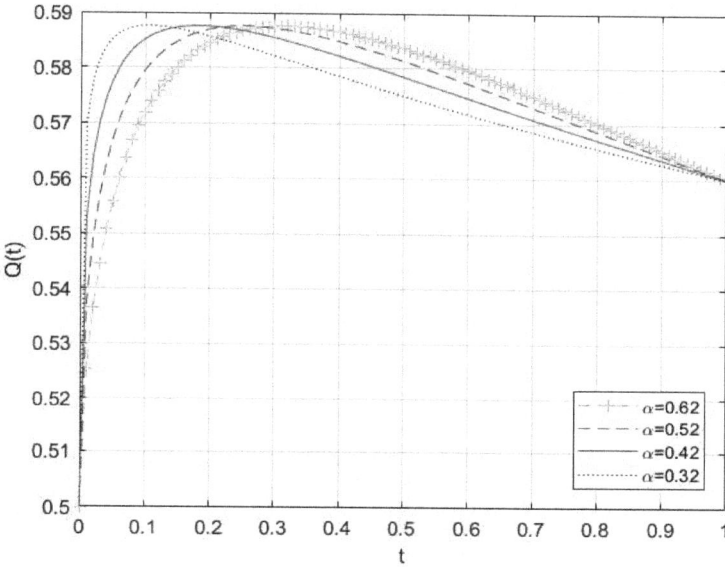

Fig. 8.16: Graph of Eq.(5.68) using $A = 0.5$, $R = 3$, $L = 0.8$, $C = 2$

using the initial conditions $i(0) = A$, and $D_F^\alpha i(t)|_{t=0} = A$. and applying the fractal Laplace transform to both sides of Eq.(8.188), then it turns into

$$i(\omega) = A \left(\frac{S_F^\alpha(\omega)}{S_F^\alpha(\omega)^2 + \frac{S_F^\alpha(\omega)}{RC} + \frac{1}{LC}} + \frac{1}{S_F^\alpha(\omega)^2 + \frac{S_F^\alpha(\omega)}{RC} + \frac{1}{LC}} \right.$$
$$\left. + \frac{1}{RC} \frac{1}{S_F^\alpha(\omega)^2 + \frac{S_F^\alpha(\omega)}{RC} + \frac{1}{LC}} \right). \tag{8.189}$$

Taking the inverse Laplace of Eq.(8.189), we get

$$i(t) = \frac{AL(1 + 2CR)}{2\sqrt{L(L - 4CR^2)}} \left(-1 + \exp\left(\frac{\sqrt{L(L - 4CR^2)}}{CLR} S_F^\alpha(t) \right) \right)$$
$$* \exp(-\frac{L + \sqrt{L(L - 4CR^2)}}{2CLR} S_F^\alpha(t))$$
$$+ \frac{A\sqrt{L(L - 4CR^2)}}{2\sqrt{L(L - 4CR^2)}} \left(1 + \exp\left(\frac{\sqrt{L(L - 4CR^2)}}{CLR} S_F^\alpha(t) \right) \right)$$
$$* \exp\left(-\frac{L + \sqrt{L(L - 4CR^2)}}{2CLR} S_F^\alpha(t) \right). \tag{8.190}$$

By the upper bound of the staircase function $(S_F^\alpha(t) \leq t^\alpha)$, we get

$$i(t) \sim \frac{AL(1+2CR)}{2\sqrt{L(L-4CR^2)}}\left(-1+\exp\left(\frac{\sqrt{L(L-4CR^2)}}{CLR}\right)t^\alpha\right)$$

$$*\exp\left(-\frac{L+\sqrt{L(L-4CR^2)}}{2CLR}t^\alpha\right)$$

$$+\frac{A\sqrt{L(L-4CR^2)}}{2\sqrt{L(L-4CR^2)}}\left(1+\exp\left(\frac{\sqrt{L(L-4CR^2)}}{CLR}\right)t^\alpha\right)$$

$$*\exp\left(-\frac{L+\sqrt{L(L-4CR^2)}}{2CLR}t^\alpha\right). \tag{8.191}$$

In Figures 8.17 and 8.18, we have sketched Eqs.(8.190) and (8.191) that are solutions of the fractal differential equation that models parallel RLC circuit which evolves with the fractal time.

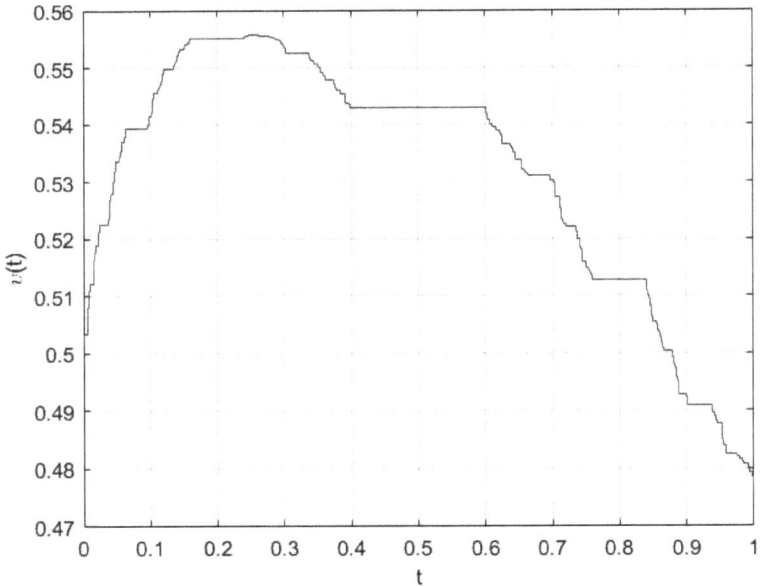

Fig. 8.17: Graph of Eq.(8.190) using $A = 0.5$, $R = 2$, $L = 7$, $C = 0.1$

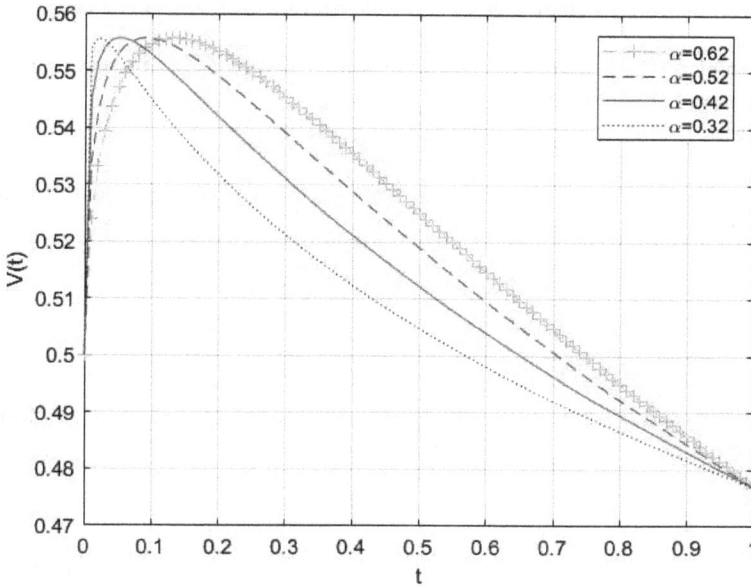

Fig. 8.18: Graph of Eq.(8.191) using $A = 0.5$, $R = 2$, $L = 7$, $C = 0.1$

8.17 Kepler's third law on fractal time

By Eq.(8.175) the mathematical model of the moving particle under force $f = -\kappa x(t)^{-2}$ in the fractal time is [Golmankhaneh (2018)]

$$m_K (D_{F,t}^\alpha)^2 x(t) + \frac{\kappa}{x(t)^2} = 0. \qquad (8.192)$$

where κ is a constant and $m_K = 1$. Scale transform of Eq.(8.192), gives

$$t \rightarrow Tt, \quad x \rightarrow \lambda x, \quad S_F^\alpha(Tt) \rightarrow T^\alpha S_F^\alpha(t)$$

$$D_F^\alpha \lambda x(Tt) \rightarrow \frac{\lambda}{T^\alpha} D_F^\alpha x(t). \qquad (8.193)$$

Since Eq.(8.192) must be invariance under scale transform then we get

$$T = \lambda^{3/2\alpha}, \qquad (8.194)$$

where T is the orbital period and λ the semi-major axis of the particle moves on a elliptic orbit. The Eq.(8.194) is called fractal Kepler's third law.

8.18   Lagrangian and Hamiltonian mechanics on fractal

In this section we preset the Lagrange and Hamilton equations on fractal time [Golmankhaneh *et al.* (2013c)].
The action functional of a given system is defined by

$$A^\alpha = \int_{t_1}^{t_2} L^\alpha(t, x, \ D_t^\alpha x) d_F^\alpha t \qquad L^\alpha : F \times R \times R \to R, \qquad (8.195)$$

where the function L_F^α is its Lagrangian. Then fractal Hamilton's principle is

$$\delta A^\alpha = \int_{t_1}^{t_2} \left(\frac{\partial L}{\partial x} - D_t^\alpha \frac{\partial L}{\partial \rho} \right) \delta x d_F^\alpha t = 0, \qquad (8.196)$$

where $D_t^\alpha x = \varrho$. Extremizing A^α leads to the Euler-Lagrange equation on fractal time as

$$\frac{\partial L}{\partial x} - D_t^\alpha \frac{\partial L}{\partial \rho} = 0. \qquad (8.197)$$

Let the fractal generalized momentum denoted by $p^\alpha = \frac{\partial L}{\partial \rho}$, then the Hamiltonian of system is

$$H^\alpha = D_t^\alpha x \ p^\alpha - L^\alpha. \qquad (8.198)$$

Conjugacy of fractal calculus and ordinary calculus gives the fractal Hamilton equation as

$$D_t^\alpha x = \frac{\partial H^\alpha}{\partial p^\alpha}, \qquad D_t^\alpha p^\alpha = -\frac{\partial H^\alpha}{\partial x}. \qquad (8.199)$$

Example 8.5. If the Lagrangian of a particle be as the following form

$$L^\alpha(t, x, D_t^\alpha x) = a(D_t^\alpha x)^2 - bx^2 \qquad (8.200)$$

where c, e, are constant and $x : F \to R$. Then, using Eq.(8.197) the Lagrange equation is

$$-2b(x) = 2a(D_t^\alpha)^2 x \qquad (8.201)$$

Example 8.6. By Eq.(8.198) and Eq.(8.200) the Hamiltonian of the particle is

$$H^\alpha = c(p^\alpha)^2 + e(x)^2 \qquad (8.202)$$

So the Hamilton's equations will be

$$D_t^\alpha p_\alpha = -2ex(t), \qquad D_t^\alpha x = 2cp^\alpha. \qquad (8.203)$$

8.18.1 *Poisson bracket on fractals*

Let $F(t, x(t), p^\alpha)$ and $G(t, x(t), p^\alpha)$ be any two functions of dynamical variables, then the generalized Poisson bracket is defined by [Golmankhaneh *et al.* (2013c)]

$$[F, G]^\alpha = \frac{\partial F}{\partial x} \frac{\partial G}{\partial p^\alpha} - \frac{\partial F}{\partial p^\alpha} \frac{\partial G}{\partial x}. \tag{8.204}$$

Therefore the Hamiltonian equation using Poisson brackets will be

$$[p^\alpha, H^\alpha]^\alpha = D_t^\alpha p^\alpha \qquad [x, H^\alpha]^\alpha = D_t^\alpha x. \tag{8.205}$$

Poisson brackets are invariant under canonical transformation.

8.19 Gradient, divergent, curl and Laplacian on fractal curves

In this section we generalize the F^α-calculus by defining the gradient, divergent, curl and Laplacian on fractal curves imbedding in R^3. Let $f \in B(F)$ as an F-continuous function on $C(a, b) \subset F$ [Golmankhaneh *et al.* (2013b)].

Definition 8.4. The gradient of a scalar valued function $f : F \to R$ is defined by

$$\nabla_F^\alpha f = \hat{e}^i D_{x_i}^\alpha f \quad i = 1, 2, 3, \tag{8.206}$$

where the $\hat{e}^i$ is the basis of R^n.

Definition 8.5. The divergent of a vector-valued function $\mathbf{f} : F \to R^n$ is defined by:

$$\nabla_F^\alpha.\mathbf{f} = \sum_{i=1}^{3} D_{x_i}^\alpha f_i, \tag{8.207}$$

where $\mathbf{f} = f_i \hat{e}^i$ is a vector field.

Definition 8.6. The Laplacian of a scalar valued function $f : F \to R$ is defined as

$$\triangle_F^\alpha f = (\nabla_F^\alpha)^2 f = \sum_{i=1}^{3} (D_{x_i}^\alpha)^2 f, \tag{8.208}$$

where the $\triangle_F^\alpha$ is called Laplacian on fractal curve.

8.20 Fractal differential forms

In this section we study the F^ξ-fractional form on fractals [Golmankhaneh *et al.* (2013b)].

Definition 8.7. A fractal 1-form is $H(x,y,z)d_F^\alpha x + G(x,y,z)d_F^\beta y + N(x,y,z)d_F^\epsilon z$ where H, G, N are functions on F.

Definition 8.8. Let $f(x,y,z) : F \to R^3$, then its exterior derivative is

$$d_F^\xi f(x,y,z) = D_x^\alpha f d_F^\alpha x + D_y^\beta f d_F^\beta y + D_z^\epsilon f d_F^\epsilon z \qquad (8.209)$$

where $\xi = \alpha + \beta + \epsilon$.

Definition 8.9. A fractal 1-form $Hd_F^\alpha x + Gd_F^\beta y + Nd_F^\epsilon z$ is called exact if $d_F^\xi f = Hd_F^\alpha x + Gd_F^\beta y + Nd_F^\epsilon z$ or

$$D_x^\alpha f = H, \qquad D_y^\beta f = G, \qquad D_z^\epsilon f = N. \qquad (8.210)$$

Definition 8.10. A fractal 1-form $Hd_F^\alpha x + Gd_F^\beta y + Nd_F^\epsilon z$ is closed if we have

$$D_y^\beta N = D_z^\epsilon G, \qquad D_x^\alpha G = D_y^\beta H, \qquad D_z^\epsilon H = D_x^\alpha N. \qquad (8.211)$$

Definition 8.11. A fractal 2-forms is defined by

$$M(x,y,z)d_F^\alpha x \bigwedge d_F^\beta y + W(x,y,z)d_F^\beta y \bigwedge d_F^\epsilon z + L(x,y,z)d_F^\epsilon z \bigwedge d_F^\alpha x \qquad (8.212)$$

where $\bigwedge$ is wedge product of two fractal 1-forms with following properties

$$d_F^\alpha x \bigwedge d_F^\beta y = -d_F^\beta y \bigwedge d_F^\alpha x \qquad d_F^\alpha x \bigwedge d_F^\alpha x = 0 \qquad (8.213)$$

Definition 8.12. Fractal gradient, divergence, and curl are defined by, respectively:

$$\nabla^\xi f = \hat{i}D_x^\alpha f + \hat{j}D_y^\beta f + \hat{k}D_z^\epsilon f \qquad \xi = \alpha + \beta + \epsilon \qquad (8.214)$$

$$\nabla^\xi . \vec{V} = D_x^\alpha V_x + D_y^\beta V_y + D_z^\epsilon V_z \qquad (8.215)$$

$$\nabla^\xi \times \vec{V} = \hat{i}(D_y^\beta V_z - D_z^\epsilon V_y) + \hat{j}(D_z^\epsilon V_x - D_x^\alpha V_z) + \hat{k}(D_x^\alpha V_y - D_y^\beta V_x). \qquad (8.216)$$

8.21 Maxwell's equations on fractals

In this section we obtain Maxwell's equations on fractals using fractal forms [Golmankhaneh *et al.* (2013b)].

Suppose a fractal 2-form as

$$\omega^\xi = (E_1 d_F^\alpha x_1 + E_2 d_F^\beta x_2 + E_3 d_F^\epsilon x_3) \bigwedge d_F^\kappa t$$

$$+ B_1 d_F^\beta x_2 \bigwedge d_F^\epsilon x_3 + B_2 d_F^\epsilon x_3 \bigwedge d_F^\alpha x_1 + B_3 d_F^\beta x_1 \bigwedge d_F^\alpha x_2. \quad (8.217)$$

where E_i, B_j, $i, j = 1, 2, 3$ are components of electromagnetic fields . If we define d_F^ξ by

$$d_F^\xi = D_{x_1}^\mu d_F^\mu x_1 + D_{x_2}^\mu d_F^\mu x_2 + D_{x_3}^\mu d_F^\mu x_3 + D_t^\mu d_F^\mu t, \quad (8.218)$$

then Eq.(8.217) seting $\alpha = \beta = \epsilon = \kappa = \mu$, and using $d_F^\xi \omega^\xi = 0$, becomes

$$\{D_{x_1}^\mu E_2 - D_{x_2}^\mu E_1\} d_F^\mu x_1 \bigwedge d_F^\mu x_2 \bigwedge d_F^\mu t$$

$$+ \{D_{x_1}^\mu E_3 - D_{x_3}^\mu E_1\} d_F^\mu x_1 \bigwedge d_F^\mu x_3 \bigwedge d_F^\mu t$$

$$+ \{D_{x_1}^\mu E_2 - D_{x_2}^\mu E_3\} d_F^\mu x_3 \bigwedge d_F^\mu x_2 \bigwedge d_F^\mu t$$

$$- D_t^\mu B_3 d_F^\mu x_1 \bigwedge d_F^\mu x_2 \bigwedge d_F^\mu t - D_t^\mu B_2 d_F^\mu x_1 \bigwedge d_F^\mu x_3 \bigwedge d_F^\mu t$$

$$- D_t^\mu B_1 d_F^\mu x_3 \bigwedge d_F^\mu x_2 \bigwedge d_F^\mu t = 0 \quad (8.219)$$

and,

$$(D_{x_1}^\mu B_1 + D_{x_2}^\mu B_2 + D_{x_3}^\mu B_3) d_F^\mu x_1 d_F^\mu x_2 d_F^\mu x_3 = 0. \quad (8.220)$$

Using the vector notation we have

$$\nabla^\mu \times \vec{E} = -D_t^\mu \vec{B} \quad (8.221)$$

$$\nabla^\mu . \vec{B} = 0. \quad (8.222)$$

which are called Maxwell's equations on fractals.
Now let define a fractal 2-form as

$$\pi^\mu = (J_1 d_F^\mu x_2 \bigwedge d_F^\mu x_3 + J_2 d_F^\mu x_3 \bigwedge d_F^\mu x_1 + J_3 d_F^\mu x_1 \bigwedge d_F^\mu x_2) \bigwedge d_F^\mu t$$

$$- \rho d_F^\mu x_1 \bigwedge d_F^\mu x_2 \bigwedge d_F^\mu x_3. \quad (8.223)$$

where J_i, $i = 1, 2, 3$ is components of current and ρ density of charge . Applying d_F^μ to Eq.(8.223) and using $d_F^\mu \pi^\mu = 0$ we obtain

$$(D_{x_1}^\mu J_1 + D_{x_2}^\mu J_2 + D_{x_3}^\mu J_3 + D_t^\mu \rho) d_F^\mu x_1 \bigwedge d_F^\mu x_2 \bigwedge d_F^\mu x_3 \bigwedge d_F^\mu t = 0, \quad (8.224)$$

and

$$\nabla^\mu . \vec{J} + D_t^\mu \rho = 0, \quad (8.225)$$

which is called fractal conservation of charge . Let consider fractal 1-form as

$$\zeta^\mu = A_1 d_F^\mu x_1 + A_2 d_F^\mu x_2 + A_3 d_F^\mu x_3 + \varphi d_F^\mu t, \qquad (8.226)$$

where $A_i, i = 1, 2, 3$ is vector potential . Using $d_F^\mu \zeta^\mu = 0$, one can obtain

$$\nabla^\mu \times \vec{A} = \vec{B} \qquad \nabla^\mu \varphi - D_t^\mu \vec{A} = \vec{E}. \qquad (8.227)$$

It is straightforward to get

$$c^2 \nabla^\mu \times \vec{B} = D_t^\mu \vec{E}, \qquad (8.228)$$

where c is speed of light.

8.22   Anomalous diffusion on fractal Cantor tartan

Consider diffusion equation on Cantor tartan $\mathbb{F}$ as the following [Golmankhaneh and Fernandez (2018)]:

$$\substack{C\\0}\mathcal{D}_t^\beta u(x, y, t) = \mathbf{K}\left((D_x^\eta)^2 u(x, y, t) + (D_y^\eta)^2 u(x, y, t)\right), \qquad (8.229)$$

for $(x, y) \in \mathbb{F}$ and $t \in F$. Here $\substack{C\\0}\mathcal{D}_t^\beta$ is the fractal left-sided Caputo derivative of order β on a fractal set F and $\mathbf{K}$ is the fractal diffusion coefficient. We also impose the following initial and boundary conditions:

$$u(x, y, t) = 0, \quad (x, y) \in \partial\mathbb{F}, \ t > 0, \qquad (8.230)$$

$$u(x, y, 0) = \sin(x\pi)\sin(y\pi), \quad (x, y) \in \mathbb{F} \cup \partial\mathbb{F}. \qquad (8.231)$$

The solution of the differential Eq.(8.229), using conjugacy of fractal calculus and standard calculus will be

$$u(x, y, t) = \sin(S_F^\eta(x)\pi)\sin(S_F^\eta(y)\pi)E_\beta^\eta\left(-2(S_F^\eta(t))^\epsilon\right), \quad \mathbf{K} = \frac{1}{\pi}, \qquad (8.232)$$

where $E_\beta^\eta(.)$ is the fractal Mittag-Leffler function of one parameter.

8.23   Brownian motion on fractal sets

In this section we present Brownian motion and fractional Brownian motion on fractal sets [Golmankhaneh and Sibatov (2021)].

The normalized Brownian motion on fractal sets is a random process that is denoted by $B_F^\zeta(t)$, where $t \in F$ with the following properties:
Its mean is given by

$$E[B_F^\zeta(t)] = 0, \qquad (8.233)$$

and its autocorrelation is defined by

$$E[B_F^\zeta(t)B_F^\zeta(s)] = \frac{1}{2}(|S_F^\zeta(t)| + |S_F^\zeta(t)| - |S_F^\zeta(t) - S_F^\zeta(s)|). \qquad (8.234)$$

By using the upper bound $S_F^\zeta(t) < t^\zeta$ we obtain

$$E[B_F^\zeta(t)B_F^\zeta(s)] \approx \frac{1}{2}(|t^\zeta| + |s^\zeta| - |t^\zeta - s^\zeta|), \qquad (8.235)$$

where $0 < \zeta \le 1$ is the dimension of fractal time set, which is the fractal parameter space of the random process. In Figure 8.19, we plotted a Brownian motion on a real line and on fractal sets.

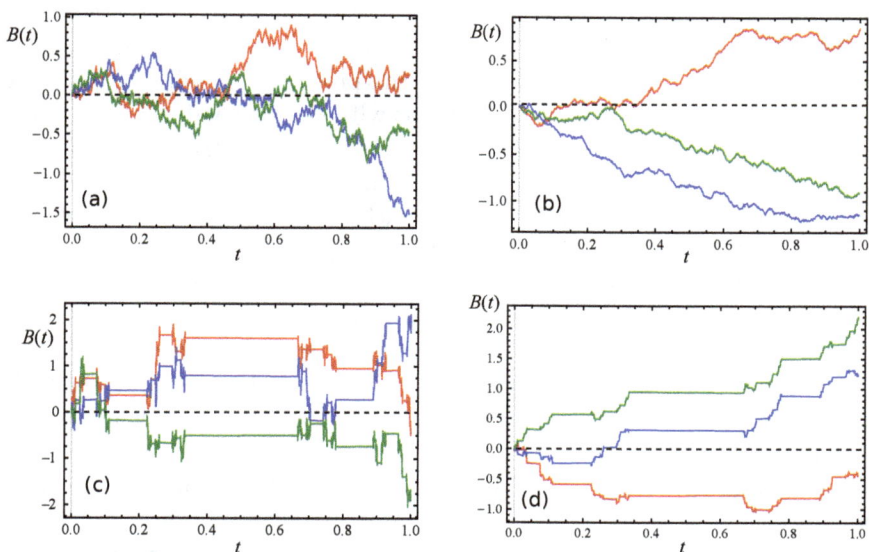

Fig. 8.19: Trajectories of the stochastic processes under consideration, (a) The Brownian motion on a real line with γ-dimension=1, $H = 0.5$, (b) The fractional Brownian motion on a real line with γ-dimension=1, $H = 0.75$, (c) The Brownian motion on a Cantor-like set with γ-dimension= $\alpha = 0.5$, $H = 0.25$, (d) The fractional Brownian motion on a Cantor-like set with γ-dimension=$\alpha = 0.5$, $H = 0.375$.

8.24   Fractional Brownian motion on fractal sets

In this section, we define the Fractional Brownian Motion (FBM) with
fractal support in terms of probability density function, stochastic process,
and fractal stochastic integral [Golmankhaneh and Sibatov (2021)]. It is
known that the FBM is a random walk with a continuous time parameter
space, which provides useful models for many physical phenomena for which
the empirical spectral power is given by a fractional law. FBM is defined
by the so-called Hurst exponent $H \in (0,1)$. The increments of FBM are
correlated (except the case $H = 1/2$) and obtained from Gaussian noise via
fractional integral

$$B_H(t) = \frac{1}{\Gamma(H+1/2)} \int_{-\infty}^{\infty} \left[(t-t')_+^{H-1/2} - (-t')_+^{H-1/2} \right] dB(t'), \quad (8.236)$$

that produces an auto-correlation function as

$$\langle B_H(t_1)B_H(t_2)\rangle = C \left[|t_1|^{2H} + |t_2|^{2H} - |t_1 - t_2|^{2H} \right], \quad (8.237)$$

where C is a constant. Here, $(t-t')_+ = 0$, if $(t-t') < 0$. When $H = 1/2$,
FBM becomes the ordinary BM. If $1/2 < H < 1$, we have super-diffusion
(enhanced diffusion), and if $H < 1/2$, the process is sub-diffusive. FBM
is non-stationary and non-Markovian but with stationary-dependent increments with normal distribution. It has a self-similar structure, and it is
defined as a stochastic integral of white noise . Its transition probability function is a normal distribution, which is the solution for heat equation. FBM is the generalization of the BM, which allows its increment
to be correlated over time which can be positive/persistent correlation or
negative/anti-persistent correlation . On one hand, the fractional Gaussian noise (FGN) is defined as the stochastic derivative of FBM. On the
other hand, the stochastic integration of FGN gives FBM; therefore, both
of them are characterized by Hurst parameter $(0 < H < 1)$, which is de-
fined by $Q \propto (\Delta t)^H$, where Q is standard derivation and Δt indicates lag
time. FBMs are characterized by H in this way if $H < 0.5$. it is called
anti-persistent, persistent if $0.5 < H$, and Brownian motion if $H = 0.5$.
FBM is called a non-stationary stochastic process with time-dependence
variance, while FGN is stationary. Fractional derivatives (FDs) were used
to define FBM and show its properties; hence, both of them are associated
with anomalous diffusion. The fractional models have proven a linear relation with the order of derivatives and fractal dimensions. FBMs trajectories are continuous but non-differentiable in the sense of standard calculus.
Anomalous diffusion in strong disorder media was studied by using FBM

and the corresponding mean square passion that does not increase linearly with time. The DNA sequence was converted into a format capable of being processed with the information theory and signal processing technique results with their interpretations, which lead to FBM.

8.24.1 *First representation*

To model anomalous sub-diffusion using non-local derivatives, which is called fractal fractional Brownian motion, we have [Golmankhaneh and Sibatov (2021)]

$$D_t^\zeta p(x,t) = {}_0\mathcal{D}_t^{1-\beta}(\Gamma_F^\zeta(\beta)\mathbf{D})(D_x^\alpha)^2 p(x,t)$$

$$= \left(S_F^\zeta(t)\right)^{\beta-1}\mathbf{D}\,(D_x^\alpha)^2 p(x,t), \qquad (8.238)$$

where $\mathbf{D}$ is constant and $p(x,t)$ is the probability density function of fractal fractional Brownian motion. The solution of Eq.(8.238) is

$$p(x,t) = \frac{1}{\sqrt{4\pi D(S_F^\zeta(t))^\beta}}\exp\left[-\frac{S_F^\alpha(x)^2}{4\pi D(S_F^\zeta(t))^\beta}\right]. \qquad (8.239)$$

By using the upper bound of the staircase function,

$$0 < S_F^\zeta(t) \le t^\zeta, \qquad (8.240)$$

Then, we have

$$p(x,t) \approx \frac{1}{\sqrt{4\pi Dt^{\zeta\beta}}}\exp\left[-\frac{x^{2\alpha}}{4\pi D\,t^{\zeta\beta}}\right], \quad 0 < \beta < \zeta, \qquad (8.241)$$

where ζ and α are fractal dimensions of the time and space, respectively, and β is a free parameter, which is called the non-local order derivatives on fractal time space. Using Eq.(8.241), the mean square displacement is obtained as follows:

$$< S_F^\alpha(x)^2 > 2D(S_F^\zeta(t))^\beta. \qquad (8.242)$$

Using Eq.(8.240), we obtain

$$< S_F^\alpha(x)^2 > \approx 2Dt^{\zeta\beta}. \qquad (8.243)$$

The mean square displacement in terms of Hurst parameter H is defined as

$$< S_F^\alpha(x)^2 > = 2DS_F^\zeta(t)^{2H}. \qquad (8.244)$$

Utilizing Eq.(8.240), we have

$$< S_F^\alpha(x)^2 > \approx 2Dt^{2\zeta H}. \qquad (8.245)$$

Hence, we get

$$\beta = 2H, \tag{8.246}$$

and

$$H_F^\zeta = \zeta H, \tag{8.247}$$

where H_F^ζ is the fractal Hurst exponent. We note that the fractal power law diffusivity is related to fractional derivative.

8.24.2 *Second representation*

The fractional Brownian motion on F is defined as a stochastic process with the following properties [Golmankhaneh and Sibatov (2021)].
Let $B^{H_F^\zeta}(t) = B_F^\zeta(t)$ and $H_F^\zeta \in [0, \zeta]$, then its correlations function and self similarity properties can be expressed as follows:

(1) Correlations :

$$E(B_F^\zeta(t)B_F^\zeta(s)) = \mathrm{Corr}(t,s)$$
$$= \frac{1}{2}\left(|S_F^\zeta(t)|^{2H} + |S_F^\zeta(s)|^{2H} - |S_F^\zeta(t) - S_F^\zeta(s)|^{2H}\right), \tag{8.248}$$

or its approximation

$$\mathrm{Corr}(t,s) \approx \frac{1}{2}\left(|t^\zeta|^{2H} + |s^\zeta|^{2H} - |t^\zeta - s^\zeta|^{2H}\right), \quad t, \ s \in F. \tag{8.249}$$

(2) Self Similarity:

$$B_F^\zeta(at) \sim a^{H_F^\zeta} B_F^\zeta(t). \tag{8.250}$$

(3) Second moment:

$$E(B_F^\zeta(t) - B_F^\zeta(s))^2 = E(B_F^\zeta(t))^2 + E(B_F^\zeta(s))^2 - 2E(B_F^\zeta(t)B_F^\zeta(s))$$
$$= |S_F^\zeta(t)|^{2H} + |S_F^\zeta(s)|^{2H} - |S_F^\zeta(t)|^{2H} - |S_F^\zeta(s)|^{2H} + |S_F^\zeta(t) - S_F^\zeta(s)|^{2H}$$
$$= |S_F^\zeta(t) - S_F^\zeta(s)|^{2H}. \tag{8.251}$$

8.24.3 *Third representation*

The fractal fractional Brownian motion is defined by [Golmankhaneh and Sibatov (2021)]

$$B^{H_F^\zeta}(t) = \frac{1}{\Gamma_F^\alpha(H_F^\zeta + \frac{1}{2})} \int_0^t (t-s)^{(H_F^\zeta - 1/2)} d_F^\alpha B(s), \tag{8.252}$$

where we suppose

$$\phi(s) = (t - s)^{(H_F^\zeta - 1/2)}. \tag{8.253}$$

To interpret Eqs.(8.252) and (8.253), we define them for the partition $t_1 = a, \ldots, t_l = b$ and $s \in (t_j, t_{j+1}], s \in F$ as follows:

$$\int_a^b \phi(s) d_F^\alpha B(s) = \sum_{n=1}^l \phi(s) \Gamma(\alpha + 1) \left(B^{H_F^\zeta}(t_{j+1}) - B^{H_F^\zeta}(t_j) \right)^\alpha$$

$$= \sum_{n=1}^l \phi(j) \Gamma(\alpha + 1) \left(B^{H_F^\zeta}(t_{j+1}) - B^{H_F^\zeta}(t_j) \right)^\alpha. \tag{8.254}$$

In Figure 8.19, we sketched the fractional Brownian motion on a real line and on fractal sets.

8.25 Spectral density of fractional Brownian motion on fractal sets

In this section, we present the spectral density FBM on fractal sets [Golmankhaneh and Sibatov (2021)]. In view of the conjugacy between the fractal calculus and standard calculus, we suggest the following spectral density:

$$p_{B_F^\xi}(\omega) = F_-^\alpha \lim_{T \to \infty} p_{B_{F,T}^\xi}(\omega) = F_-^\alpha \lim_{T \to \infty} \frac{1}{T} E|\hat{B}(\omega)|^2, \tag{8.255}$$

where

$$\hat{B}(\omega) = \int_{t_0}^{t_0+T} B_F^\zeta(t) d_F^\alpha t, \tag{8.256}$$

is the fractal Fourier of $B_F^\zeta(t)$ and

$$E|\hat{B}(\omega)|^2$$
$$= \int_{t_0}^{t_0+T} \int_{t_0}^{t_0+T} E(B_F^\zeta(t_1) B_F^\zeta(t_2)) e^{-i S_F^\zeta(\omega)(S_F^\zeta(t_1) - S_F^\zeta(t_2))} d_F^\alpha t_1 d_F^\alpha t_2, \tag{8.257}$$

where

$$E(B_F^\zeta(t_1) B_F^\zeta(t_2))$$
$$= \frac{1}{2} (|S_F^\zeta(t_1) - S_F^\zeta(t_0)|^{2H} + |S_F^\zeta(t_2) - S_F^\zeta(t_0)|^{2H} - |S_F^\zeta(t_2) - S_F^\zeta(t_1)|^{2H}). \tag{8.258}$$

By virtue of the conjugacy of fractal calculus with the ordinary one, we can write the following:

$$p_{B_{F,T}^\xi}(\omega) = \begin{cases} \frac{S_F^\zeta(T)^{2H+1}}{2(H+1)}, & S_F^\zeta(\omega) = 0; \\ \frac{\pi}{4}\left(2H\frac{\pi^{2H-1}}{(S_F^\zeta(\omega))^{2H+1}} + \frac{S_F^\zeta(T)^{2H-1}}{S_F^\zeta(\omega)^2}\right), & S_F^\zeta(\omega) \neq 0. \end{cases}$$

$$\propto \begin{cases} \frac{T^{\zeta(2H+1)}}{2(H+1)}, & S_F^\zeta(\omega) = 0; \\ \frac{\pi}{4}(2H\frac{\pi^{2H-1}}{(\omega)^{\zeta(2H+1)}} + \frac{T^{\zeta(2H-1)}}{(\omega)^{2\zeta}}), & S_F^\zeta(\omega) \neq 0. \end{cases} \quad (8.259)$$

and

$$p_{B_F^\xi}(\omega) = \begin{cases} \infty, & S_F^\zeta(\omega) = 0, \quad 0 < H < 1, \\ \frac{H\pi^{2H}}{2}\frac{1}{S_F^\zeta(\omega)^{(2H+1)}}, & S_F^\zeta(\omega) \neq 0, \quad 0 < H < 1/2, \\ \frac{\pi}{2}\frac{1}{S_F^\zeta(\omega)^2}, & S_F^\zeta(\omega) \neq 0, \quad H = 1/2, \\ \infty, & S_F^\zeta(\omega) \neq 0, \quad 1/2 < H < 1. \end{cases}$$

$$\propto \begin{cases} \infty, & S_F^\zeta(\omega) = 0, \quad 0 < H < 1, \\ \frac{H\pi^{2H}}{2}\frac{1}{\omega^{\zeta(2H+1)}}, & S_F^\zeta(\omega) \neq 0, \quad 0 < H < 1/2, \\ \frac{\pi}{2}\frac{1}{\omega^{2\zeta}}, & S_F^\zeta(\omega) \neq 0, \quad H = 1/2, \\ \infty, & S_F^\zeta(\omega) \neq 0, \quad 1/2 < H < 1. \end{cases} \quad (8.260)$$

Remark 8.5. For example, if we consider FBM on the triadic Cantor set ($\zeta = 0.6$), then the Hurst parameter $0 < H < 0.6$; in particular, we can characterize it as follows: if $H < 0.3$, then we have anti-persistent FBM, FBM persistent if $0.3 < H$, and BM if $H = 0.3$. Using the virtue of spectral density, we can also categorize FBM on the triadic Cantor set as follows: if $0 < H \leq 0.3$, then by using Eq.(8.260), we have $p_{B_F^\xi}(\omega) \propto \omega^{-0.6(2H+1)}$. In addition, from Eq.(8.259), if $0 < H \leq 0.3 \Rightarrow p_{B_{F,T}^\xi}(\omega) \propto \omega^{-0.6(2H+1)}$, but $0.3 < H < 0.6 \Rightarrow p_{B_{F,T}^\xi}(\omega) \propto \omega^{-1.2}$.

8.26 Fractal logistic equation

In this section we give the fractal logistic equation and its solution [Golmankhaneh and Cattani (2019)].
Let us consider fractal logistic equation as follows:

$$D_t^\alpha z(t) = r_K z(t)\left(1 - \frac{z(t)}{r_K'}\right), \quad (8.261)$$

where r_F is called fractal growth parameter and r_F' is called fractal carrying capacity. Applying conjugacy of fractal calculus and standard calculus we obtain the solution of Eq.(8.261) as follows:

$$z(t) = \frac{z(0)r_F}{z(0) + (r_F - z(0))\exp(-r'_F S_F^\alpha(t))}, \tag{8.262}$$

where $z(0)$ is called initial population. If we consider upper limit of $S_F^\alpha(t) < t^\alpha$ we get

$$z(t) \simeq \frac{z(0)r_K}{z(0) + (r_F - z(0))\exp(-r'_F t^\alpha)}, \tag{8.263}$$

where $z(t)$ is called fractal Logistic function. In Figure 8.20 we have plotted Eq.(8.262). The fractal derivative of Eq.(8.262) is

$$D_t^\alpha z(t) = \frac{z(0)r_F(r_F - z(0))r'_F \exp(-r'_F S_F^\alpha(t))}{(z(0) + (r_F - z(0))\exp(-r'_F S_F^\alpha(t)))^2}, \tag{8.264}$$

and its upper bound is

$$D_t^\alpha z(t) \simeq \frac{z(0)r_F(r_F - z(0))r'_F \exp(-r'_F t^\alpha)}{(z(0) + (r_F - z(0))\exp(-r'_F t^\alpha))^2}. \tag{8.265}$$

In Figures 8.21 and 8.22, we present the upper bound of fractal logistic function and its fractal derivatives by setting different value of α.

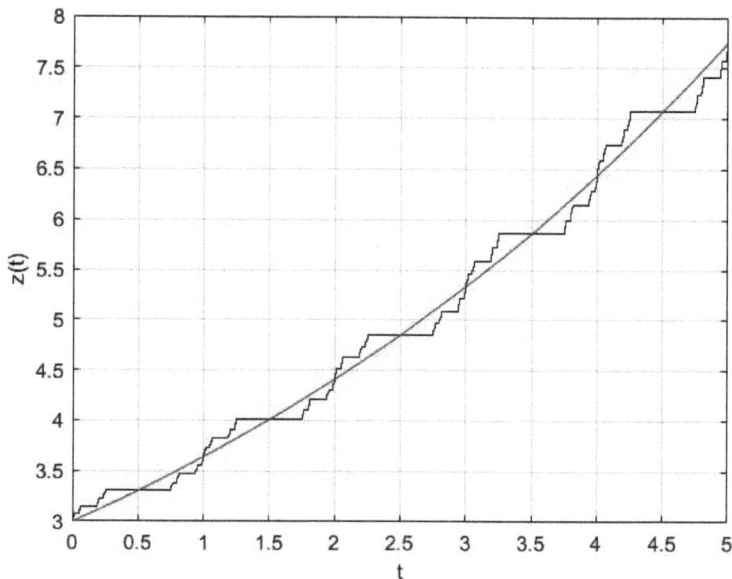

Fig. 8.20: Graph of fractal logistic curve with $\alpha = 0.5$ where $\mu = 1/2$, $z(0) = 3$, $r_F = 100$, $r'_F = 0.2$

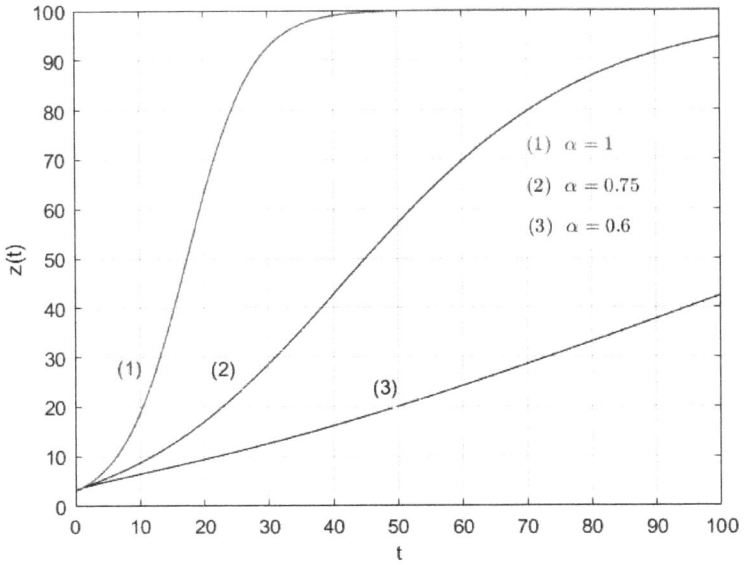

Fig. 8.21: Upper bound function for the value $z(0) = 3$, $r_F = 100$, $r'_F = 0.2$

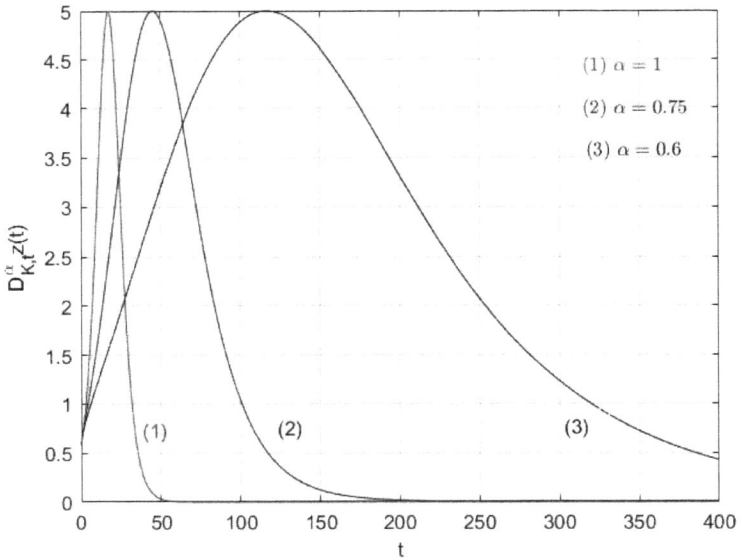

Fig. 8.22: Upper bound of fractal derivative for the values $z(0) = 3$, $r_F = 100$, $r'_F = 0.2$

The inflection point $(t = t_{ip})$ of fractal Logistic function is defined $z(t_{ip}) = r_F/2$, then we have

$$S_F^\alpha(t_{ip}) = \frac{\ln(r_F/z(0) - 1)}{r_F'}. \tag{8.266}$$

Using the upper bound of $S_F^\alpha(t) \leq t^\alpha$, we get

$$t_{ip}(\alpha) = \left(\frac{\ln(r_F/z(0) - 1)}{r_F'}\right)^\alpha. \tag{8.267}$$

In Figure 8.23 we have sketched Eq.(8.267) which indicates inflection points versus α.

Remark 8.6. In Figures 8.21, 8.22 and 8.23, we have shown that the power law model for the processes with the fractal structure by using fractal calculus leads to ordinary cases by letting $\alpha = 1$.

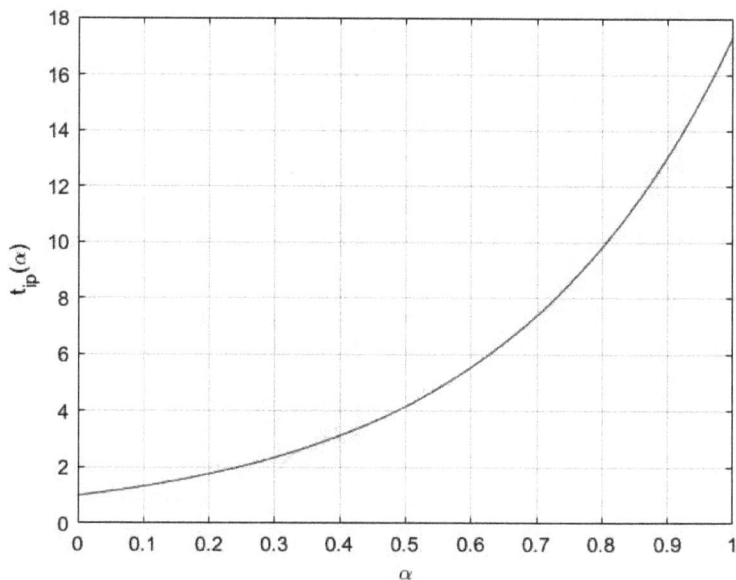

Fig. 8.23: Graph of inflection points as function of α by choosing $z(0) = 3$, $r_F = 100$, $r_F' = 0.2$

8.27 Distribution on fractals

In this section we defined the random variables and their feathers on fractals [Golmankhaneh and Fernandez (2019)].

Example 8.7. Consider fractal Cauchy random variable with probability density function with parameter a as follows

$$f_X(y) = \frac{a}{\pi((y\chi_F)^2 + a^2)}, \quad -\infty < y < +\infty \qquad (8.268)$$

where a is constant.

In Figure 8.24 we have plotted Eq.(8.268).

Distribution function of fractal Cauchy random variable by using Eq.(7.92) is

$$F_X(y) = \frac{\Gamma(1+\alpha)}{2} + \frac{\Gamma(1+\alpha)}{\pi} \arctan\left(\frac{S_F^\alpha(y)}{a\Gamma(1+\alpha)}\right) \qquad (8.269)$$

In Figure 8.25 we have sketched Eq.(8.269).

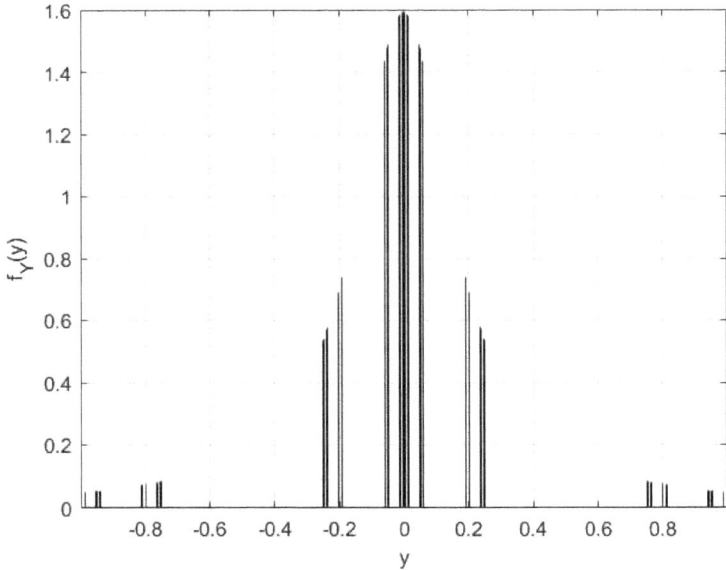

Fig. 8.24: Probability density function of fractal Cauchy random choosing, $\alpha = 0.5$, $a = 0.2$

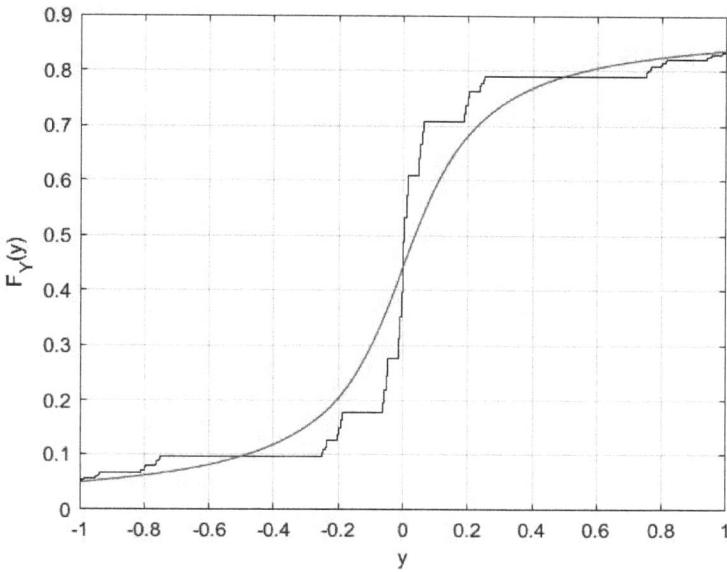

Fig. 8.25: Distribution function of fractal Cauchy random, $\alpha = 0.5$, $a = 0.2$

Using Eq.(7.98) we obtain characteristic function of fractal Cauchy as follows

$$\Psi_X(\omega) = \int_{-\infty}^{\infty} e^{i\omega y} \frac{a}{\pi((y\chi_F)^2 + a^2)} d_F^\alpha y = \Gamma(1+\alpha) e^{-a \left| \frac{S_F^\alpha(\omega)}{\Gamma(1+\alpha)} \right|} \quad (8.270)$$

In Figure 8.26 we have shown Eq.(8.270).
The fractal Shannon entropy of fractal Cauchy random variable utilizing Eq.(7.99) is

$$H_Y(y) = -\int_{-\infty}^{\infty} \frac{a}{\pi((y\chi_F)^2 + a^2)} \log\left(\frac{a}{\pi((y\chi_F)^2 + a^2)}\right) d_F^\alpha y$$
$$= \Gamma(\alpha + 1) \ln(4\pi a) \quad (8.271)$$

In Figure 8.27 we have drawn Eq.(8.271).

Example 8.8. Consider fractal Laplace random variable Y with following probability density function

$$f_X(t) = \frac{1}{2c} \exp\left(-\frac{|t\chi_F - \nu|}{c}\right), \quad (8.272)$$

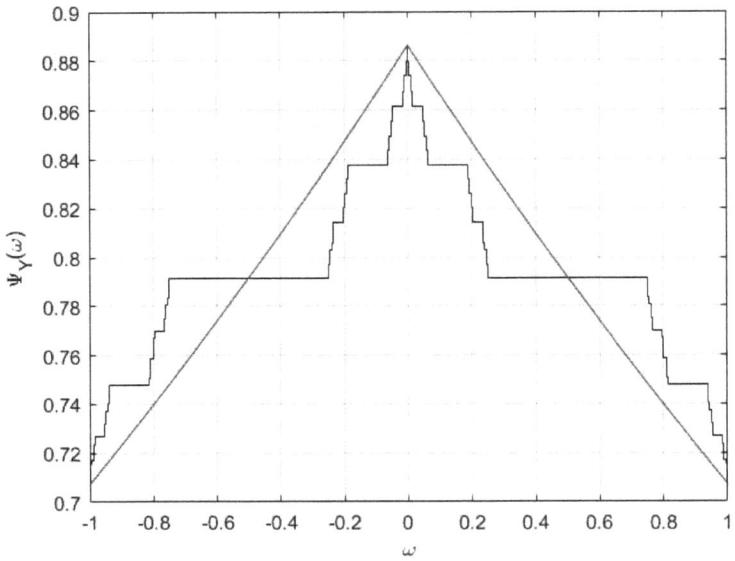

Fig. 8.26: Characteristic function of fractal Cauchy random, $\alpha = 0.5$, $a = 0.2$

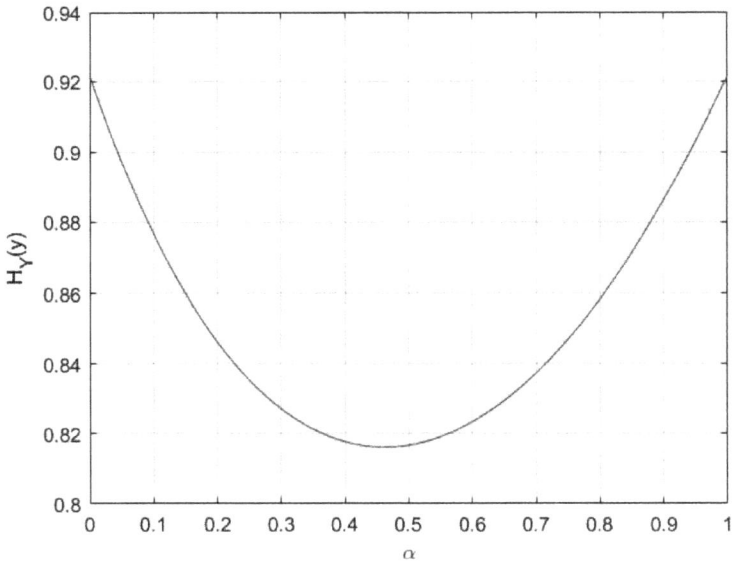

Fig. 8.27: Entropy of fractal Cauchy random, $a = 0.2$

Distribution function of fractal Laplace random variable is

$$F_X(t) = \frac{\Gamma(1+\alpha)}{2} + \frac{\Gamma(1+\alpha)}{2}\mathrm{sgn}(t-\nu)\left(1 - \exp(\frac{S_F^\alpha(t) - \nu\Gamma(1+\alpha)}{c\Gamma(1+\alpha)})\right),$$

(8.273)

The mean fractal Laplace random variable is

$$E[X] = \Gamma(1+\alpha)\nu \tag{8.274}$$

The variance of the fractal Laplace random variable is

$$Var[X] = 2(c\Gamma(1+\alpha))^2 \tag{8.275}$$

The fractal Shannon entropy of fractal Laplace random variable is

$$H_X(t) = \log(2c\Gamma(1+\alpha)e) \tag{8.276}$$

The characteristic function of the fractal Laplace random variable is

$$\Psi_X(\omega) = \frac{\Gamma(1+\alpha)\exp\left(\Gamma(1+\alpha)\nu i S_F^\alpha(\omega)\right)}{1 + (c\Gamma(1+\alpha))^2 S_F^\alpha(\omega)^2} \tag{8.277}$$

In Figures 8.28, 8.29, 8.30, and 8.31 we have plotted the graphs of fractal probability density function, fractal distribution function, fractal characteristic function, fractal Shannon entropy, of Laplace random variable.

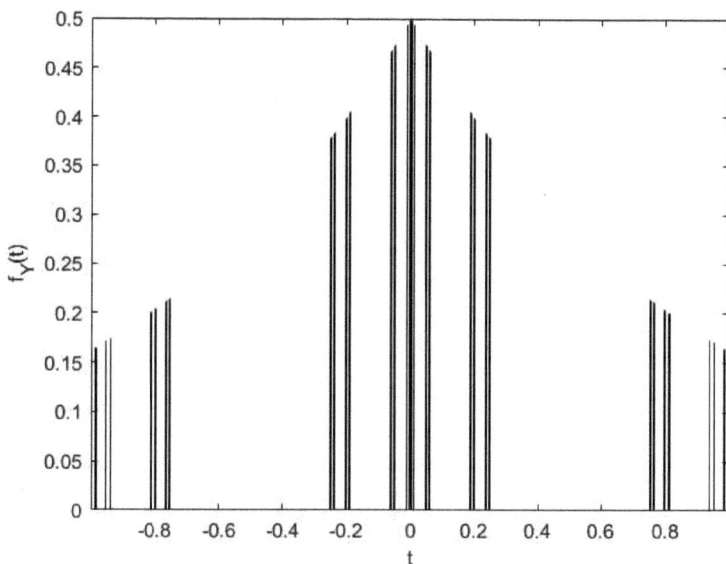

Fig. 8.28: Probability density function of fractal Laplace random variable supposing, $c = 1$, $\alpha = 0.5$, $\nu = 0$

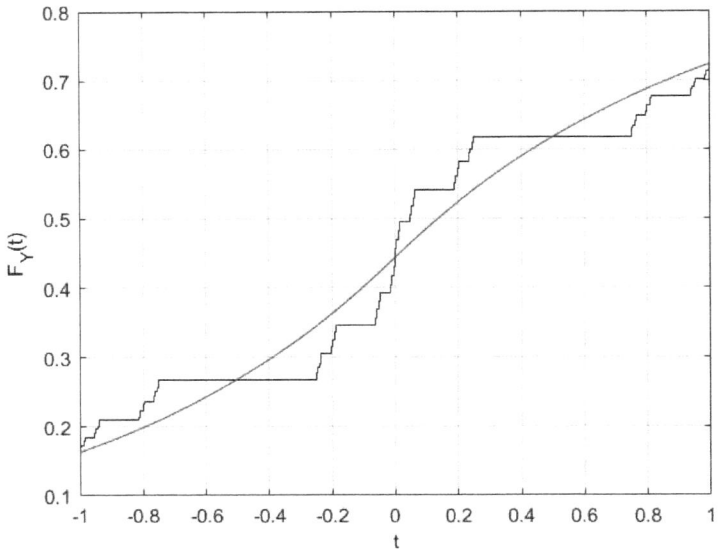

Fig. 8.29: Distribution function of fractal Laplace random variable Cauchy letting, $c = 1$, $\alpha = 0.5$, $\nu = 0$

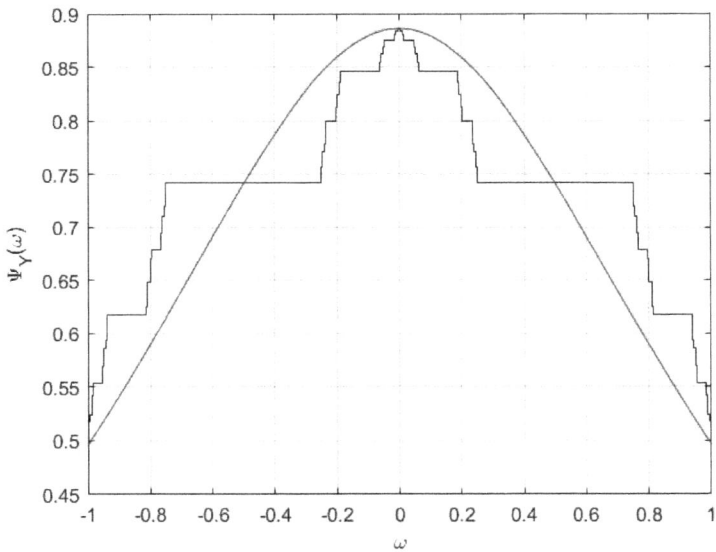

Fig. 8.30: Characteristic function of fractal Laplace random letting, $c = 1$, $\alpha = 0.5$, $\nu = 0$

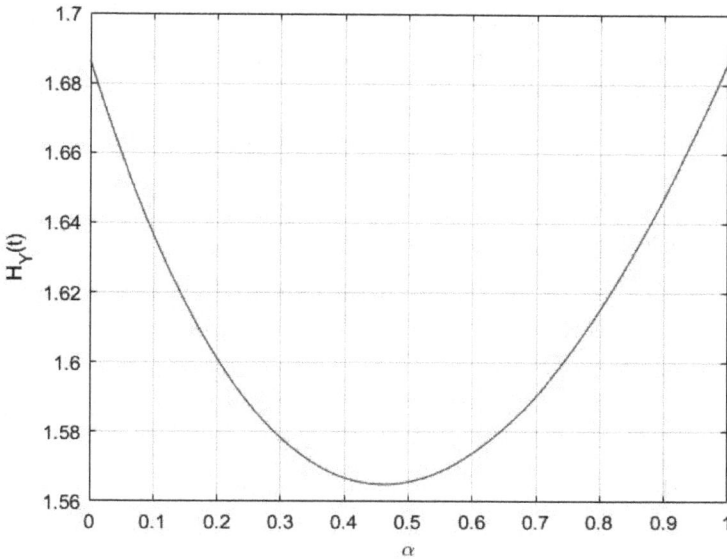

Fig. 8.31: Entropy of fractal Laplace random setting, c=1

Remark 8.7. The smooth line in the graphs show the case with continuous support and $\alpha = 1$ we recall standard result and the examples work out using conjugacy of fractal calculus and standard calculus.

8.28 Hierarchy of stable distributions on fractals

In this section, we define the stable distribution on fractals and present examples such as fractal Cauchy and Gaussian stable distribution [Golmankhaneh and Fernandez (2019)].

If $X_1(t)$ and $X_2(t)$ are independent and identically distributed (*iid*) on the Cantor-like sets ($t \in F$) with means $\mu \in (-\infty, +\infty)$ and variances $\sigma \in (0, +\infty)$. A fractal random variables $X(t)$ with cumulative distribution function (CDF) $F_X(t)$ on fractal sets is defined by

$$X(t) = \frac{a_1 X_1 + a_2 X_2 - b}{a}. \tag{8.278}$$

X has same CDF for given positive a_1, a_2 if we have following condition

$$F_X\left(\frac{t}{a_1}\right) F_X\left(\frac{t}{a_2}\right) = F_X\left(\frac{t-b}{a}\right), \quad t \in F. \tag{8.279}$$

If a, b exist, hence $F_X(t)$ is called fractal stable distribution. In generally, there is no solution to Eq.(8.279). Another definition of the stable fractal random variable X if its characteristic function can be written as

$$\Psi_X(\omega; \lambda, c, \nu, \mu) = \exp(i\mu S_F^\alpha(\omega) - \Phi). \qquad (8.280)$$

$$\Phi = \begin{cases} |cS_F^\alpha(\omega)|^\lambda (1 - i\nu \, \text{sgn}(S_F^\alpha(\omega)) \tan \frac{\pi\lambda}{2}), & \lambda \neq 1; \\ |cS_F^\alpha(\omega)|^\lambda (1 + i\nu \frac{2}{\pi} \, \text{sgn}(S_F^\alpha(\omega)) \ln |S_F^\alpha(\omega)|), & \lambda = 1, \end{cases} \qquad (8.281)$$

where $i = \sqrt{-1}$, $\lambda \in (0, 2]$ stability parameter, $\nu \in [-1, 1]$ skewness parameter, $c \in (0, \infty)$ scale parameter, $\mu \in (-\infty, +\infty)$ location

$$\text{sgn}(S_F^\alpha(\omega)) = \begin{cases} 1, & \omega > 0; \\ 0, & \omega = 0; \\ -1, & \omega < 0. \end{cases} \qquad (8.282)$$

in view of the upper bound $S_F^\alpha(\omega) \leq \omega^\alpha$ we have

$$\Psi_X(\omega; \lambda, c, \nu, \mu) \approx \begin{cases} \exp\left(i\mu\omega^\alpha - |(c\omega)^\alpha|^\lambda (1 - i\nu \, \text{sgn}(\omega^\alpha) \tan \frac{\pi\lambda}{2})\right), & \lambda \neq 1/\alpha; \\ \exp\left(i\mu\omega^\alpha - |(c\omega)^\alpha|^\lambda (1 + i\nu(\frac{2}{\pi} \, \text{sgn}(\omega^\alpha) \ln |\omega^\alpha|)\right), & \lambda = 1/\alpha. \end{cases} \qquad (8.283)$$

If we choose $\mu = 0$, and $\nu = 0$, then we lead to

$$\Psi_X(\omega; \lambda, c, 0, 0) \approx \exp\left[-|(c\omega)^\alpha|^\lambda\right], \quad 0 < \lambda \leq \frac{2}{\alpha}; \qquad (8.284)$$

where if $\lambda < 0$ then it rules out means $\Psi_X(\omega) = 1$ as $|\omega| \to \infty$ then there is no fractal inverse Fourier transformation. Also, $\lambda > \frac{2}{\alpha}$ leads to the negative $f_X(t)$.

Here, we define some of fractal stable distribution:

(1) Gaussian stable distribution on fractal sets: In Eq.(8.280) if we choose $\lambda = 2$, $\nu = 0$ and $\sigma^2 = 2c^2$ then we have $\Psi_X(\omega)$ and its asymptotic expression using upper bound gives

$$\Psi_X(\omega, 2, c, 0, \mu) = \exp\left[i\mu S_F^\alpha(\omega) - \frac{1}{2}\sigma^2 |S_F^\alpha(\omega)|^2\right]$$

$$\approx \exp\left[i\mu\omega^\alpha - \frac{1}{2}\sigma^2\omega^{2\alpha}\right] \qquad (8.285)$$

where σ and μ are variance and mean square, respectively. The corresponding probability distribution function which is inverse Fourier transformation of Eq.(8.285) is as follows:

$$p(t, 2, c, 0, \mu) = \frac{1}{\sqrt{2\pi\sigma^2}} \exp\left[\frac{(t\chi_F - \mu)^2}{2\sigma^2}\right]$$

(2) Cauchy stable distribution on fractal sets: If we choose $\lambda = 1$ therefore Eq.(8.280) gives

$$\Psi_X(\omega, 1, 1, 0, \mu) = \exp\left[i\mu S_F^\alpha(\omega) - |S_F^\alpha(\omega)|\right] \qquad (8.286)$$

The corresponding probability distribution function is as

$$p(t, 1, 1, 0, \mu) = \frac{1}{\pi(t\chi_F - \mu)^2 + 1)} \qquad (8.287)$$

(3) Levy α-Stable distribution on fractal sets: If we choose $\lambda = 1/2$ therefore Eq.(8.280) gives

$$\Psi_X(\omega, 1/2, 1, 1, \mu) = \exp\left[i\mu S_F^\alpha(\omega) - |S_F^\alpha(\omega)|^{1/2}(1 - i \operatorname{sgn} S_F^\alpha(\omega))\right]$$
$$\approx \exp\left[i\mu\omega^\alpha - |\Gamma(1 + \alpha)\omega^\alpha|^{1/2}(1 - i \operatorname{sgn} \omega^\alpha)\right] \qquad (8.288)$$

The corresponding probability distribution function is as

$$p(t, 1/2, 1, 1, \mu) = \sqrt{\frac{1}{2\pi(t\chi_F - \mu)^3}} \exp\left[\frac{-1}{2(t\chi_F - \mu)}\right] \qquad (8.289)$$

Remark 8.8. We can summarized the above results as follows:

(1) If we set $\lambda = 2/\alpha$ the stable distribution has the mean and the variance limited.
(2) If we set $0 < \lambda < 1/\alpha$ mean and variance is infinite.
(3) If we set $1/\alpha < \lambda < 2/\alpha$ mean is finite and variance is infinite (it is called Heavy-tailed).

Remark 8.9. If X is a random variable on fractal sets has stable distribution with exponent $1/\alpha < \lambda < 2/\alpha$ then $1/X^\lambda$ has also stable distribution with exponent $1/\lambda$.

Example 8.9. Consider random walk on fractal set the first passing time of point $b \in C^\beta$ conditional density probability distribution between time $(t, t + dt)$ has the following form

$$p(t, b|0) = \frac{S_F^\alpha(b)}{\sqrt{4\pi Dt^3}} \exp\left[\frac{-S_F^\alpha(b)^2}{4Dt}\right], \qquad (8.290)$$

where D is constant.

Example 8.10. Two particles have random walks on a fractal set with positions X_1, $X_2 \in F$ the new fractal random variable $Y = X_1/X_2$ has fractal

Cauchy distribution. Here each X_1, X_2 has fractal Gaussian distribution. Since If we suppose probability distribution of Y is denoted by $\rho(y,t)$, then

$$\rho(y,t) = \int_{-\infty}^{\infty} d_F^\alpha x_1 \int_{-\infty}^{\infty} d_F^\alpha x_2 \frac{1}{4\pi Dt} \exp\left[\frac{(-x_1^2 - x_2^2)\chi_F}{4Dt}\right] \delta_F^\alpha\left((y - \frac{x_1}{x_2})\chi_F\right)$$

$$= \int_{-\infty}^{\infty} d_F^\alpha x_2 \frac{|x_2\chi_F|}{4\pi Dt} \exp\left[\frac{-x_2^2(1+y^2)\chi_F}{4Dt}\right]$$

$$= 2\int_0^{\infty} d_F^\alpha x_2 \frac{x_2\chi_F}{4\pi Dt} \exp\left[\frac{-x_2^2(1+y^2)\chi_F}{4Dt}\right]$$

$$= \frac{1}{\pi(1 + S_F^\alpha(y)^2)}$$

$$\approx \frac{1}{\pi(1 + y^{2\alpha})}. \tag{8.291}$$

In Figure 8.32, we have plotted $\rho(y,t)$ for different dimensions.

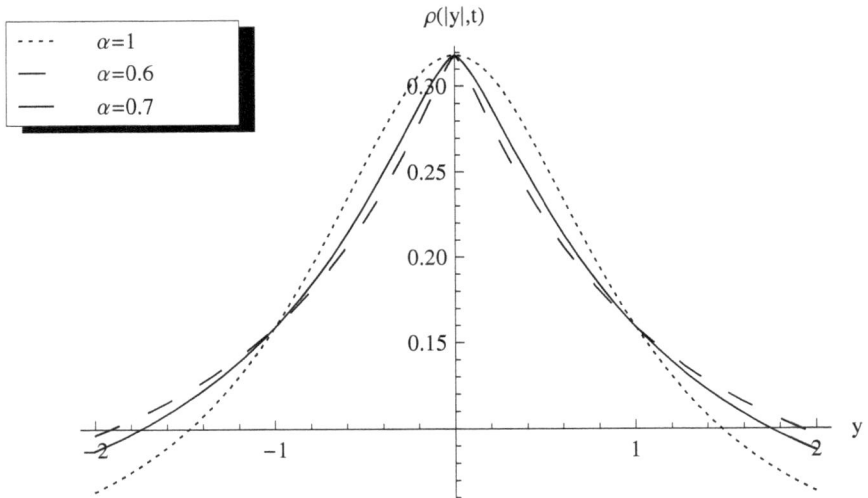

Fig. 8.32: Graph of probability distribution $\rho(y,t)$

8.29 Kronig-Penney model on fractals

In this section, we define α-dimensional Schrödinger equation to generalize the Kronig-Penney model [Golmankhaneh and Kamal Ali (2021)].

A fractal crystal lattice with α-dimension, and period a is built by lattice translation vector $T = ma\hat{i}$, $m = \pm 1, \pm 2, \ldots$ that operates on fractal.

A fractal potential energy function on fractal is defined by

$$U(x) = \begin{cases} U_0, \ -b < x < 0 \quad x \in F; \\ 0, \quad 0 < x < a; \\ U_0, \ a < x < b, \quad x \in F. \end{cases} \tag{8.292}$$

where U_0 is constant, and $U(x) = U(x + a)$ is called the potential energy function of the fractal Kronig-Penney model or the fractal potential energy comb.

In Figure 8.33 we have plotted the potential of the fractal Kronig- Penney model on fractal.

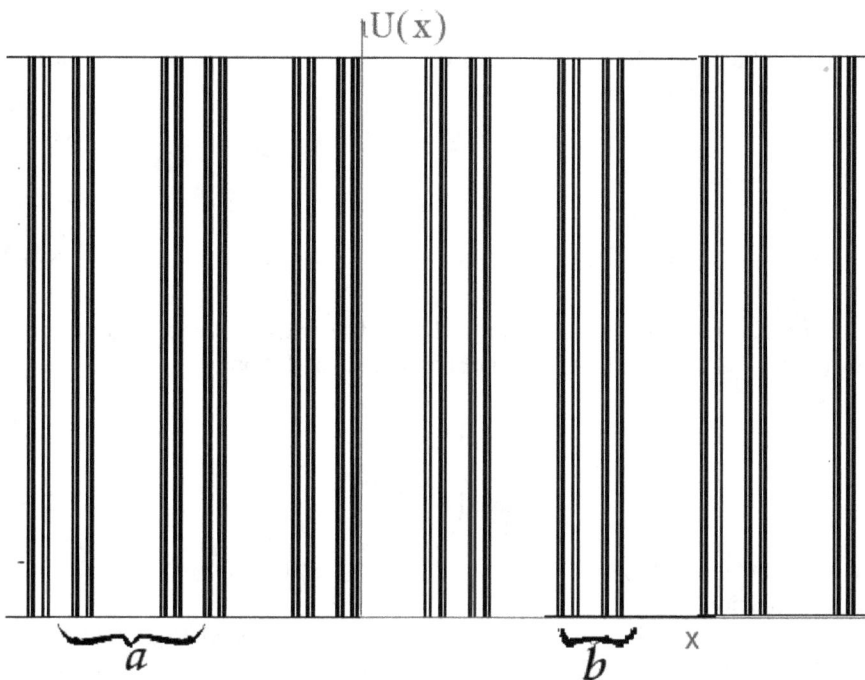

Fig. 8.33: Graph of potential of the fractal Kronig-Penney model

Consider the α-dimensional Schrödinger equation is given by

$$-\frac{\hbar^2}{2m} D_x^{2\alpha} \Psi(x,t) + U(x)\Psi(x,t) = i\hbar D_t^\alpha \Psi(x,t), \quad x,t \in F, \quad (8.293)$$

where $i = \sqrt{-1}$, $\hbar$ is the reduced Planck's constant, t is fractal time [Shlesinger (1988); Vrobel (2011); Welch (2020)], x is fractal position, $\Psi(x,t)$ is called fractal wave function , and the left hand side of Eq.(5.68) is equivalent to the fractal Hamiltonian energy operator acting on $\Psi(x,t)$. Eq.(5.68) is called the general form of the Schrödinger equation on fractal time-space. We will investigate the fractal Schrödinger time-dependent equation and time-independent for the given fractal potential energy. Considering that the fractal wave function $\Psi(x,t)$ is separable. In addition, the fractal wave function of two variables are represented as the product of two fractal separate functions of a fractal single variable:

$$\Psi(x,t) = \psi(x)T(t). \quad (8.294)$$

Then, in view of conjugacy fractal calculus with ordinary calculus and by using mathematical methods for solving partial differential equations, it is noticeable that the fractal wave equation is written into two separate differential equations as follows:

$$i\hbar D_t^\alpha T(t) = E_F^\alpha T(t), \quad (8.295)$$

$$-\frac{\hbar^2}{2m} D_x^{2\alpha} \psi(x) + U(x)\psi(x) = E_F^\alpha \psi(x), \quad (8.296)$$

which is called the fractal Schrödinger time-independent equation. Note that Eq.(8.295) depends on the fractal time $T(t)$ while Eq.(8.296) depends only on the fractal position $\psi(x)$. E_F^α is the fractal energy eigenvalue . Eq.(8.295) can be solved immediately to give the following:

$$T(t) = \exp\left(\frac{-iE_F^\alpha S_F^\alpha(t)}{\hbar}\right), \quad (8.297)$$

$$\approx \exp\left(\frac{-iE_F^\alpha t^\alpha}{\hbar}\right). \quad (8.298)$$

In the above solution, we have used $c_1 t^\alpha \leq S_F^\alpha(t) \leq c_2 t^\alpha$ to write the approximation.

In Figure 8.34, 8.35, 8.36, and 8.37, we have sketched Eqs.(8.297) and (8.298). Solutions of Eq.(8.295) depend on the fractal potential energy function $U(x)$. In the following part, we present an approach to solve Eq.(8.295) for the fractal potential energy comb.

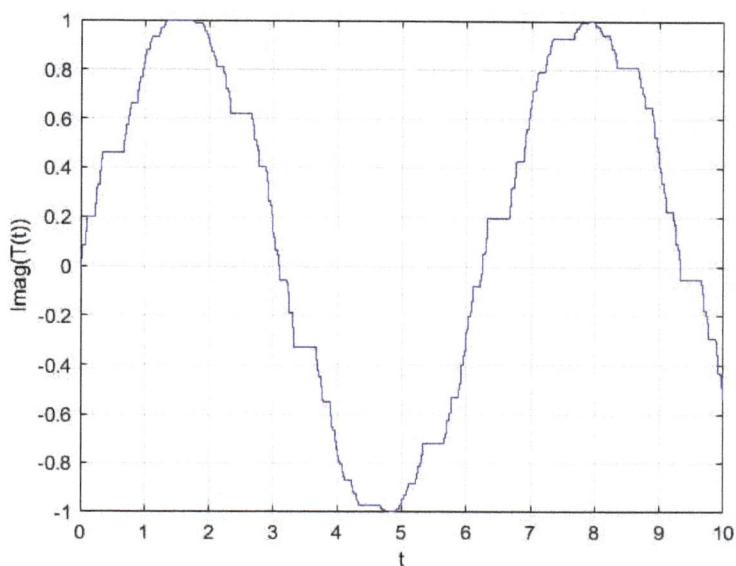

Fig. 8.34: Graph of imaginary part of Eq.(8.297)

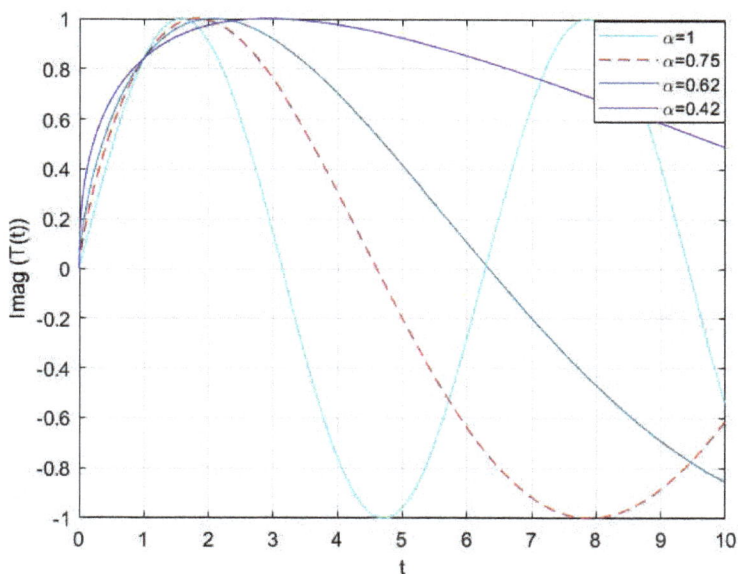

Fig. 8.35: Graph of imaginary part of Eq.(8.298)

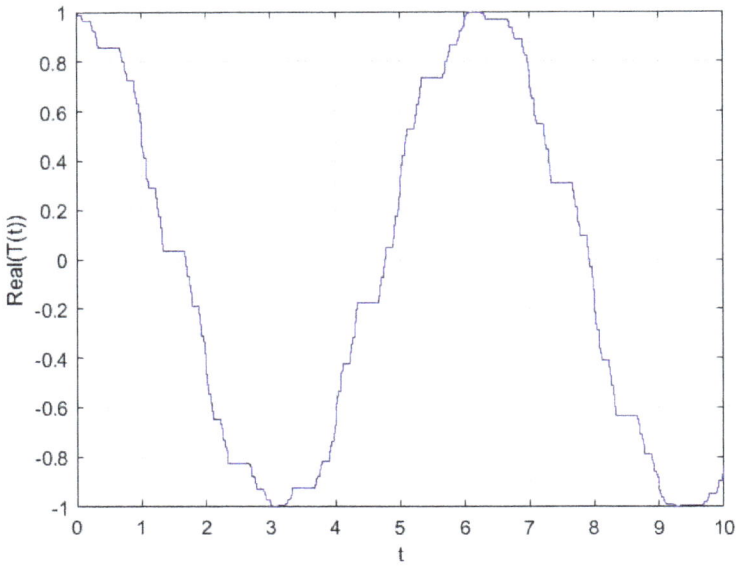

Fig. 8.36: Graph of real part of Eq.(8.297)

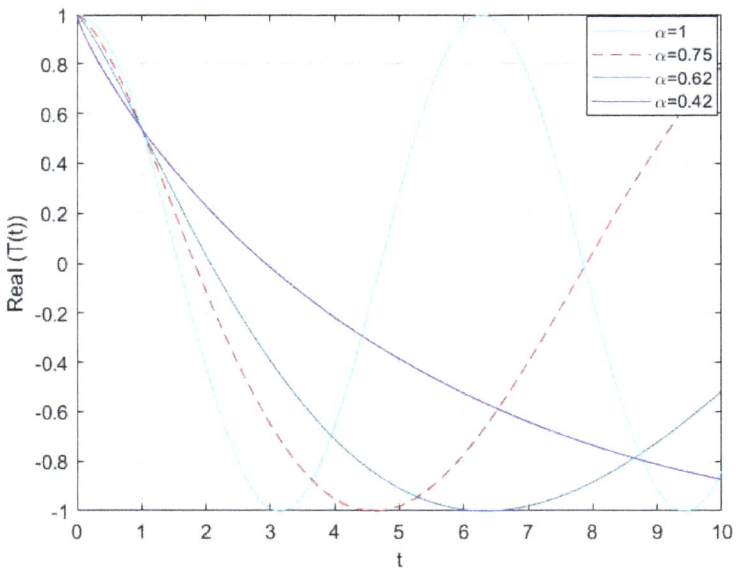

Fig. 8.37: Graph of real part Eq.(8.298) for different α

Theorem 8.2. *The Fractal Bloch theorem states that the solutions of the fractal Schrödinger equation Eq.(8.293) for a fractal periodic potential Eq.(8.292) can be expressed in the following form:*

$$\psi\left(x\right) = u_k\left(x\right) e^{\left(iS_F^\alpha(k)S_F^\alpha(x)\right)}, \tag{8.299}$$

where $u_k\left(x\right)$ has a period of the fractal crystal lattice with $u_k(x) = u_k(x+a)$, and k is a fractal crystal momentum of a particle.

Proof. For the fractal lattice with the period a, we have

$$|\psi(x)|^2 = |\psi(x+a)|^2, \tag{8.300}$$

where $p(x, t) = |\psi(x)|^2$ is called the fractal probability density function of the particle moving under the applied fractal force corresponding to the fractal potential function $U(x)$. Then Eq.(8.300) refers to the following:

$$\psi(x+a) = \mathcal{C}\psi(x), \tag{8.301}$$

where $|\mathcal{C}|^2 = 1$. Then, one may write $\mathcal{C} = \exp(iS_F^\alpha(a)S_F^\alpha(k))$, where k is an arbitrary parameter. Then, we can write

$$\psi(x) = \exp(-iS_F^\alpha(a)S_F^\alpha(k))\psi(x+a). \tag{8.302}$$

Multiplying both sides of Eq.(8.302) by $\exp(-iS_F^\alpha(x)S_F^\alpha(k))$, we get

$$\exp(-iS_F^\alpha(x)S_F^\alpha(k))\psi(x) = \exp(-iS_F^\alpha(k)(S_F^\alpha(x)+S_F^\alpha(a))\psi(x+a). \tag{8.303}$$

Eq.(8.303) shows that

$$u_k(x) = \exp(-iS_F^\alpha(x)S_F^\alpha(k))\psi(x), \tag{8.304}$$

which is a fractal periodic function with period a. By rewriting Eq.(8.304), we obtain

$$\psi(x) = u_k(x)\exp(iS_F^\alpha(k)S_F^\alpha(x)), \tag{8.305}$$

which completes the proof. $\square$

The fractal Kronig-Penney model is suggested by considering the fractal Schrödinger time-independent equation Eq.(8.296), involving the fractal potential energy comb Eq.(8.292). We preset the following solution of the model:

In the region $0 < x < a$ in which $U(x) = 0$, the eigenfunction is a linear combination of plane waves traveling to the right and to the left, namely

$$\psi\left(x\right) = Ae^{iS_F^\alpha(K)S_F^\alpha(x)} + Be^{-iS_F^\alpha(K)S_F^\alpha(x)}, \tag{8.306}$$

where A, B, are constant and

$$E_F^\alpha = \frac{\hbar^2 S_F^\alpha(K)^2}{2m},$$ (8.307)

which is called the fractal energy. In the region $-b < x < 0$ with the fractal barrier the solution is of the following form:

$$\psi(x) = Ce^{S_F^\alpha(Q)S_F^\alpha(x)} + De^{S_F^\alpha(Q)S_F^\alpha(x)},$$ (8.308)

where C, D are constants, and

$$E_F^\alpha = U_0 - \frac{\hbar^2 S_F^\alpha(Q)^2}{2m}.$$ (8.309)

Consequently, using the fractal Bloch theorem Eq.(8.299), the solution in the region $a < x < a + b$ is related to the solution of Eq.(8.308) in the region $-b < x < 0$ as follows:

$$\psi(a < x < a + b) = \psi(-b < x < 0)\,e^{iS_F^\alpha(k)(a+b)},$$ (8.310)

where k is index to label the solution. By boundary conditions, ψ and $D_{F^\mu,x}^\alpha \psi$ are continuous at $x = 0$ and $x = a$. We can find A, B, C, D as follows:

Using Eq.(8.306), Eq.(8.308) with Eq.(8.310), we have

$$Ae^{iS_F^\alpha(K)S_F^\alpha(x)} + Be^{-iS_F^\alpha(K)S_F^\alpha(x)}$$
$$= \left(Ce^{S_F^\alpha(Q)S_F^\alpha(x)} + De^{-S_F^\alpha(Q)S_F^\alpha(x)}\right)e^{iS_F^\alpha(k)(a+b)},$$ (8.311)

Derivative Eq.(8.311) with respect to x, we get

$$iS_F^\alpha(K)\chi_F\left(Ae^{iS_F^\alpha(K)}S_F^\alpha(x) - Be^{-iS_F^\alpha(K)S_F^\alpha(x)}\right)$$
$$= S_F^\alpha(Q)\chi_F Ce^{S_F^\alpha(Q)S_F^\alpha(x)}e^{iS_F^\alpha(k)(a+b)}$$
$$- S_F^\alpha(Q)\chi_F De^{-S_F^\alpha(Q)S_F^\alpha(x)}e^{iS_F^\alpha}(k)(a+b),$$ (8.312)

at $x = 0$, Eq.(8.311) and Eq.(8.312) respectively, becomes

$$A + B - C - D = 0,$$ (8.313)

and

$$iS_F^\alpha(K)\chi_F(x)A$$
$$- iS_F^\alpha(K)\chi_F(x)B - S_F^\alpha(Q)\chi_F(x)C + S_F^\alpha(Q)\chi_F(x)D = 0.$$ (8.314)

At $x = a$, by using Eq.(8.311) and Eq.(8.312), respectively, we obtain the following:

$$Ae^{iS_F^\alpha(K)S_F^\alpha(a)} + Be^{-iS_F^\alpha(K)S_F^\alpha(a)}$$
$$- Ce^{-S_F^\alpha(Q)S_F^\alpha(b)+iS_F^\alpha(k)(a+b)} - De^{S_F^\alpha(Q)S_F^\alpha(b)+iS_F^\alpha(k)(a+b)} = 0, \quad (8.315)$$

and

$$iS_F^\alpha(K)\chi_F(x)\left(Ae^{iS_F^\alpha(K)S_F^\alpha(b)} - Bie^{-iS_F^\alpha(K)S_F^\alpha(b)}\right) -$$
$$S_F^\alpha(Q)\chi_F(x)\left(Ce^{-S_F^\alpha(Q)S_F^\alpha(b)+iS_F^\alpha(k)(a+b)} + De^{S_F^\alpha(Q)S_F^\alpha(x)+iS_F^\alpha(k)(a+b)}\right) = 0,$$
$$(8.316)$$

The Eqs.(8.313), (8.314), (8.315), and (8.316) are written in the form of a matrix as follows:

$$\begin{pmatrix} 1 & 1 & -1 & -1 \\ \mathcal{H} & \mathcal{I} & \mathcal{J} & \mathcal{L} \\ H & I & J & L \\ M & N & O & P \end{pmatrix} \begin{pmatrix} A \\ B \\ C \\ D \end{pmatrix} = \begin{pmatrix} 0 \\ 0 \\ 0 \\ 0 \end{pmatrix} \quad (8.317)$$

where

$$\mathcal{H} = iS_F^\alpha(K)\chi_F(a),$$
$$\mathcal{I} = -iS_F^\alpha(K)\chi_F(a),$$
$$\mathcal{J} = -S_F^\alpha(Q)\chi_F(a)S_F^\alpha(Q)\chi_F(a),$$
$$H = e^{iS_F^\alpha(K)S_F^\alpha(a)},$$
$$I = e^{-iS_F^\alpha(K)S_F^\alpha(a)},$$
$$J = -e^{-S_F^\alpha(Q)S_F^\alpha(a)+iS_F^\alpha(k)(a+b)},$$
$$L = -e^{S_F^\alpha(Q)S_F^\alpha(a)+iS_F^\alpha(k)(a+b)},$$
$$M = iS_F^\alpha(K)\chi_F(a)e^{iS_F^\alpha(K)S_F^\alpha(a)},$$
$$N = -iS_F^\alpha(K)\chi_F(a)e^{-iS_F^\alpha(k)S_F^\alpha(a)},$$
$$O = -S_F^\alpha(Q)\chi_F(a)e^{-S_F^\alpha(Q)S_F^\alpha(b)+iS_F^\alpha(k)(a+b)},$$
$$P = S_F^\alpha(Q)\chi_F(a)e^{S_F^\alpha(Q)S_F^\alpha(b)+iS_F^\alpha(k)(a+b)}. \quad (8.318)$$

Eq.(8.317) has a solution that exists only if the determination of the coefficients of A, B, C, D is zero. By expanding determinant of the matrix and using trigonometric identities, we have

$$
\begin{aligned}
&S_F^\alpha(K)^2(\chi(a))^2 e^{i(a+b)S_F^\alpha(k)-iaS_F^\alpha(K)-bS_F^\alpha(Q)}\\
&- S_F^\alpha(K)^2(\chi(a))^2 e^{i(a+b)S_F^\alpha(k)+iaS_F^\alpha(K)-bS_F^\alpha(Q)}\\
&- S_F^\alpha(K)^2(\chi(a))^2 e^{i(a+b)S_F^\alpha(k)-iaS_F^\alpha(K)+bS_F^\alpha(Q)}\\
&+ (S_F^\alpha(K))^2(\chi(a))^2 e^{i(a+b)S_F^\alpha(k)+iaS_F^\alpha(K)+bS_F^\alpha(Q)}\\
&+ 4iS_F^\alpha(K)\,S_F^\alpha(Q)\,(\chi(a))^2 + 4iS_F^\alpha(K)\,S_F^\alpha(Q)\,(\chi(a))^2 e^{2i(a+b)S_F^\alpha(k)}\\
&- 2iS_F^\alpha(K)\,S_F^\alpha(Q)\,(\chi(a))^2 e^{i(a+b)S_F^\alpha(k)-iaS_F^\alpha(K)-bS_F^\alpha(Q)}\\
&- 2iS_F^\alpha(K)\,S_F^\alpha(Q)\,(\chi(a))^2 e^{i(a+b)S_F^\alpha(k)+iaS_F^\alpha(K)-bS_F^\alpha(Q)}\\
&- 2iS_F^\alpha(K)\,S_F^\alpha(Q)\,(\chi(a))^2 e^{i(a+b)S_F^\alpha(k)-iaS_F^\alpha(K)+bS_F^\alpha(Q)}\\
&- 2iS_F^\alpha(K)\,S_F^\alpha(Q)\,(\chi(a))^2 e^{i(a+b)S_F^\alpha(k)+iaS_F^\alpha(K)+bS_F^\alpha(Q)}\\
&- S_F^\alpha(Q)^2(\chi(a))^2 e^{i(a+b)S_F^\alpha(k)-iaS_F^\alpha(K)-bS_{F\mu}^\alpha(Q)}\\
&+ S_F^\alpha(Q)^2(\chi(a))^2 e^{i(a+b)S_{F\mu}^\alpha(k)+iaS_F^\alpha(K)-bS_F^\alpha(Q)} +\\
&S_F^\alpha(Q)^2(\chi(a))^2 e^{i(a+b)S_F^\alpha(k)-iaS_F^\alpha(K)+bS_F^\alpha(Q)}\\
&- S_F^\alpha(Q)^2(\chi(a))^2 e^{i(a+b)S_F^\alpha(k)+iaS_F^\alpha(K)+bS_F^\alpha(Q)} = 0.
\end{aligned}
\tag{8.319}
$$

Dividing Eq.(8.319) by $\left(-i8(\chi(a))^2 S_F^\alpha(K)\,S_F^\alpha(Q)\,e^{i(a+b)S_F^\alpha(k)}\right)$ and after some simplifications, we have

$$
\begin{aligned}
&\left(\frac{S_F^\alpha(Q)^2 - S_F^\alpha(K)^2}{2S_F^\alpha(K)S_F^\alpha}\right)\\
&\left(\frac{e^{iaS_F^\alpha(K)+bS_F^\alpha(Q)} - e^{-iaS_F^\alpha(K)+bS_F^\alpha(Q)} - e^{iaS_F^\alpha(K)-bS_F^\alpha(Q)}+e^{-iaS_F^\alpha(K)-bS_F^\alpha(Q)}}{4i}\right)\\
&+ \left(\frac{e^{iaS_F^\alpha(K)+bS_F^\alpha(Q)}+e^{-iaS_F^\alpha(K)+bS_F^\alpha(Q)}+e^{iaS_F^\alpha(K)-bS_F^\alpha(Q)}+e^{-iaS_F^\alpha(K)-bS_F^\alpha(Q)}}{4}\right)\\
&- \left(\frac{e^{i(a+b)S_F^\alpha(k)}+e^{-i(a+b)S_F^\alpha(k)}}{2}\right) = 0.
\end{aligned}
\tag{8.320}
$$

Using conjugacy of fractal calculus with ordinary calculus we have

$$
\left(\frac{S_F^\alpha(Q)^2 - S_F^\alpha(K)^2}{2S_F^\alpha(K)\,S_F^\alpha(Q)}\right) \times
$$
$$
\sinh\left(bS_F^\alpha(Q)\right)\sin\left(aS_F^\alpha(K)\right) + \cosh\left(bS_F^\alpha(Q)\right)\cos\left(aS_F^\alpha(K)\right)
$$
$$
= \cos\left((a+b)S_F^\alpha(k)\right).
\tag{8.321}
$$

The result is simplified if the potential is presented via the fractal periodic Dirac delta function which is obtained when we pass to the limit $b = 0$ and $U_0 = \infty$ in such a way that $\frac{baS_F^\alpha(Q)^2}{2} = P$ is a finite quantity. In this limit, $S_F^\alpha(Q) \gg S_F^\alpha(K)$ and $bS_F^\alpha(Q) \ll 1$. Then, Eq.(8.321) reduces to

$$\left(\frac{P}{aS_F^\alpha(K)}\right) \sin\left(aS_F^\alpha(K)\right) + \cos\left(aS_F^\alpha(K)\right) = \cos\left(aS_F^\alpha(k)\right), \qquad (8.322)$$

which might be called the dispersion relation for Kronig-Penney model with the fractal Dirac comb potential energy (see Figure 8.38).
Subsequently, we approximate Eq.(8.321) and Eq.(8.322) for the continuous case and the experimental data.
In view of $c_1 t^\alpha \leq S_F^\alpha(t) \leq c_2 t^\alpha$, we have

$$\left(\frac{Q^{2\alpha} - K^{2\alpha}}{2K^\alpha Q^\alpha}\right) \sinh(bQ^\alpha)\sin(aK^\alpha) + \cosh(bQ^\alpha)\cos(aK^\alpha)$$
$$\approx \cos((a+b)k^\alpha). \qquad (8.323)$$

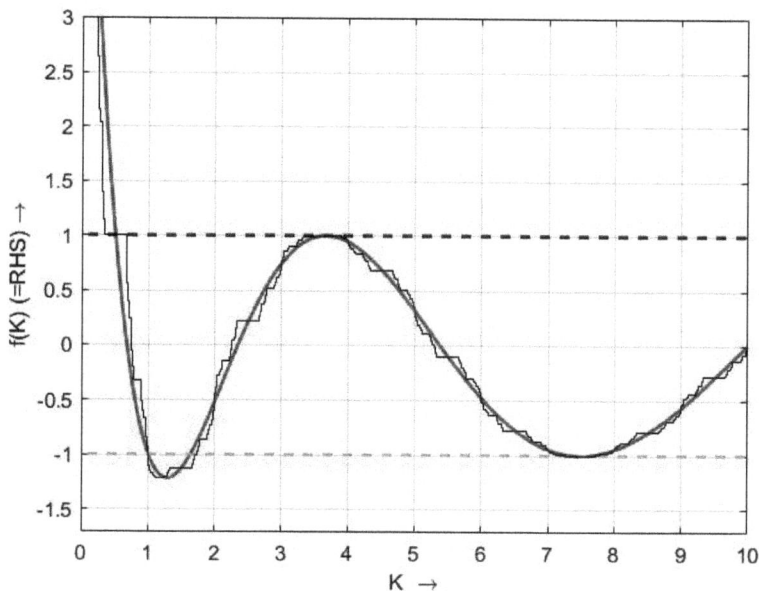

Fig. 8.38: Graph of $f(K)$ versus to K, namely Eq.(8.322)

Here, we introduce reduced variables as follows:

$$b_a = b/a, \quad \varepsilon = \frac{2ma^{2/\alpha}E_F^\alpha}{\hbar^2}, \quad q = \frac{2ma^{2/\alpha}U_0}{\hbar^2}, \quad aK^\alpha = \varepsilon^{\alpha/2},$$

$$aQ^\alpha = (q - \varepsilon)^{\alpha/2}, \tag{8.324}$$

where ε is called fractal reduced energy. Then, we obtain

$$\left(\frac{(q-\varepsilon)^\alpha - \varepsilon^\alpha}{2(\varepsilon(q-\varepsilon))^{\alpha/2}} \right) \sinh\left(b_a(q-\varepsilon)^{\alpha/2} \right) \sin\left(\varepsilon^{\alpha/2} \right)$$

$$+ \cosh\left(b_a(q-\varepsilon)^{\alpha/2} \right) \cos\left(\varepsilon^{\alpha/2} \right) \approx \cos\left((a+b)\,k^\alpha \right). \tag{8.325}$$

Eq.(8.322) can be written in the following form:

$$\left(\frac{b_a(q-\varepsilon)^\alpha}{2\varepsilon^{\alpha/2}} \right) \sin\left(\varepsilon^{\alpha/2} \right) + \cos\left(\varepsilon^{\alpha/2} \right) \approx \cos\left((a)\,k^\alpha \right), \tag{8.326}$$

which might be called fractal dispersion relation.

In Figure 8.39, we have shown the fractal lattice has wider band energies.

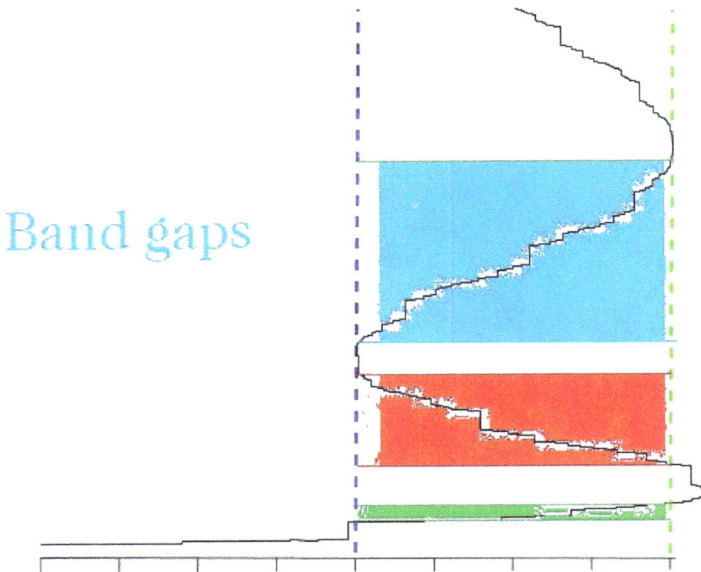

Band gaps

Fig. 8.39: Graph of band gaps for fractal lattice

8.30 Tsallis entropy on fractals

In this section, we suggest the fractal Tsallis statistics which are useful for the nonlinear, not ergodic nor close systems with fractal structure. We connect the q-parameter in Tsallis statistics to γ-dimension in fractal calculus which might be considered a model for the system with fractal structure [Golmankhaneh (2021)].

Consider the fractal differential equation as follows

$$D_x^\alpha y(x) = y^q, \quad q \in R. \tag{8.327}$$

The solution of Eq.(8.327) is given by

$$y(x) = [1 + (1-q)S_F^\alpha(x)]^{\frac{1}{1-q}},$$
$$\approx [1 + (1-q)x^\alpha]^{\frac{1}{1-q}} \tag{8.328}$$

In Figure 8.40, we have plotted Eq.(8.328) for the cases of $\alpha = 0.6$, $q = 0.8$ (below), and $q = 0.5$ (upper).

We define q-fractal exponent as follows:

$$e_q^\alpha(t) = \begin{cases} (1 + qS_F^\alpha(x))^{1/q}, & 1 + qS_F^\alpha(x) < 0; \\ 0, & 1 + qS_F^\alpha(x) < 0. \end{cases} \tag{8.329}$$

Here, if $q = 0$ we have $e_{q=0}^\alpha(t) = e^\alpha(x)$. Then we can rewrite Eq.(8.328) as follows

$$y(x) = e_{1-q}^\alpha(x). \tag{8.330}$$

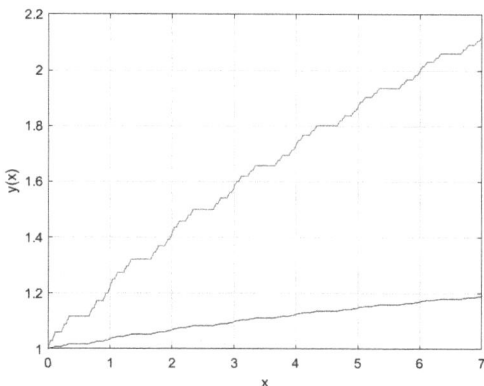

Fig. 8.40: The graph of Eq.(8.328)

The inverse function of y is denoted by $y^{-1}(x) = \ln_q^\alpha(x)$ and defined by

$$\ln_q^\alpha(x) = \frac{S_F^\alpha(x)^q - 1}{q}, \quad \ln_{q=0}^\alpha(x) = \ln^\alpha(x) \tag{8.331}$$

which is called q-fractal logarithm function. This function has the following property

$$\ln_q^\alpha(S_{F,A}^\alpha(x)S_{F,B}^\alpha(x)) = \ln_q^\alpha(S_{F,A}^\alpha(x)) + \ln_q^\alpha(S_{F,B}^\alpha(x))$$
$$+(1-q)[\ln_q^\alpha(S_{F,A}^\alpha(x))\ln_q^\alpha(S_{F,B}^\alpha(x))] \tag{8.332}$$

which is called fractal pseudo-additivity.
Fractal derivative of q-fractal exponent function is

$$D_x^\alpha e_q^\alpha(\eta x) = D_x^\alpha (1 + \eta^\alpha q S_F^\alpha(x))^{1/q}$$
$$= \eta^\alpha (1 + \eta^\alpha q S_F^\alpha(x))^{\frac{1}{q}-1}$$
$$= \eta^\alpha (1 + \frac{q}{(1-q)}\eta^\alpha(1-q)S_F^\alpha(x))^{\frac{1-q}{q}}$$
$$= \eta^\alpha e_{\frac{q}{1-q}}^\alpha(\eta x), \quad q \neq 1. \tag{8.333}$$

The extending of Eq.(8.333) to n^{th}-fractal derivative we have

$$(D_x^\alpha)^n e_q^\alpha(\eta x) = \left[\eta^{\alpha n}\prod_{i=1}^{n}(1-(i-1)q)\right]e_{\frac{q}{1-nq}}^\alpha((1-nq)\eta x), \quad q \neq \frac{1}{n}. \tag{8.334}$$

The fractal integral of the q-exponential is

$$\int e_q(x)d_F^\alpha x = \frac{1}{\eta(1+q)}\left(1 + \frac{q}{(1+q)}\eta(1+q)S_F^\alpha(x)\right)^{\frac{1+q}{q}}$$
$$= \frac{1}{\eta(1+q)}e_{\frac{q}{1+q}}((1+q)\eta x) + c_1, \quad q \neq -1. \tag{8.335}$$

Then n^{th}-fractal integral is

$$\int \cdots \left(\int e_q(x)\right)\cdots d_F^\alpha x = \left[\frac{1}{\eta^n}\prod_{i=1}^{n}\frac{1}{1+nq}\right]e_{\frac{q}{1+nq}}((1+nq)\eta x)$$
$$= +\sum_{i=1}^{n}c_i S_F^\alpha(x)^{i-1}, \quad q \neq -\frac{1}{n} \tag{8.336}$$

The fractal coupled probability density function is defined by

$$p_q^\alpha(x) = \frac{[p(x)]^{1-q}}{\int_{-\infty}^{+\infty}[p(x)]^{1-q}d_F^\alpha x} \tag{8.337}$$

The fractal Tsallis Entropy is defined on fractal sets as follows:

$$T_q^\alpha = \frac{-1 + \int p(x)^{1-q} d_F^\alpha x}{q}. \tag{8.338}$$

The fractal mean is defined by

$$E_q^\alpha[X] = \int S_F^\alpha(x) p_q^\alpha(x) d_F^\alpha x. \tag{8.339}$$

The fractal variance is defined by

$$V_q^\alpha(X) = \int \left(S_F^\alpha(x) - E_q^\alpha[X]\right)^2 d_F^\alpha x \tag{8.340}$$

The fractal maximum q-entropy with fractal finite q-variance is given by

$$G_q^\alpha(x) = \frac{\sqrt{B_q^\alpha}}{N_q^\alpha} e_q^{-B_q^\alpha (S_F^\alpha(x) - E_q^\alpha(X))^2}, \quad S_F^\alpha(x) < x^\alpha,$$

$$\approx \frac{\sqrt{B_q^\alpha}}{N_q^\alpha} e_q^{-B_q^\alpha (x^\alpha - E_q(X)^\alpha)^2} \tag{8.341}$$

where

$$B_q^\alpha = [(2+q) \, V_q^\alpha(X)^2]^{-1}, \tag{8.342}$$

and

$$N_q^\alpha = \begin{cases} \sqrt{\dfrac{\pi}{q}} \dfrac{\Gamma^\alpha(\frac{1+q}{q})}{\Gamma^\alpha(\frac{2+3q}{2q})}, & q > 0; \\[2ex] \sqrt{\pi}, & q = 0; \\[2ex] \sqrt{\dfrac{\pi}{-q}} \dfrac{\Gamma^\alpha(\frac{2+q}{-2q})}{\Gamma^\alpha(\frac{1}{-q})}, & -2 < q < 0, \end{cases} \tag{8.343}$$

where $\Gamma^\alpha(*)$ is gamma function with fractal support and $G_q^\alpha(x)$ is called the fractal q-Gaussian or generalized ζ-stable Lévy distribution on fractal sets. Eq.(8.343) indicates nonlinear fractal coupling of statistical states. For $q > 0$ nonlinear fractal coupling between the states strengthens is decayed, namely:

$$|S_F^\alpha(x) - E_q^\alpha(X)| > \sqrt{\frac{1}{q B_q^\alpha}}, \tag{8.344}$$

$$|x^\alpha - E_q(X)^\alpha| > \sqrt{\frac{1}{q B_q^\alpha}} \Rightarrow G_q^\alpha(x) \approx 0. \tag{8.345}$$

In the case of $q = 0$ we get fractal Gaussian distribution. For the $-2 < q < 0$ the decoupled states lead to a fractal heavy-tail distribution.

Now, we want to connect α-dimension of the fractal calculus to hardron system. Let us define R as follows:

$$R = \frac{<E>}{N' <U>} = \frac{(q-1)N/N'}{3 - 2q + (q-1)N}, \quad N = N' + 2/3, \quad (8.346)$$

where N is effective number of hardron subsystems and $< E >$ is mean energy of the hardron system. In view of the results in, we conclude

$$\alpha = dim_H(F) = 1 + \frac{\log N'}{\log R}, \quad (8.347)$$

where $\alpha = dim_H(F)$ which holds for fractal.

8.31   Fractal Schwarzschild metric

In this subsection, we use fractal calculus to fractalize Schwarzschild metric. The fractal the Schwarzschild metric is suggested as [El-Nabulsi and Golmankhaneh (2021b)]:

$$d_F^\beta s^2 = A(r)S_F^\beta(c)^2 d_F^\beta t - B(r)dr^2 - r^2(d\theta^2 + \sin^2\theta d\phi^2) \quad (8.348)$$

where $A(r)$ and $B(r)$ are two spatially-dependent functions to be determined. Using the fractal calculus framework, we can write:

$$d_F^\beta s^2 = \left(1 - \frac{2MG}{rS_F^\beta(c)^2}\right) S_F^\beta(c)^2 d_F^\beta t^2 - \frac{dr^2}{1 - \frac{2MG}{rS_F^\beta(c)^2}} - r^2(d\theta^2 + \sin^2\theta)d\phi^2$$

$$\approx \left(1 - \frac{2MG}{rc^{2\beta}}\right) c^{2\beta} d_F^\beta t^2 - \frac{dr^2}{1 - \frac{2MG}{rc^{2\beta}}} - r^2(d\theta^2 + \sin^2\theta)d\phi^2, \quad (8.349)$$

where M is the mass of the body, G is the gravitational constant and c is the celerity of light, and $r_{F,s}$ is called the fractal Schwarzschild radius of the massive body that is given by:

$$r = r_{F,s} = \frac{2MG}{S_F^\beta(c)^2}. \quad (8.350)$$

The fractal speed of light is given by

$$v \approx S_F^\beta(c) \left(1 - \frac{2MG}{RS_F^\beta(c)^2}\right) \quad (8.351)$$

$$\propto c^\beta \left(1 - \frac{2MG}{Rc^{2\beta}}\right), \quad (8.352)$$

where $0 < \beta < 1$ then we have $v < c$. The above expression for the deflection of light is rewritten as:

$$\varsigma = \frac{4MG}{RS_F^\beta(c)^2} \propto \frac{4MG}{Rc^{2\beta}}, \tag{8.353}$$

that is called the fractal deflection of light. Here, we define the fractal dimensionless parameter by:

$$\gamma \approx c^{2-2\beta}. \tag{8.354}$$

In Figures 8.41, and 8.42, we have illustrated the effect of fractal time dimension on Schwarzschild radius using fractional and fractal space approaches.

For $\gamma - 1 = (-2 \pm 3) \times 10^{-4}$, we get $0.999997444 < \beta < 1.0000128$, and for $\gamma - 1 = (-0.8 \pm 1.2) \times 10^{-4}$, we obtain $0.999997898 < \beta < 1.0000051$. The fractal Hawking temperature is defined by

$$T \approx \frac{c^{3\beta}}{4MGk_B} \approx c^{3\beta-3}T_H. \tag{8.355}$$

Using $0.999997444 < \beta < 1.0000128$ we have $0.99985010837\, T_H < T < 1.00074979523\, T_H$. We observe that for $0 < \beta < 1$, the fractal Schwarzschild radius of the massive body is greater than the conventional radius whereas the fractal Hawking's temperature less than the standard one.

In Table 8.2, we have summarized the main common points and dissimilarities between the fractional space and fractal approach.

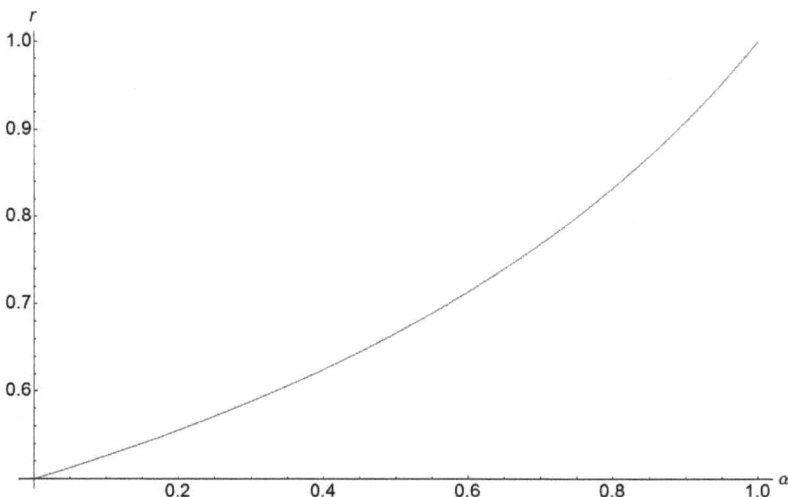

Fig. 8.41: Graph of fractional Schwarzschild radius versus to α

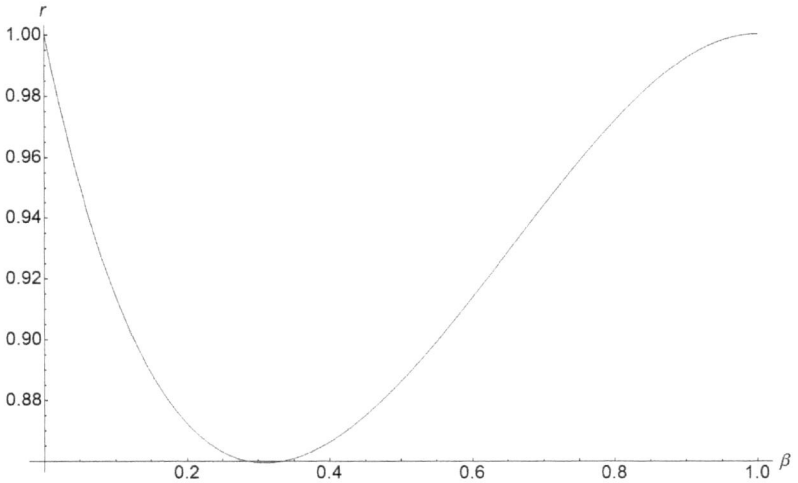

Fig. 8.42: Graph of fractal Schwarzschild radius versus to β

Table 8.2: Comparing fractional approach with fractal approach

Approaches	Fractional Approach	Fractal Approach
Lagrangian	$L_{FV} = \frac{1}{2}\Omega g_{\mu\nu}\left(\frac{x^\mu}{\tau}\right)^{\alpha-1}\left(\frac{x^\nu}{\tau}\right)^{\alpha-1}\frac{dx^\mu}{d\tau}\frac{dx^\nu}{d\tau}$	$\mathbf{L} = \frac{1}{2}\mathbf{g}_{\mu\nu}D^\beta_{F,\tau}x^\mu D^\beta_{F,\tau}x^\nu$
Euler-Lagrange Equation	$\frac{\partial L_{FV}}{\partial x^\mu} - \frac{d}{d\tau}\left(\frac{\partial L_{FV}}{\partial(\frac{dx^\mu}{d\tau})}\right) = \frac{\alpha-1}{\tau}\frac{\partial L_{FV}}{\partial(\frac{dx^\mu}{d\tau})}$	$\frac{\partial \mathbf{L}}{\partial x^\mu} - D^\beta_{F,\tau}\frac{\partial \mathbf{L}}{\partial D^\beta_{F,\tau}x^\mu} = 0$
Einstein's Field Equation	$R_{\mu\nu} - \frac{1}{2}g_{\mu\nu}R = \frac{8\pi G}{c^4}\left(T^M_{\mu\nu} - \frac{c^4(\alpha-1)}{8\pi G\mathbf{x}^2}g_{\mu\nu}\right)$	$R_{\mu\nu} - \frac{1}{2}g_{\mu\nu}R = \frac{8\pi G}{S^\beta_F(c)^4}T^M_{\mu\nu}$ $\approx \frac{8\pi G}{c^{4\beta}}T^M_{\mu\nu}$
Schwarzschild Metric	$ds^2 = c^2\left(2 - \alpha - \frac{2MG}{rc^2}\right)dt^2 - \frac{dr^2}{2 - \alpha - \frac{2MG}{rc^2}} - r^2\left(d\theta^2 + \sin^2\theta d\phi^2\right)$	$d^\beta_F s^2 \approx (1 - \frac{2MG}{rc^{2\beta}})c^{2\beta}d^\beta_F t^2 - \frac{dr^2}{1 - \frac{2MG}{rc^{2\beta}}} - r^2(d\theta^2 + \sin^2\theta d\phi^2)$
Temporal Equation	$t = c_2(c_1 + \tau^\alpha)^{1/\alpha}$	$t \approx c_1\tau^\beta + c_2$
Schwarzschild Radius	$r = \frac{2MG}{c^2(2-\alpha)}$	$r \approx \frac{2MG}{c^{2\beta}}$
Speed of Light	$v \approx c\left(1 - \frac{2MG}{(2-\alpha)Rc^2}\right)$	$v \approx c^\beta\left(1 - \frac{2MG}{Rc^{2\beta}}\right)$
Deflection of Light	$\varsigma = \frac{4MG}{(2-\alpha)Rc^2}$	$\varsigma \approx \frac{4MG}{Rc^{2\beta}}$
Hawking Temperature	$T = \frac{c^3(2-\alpha)}{4MGk_B}$	$T \approx \frac{c^{3\beta}}{4MGk_B}$
Dimensionless Parameter in PPN	$\gamma = \frac{1}{(2-\alpha)}$	$\gamma \approx c^{2-2\beta}$
Hawking Entropy	$S = \frac{k_B\pi G}{\hbar c}\frac{M^2}{(2-\alpha)}$	$S \approx \frac{k_B\pi G}{\hbar}\frac{M^2}{c^\beta}$

8.32 Fractal damped oscillators on Finsler manifold

In this section, we preset damped oscillators on fractal Finsler manifold [El-Nabulsi and Golmankhaneh (2022)]. Let $\mathbf{M}^n$ be a fractal smooth manifold and $\mathbf{TM}^n$ is the fractal analogous to the tangent bundle with 0-section removed, also $\mathbf{F} : \mathbf{TM}^n \rightarrow R$ is positive-homogeneous of degree one in $\mathbf{y} = D_\phi^\zeta \mathbf{x}$. Supposing

$$\mathbf{F}^2 = \mathbf{g}_{\alpha\beta}(D_\phi^\zeta \mathbf{x})^\alpha (D_\phi^\zeta \mathbf{x})^\beta = \mathbf{g}_{\alpha\beta} \mathbf{y}^\alpha \mathbf{y}^\beta, \qquad (8.356)$$

$$\mathbf{g}_{\alpha\beta} \equiv \mathbf{g}_{\alpha\beta}(\mathbf{x}(t), \mathbf{y}(t)), \qquad (8.357)$$

and

$$\mathbf{g}_{\alpha\beta} = D_\phi^\zeta \partial_\alpha D_\phi^\zeta \partial_\beta \left(\frac{F^2}{2}\right), \alpha, \beta = 0, 1, 2, 3, \qquad (8.358)$$

The fractal geodesic equations in $\mathbf{M}^n$ is suggested by

$$E(F(\mathbf{x}, \mathbf{y})) = 2\xi D_\phi^\zeta \mathbf{y}^\nu + 2\mathbf{G}^\nu(\mathbf{x}, \mathbf{y}) + \frac{2\xi - 1}{2} \left(\frac{\partial \ln \mathbf{g}_{\alpha\beta}}{\partial \mathbf{x}^\nu}\right) \mathbf{y}^\alpha \mathbf{y}^\beta = 0, \qquad (8.359)$$

where

$$G^\nu(\mathbf{x}, \mathbf{y}) = \frac{1}{2}\Gamma^\nu_{\alpha\beta} \mathbf{y}^\alpha \mathbf{y}^\beta, \qquad (8.360)$$

and the Finslerian metric is given by:

$$\mathbf{g}_{\alpha\beta} \equiv \frac{\partial}{\partial \mathbf{y}^\alpha} \frac{\partial}{\partial \mathbf{y}^\beta} \left(\frac{1}{2}\mathbf{F}^2\right). \qquad (8.361)$$

Let

$$D_\phi^{2\zeta} \vec{\mathbf{x}} = \frac{1}{4\xi}(\vec{\nabla}_F h_{00} + (1 - 2\xi)\vec{\nabla}(-1 + h_{00})) = \frac{1 - \xi}{2\xi}\vec{\nabla}_F h_{00}. \qquad (8.362)$$

where

$$\vec{\nabla}_F = D_\phi^{\zeta_i} \hat{e}_i, \qquad (8.363)$$

where F is a fractal manifold. The fractal Newton Equation is given by

$$D_\phi^{2\zeta} \vec{\mathbf{x}} = -\vec{\nabla}_F \Phi. \qquad (8.364)$$

The $\mathbf{g}_{00}$ is suggested by

$$\mathbf{g}_{00} = -\left(1 - \frac{2\xi}{1 - \xi}\frac{MG}{\Gamma(\zeta + 1)S_F^\zeta(R)}\right) \approx -\left(1 - \frac{2\xi}{1 - \xi}\frac{MG}{\Gamma(\zeta + 1)R^\zeta}\right). \qquad (8.365)$$

The fractal momentum is defined by

$$\mathbf{p}^\nu = mD_\varphi^\zeta \mathbf{x}^\nu. \tag{8.366}$$

8.33 Fractal electromagnetic fields in Randers spaces

In this section, we present the electromagnetic fields equation in fractal Randers spaces [El-Nabulsi and Golmankhaneh (2022)].
Let us consider

$$\mathbf{F}^{2\varepsilon(1+\xi)}(\mathbf{x}, \mathbf{y}) = \alpha(\mathbf{x}, \mathbf{y}) + \beta(\mathbf{x}, \mathbf{y}) \equiv \mathbf{L}(\alpha(\mathbf{x}, \mathbf{y}), \beta(\mathbf{x}, \mathbf{y})) \tag{8.367}$$

where

$$\alpha(\mathbf{x}, \mathbf{y}) = \sqrt{\mathbf{a}_{i,j} \mathbf{y}^i \mathbf{y}^j}, \tag{8.368}$$

and

$$\beta(\mathbf{x}, \mathbf{y}) = \mathbf{b}_i \mathbf{y}^i. \tag{8.369}$$

The corresponding fractal action is as the following

$$\mathfrak{S} = \int_{C(0,1)} F^{2\varepsilon(1+\xi)}(\mathbf{x}, \mathbf{y}) d_F^\zeta \phi. \tag{8.370}$$

It follows that

$$\frac{\partial \alpha(\mathbf{x}, \mathbf{y}) + \beta(\mathbf{x}, \mathbf{y})}{\partial \mathbf{x}^i} - D_\phi^\zeta \frac{\partial \alpha(\mathbf{x}, \mathbf{y}) + \beta(\mathbf{x}, \mathbf{y})}{\partial \mathbf{y}^i}$$
$$= \frac{2\xi - 1}{\alpha(\mathbf{x}, \mathbf{y}) + \beta(\mathbf{x}, \mathbf{y})} \frac{\partial \alpha(\mathbf{x}, \mathbf{y}) + \beta(\mathbf{x}, \mathbf{y})}{\partial \mathbf{y}^i} D_\phi^\zeta (\alpha(\mathbf{x}, \mathbf{y}) + \beta(\mathbf{x}, \mathbf{y})). \tag{8.371}$$

The fractal Lorentz equations in electrodynamics is given by:

$$2\xi D_\sigma^{\zeta i} \mathbf{y}^i + \left(\tilde{\Gamma}_{jk}^i + \frac{2\xi - 1}{2} \frac{\partial \ln a_{ij}}{\partial \mathbf{x}^k} \right) \mathbf{y}^j \mathbf{y}^k = \mathbf{F}_j^i(\mathbf{x}) \mathbf{y}^j. \tag{8.372}$$

The solution of (8.372) is

$$\mathbf{S} = \mathbf{y}^i \frac{\partial}{\partial x^i} - 2\tilde{\mathbf{G}}^i(x, y) \frac{\partial}{\partial \mathbf{y}^i}, \tag{8.373}$$

where

$$2\tilde{\mathbf{G}}^i(\mathbf{x}^i, \mathbf{y}^i) = \left(\tilde{\Gamma}_{jk}^i + \frac{2\xi - 1}{2} \frac{\partial \ln \mathbf{a}_{ij}}{\partial \mathbf{x}^k} \right) \mathbf{y}^j \mathbf{y}^k + 2\xi D_\rho^\zeta \mathbf{y}^i - \mathbf{F}_j^i(\mathbf{x}) \mathbf{y}^j. \tag{8.374}$$

Next, let us consider the fractal equation of motion as

$$mD_t^{2\zeta} \vec{\mathbf{x}} = \frac{1 - \xi}{\xi} (\vec{E} + D_t^\zeta \vec{x} \times \vec{B}). \tag{8.375}$$

For the case $\vec{E} = 0$, using the conjugacy property the solution is

$$x = x_0 + \frac{m\xi v_\perp}{1 - \xi} \frac{1}{|q|B} \sin(S_F^\zeta(\omega)S_F^\zeta(t)) \propto x_0 + \frac{m\xi v_\perp}{1 - \xi} \frac{1}{|q|B} \sin((\omega t)^\zeta)$$

$$(8.376)$$

$$y = y_0 + \frac{m\xi v_\perp}{1 - \xi} \frac{1}{|q|B} \cos(S_F^\zeta(\omega)S_F^\zeta(t)) \propto y_0 + \frac{m\xi v_\perp}{1 - \xi} \frac{1}{|q|B} \cos((\omega t)^\zeta)$$

$$(8.377)$$

where

$$S_F^\zeta(\omega) = \frac{1 - \xi}{\xi} \frac{|q|B}{m} \tag{8.378}$$

$$\omega \approx \left(\frac{1 - \xi}{\xi} \frac{|q|B}{m}\right)^{\frac{1}{\zeta}}, \tag{8.379}$$

which gives the angular frequency of motion and its relation with the dimensions of time.

In Figures 8.43, and 8.44, we have plotted Eqs.(8.376) and (8.377) for the case of $0.1 < \xi < 0.9$, $\zeta = 0.7$, and $\zeta = 0.9$ [El-Nabulsi and Golmankhaneh (2022)].

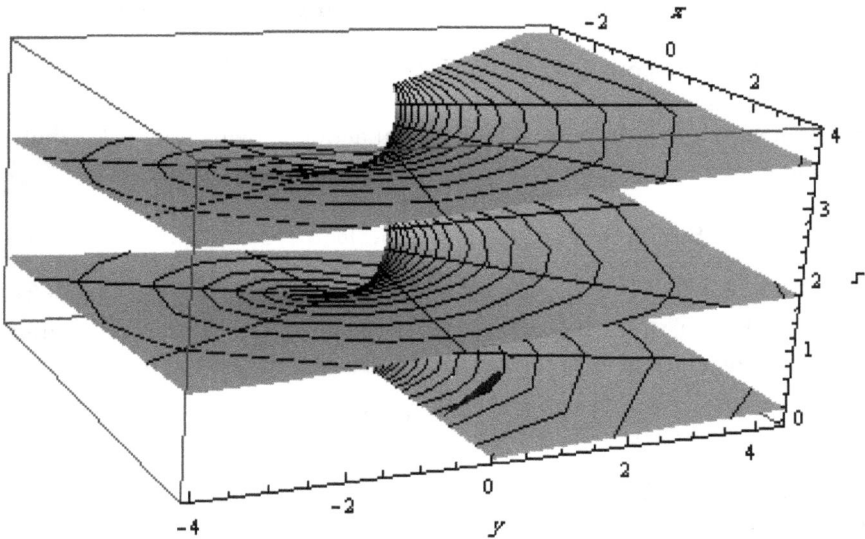

Fig. 8.43: Graph of solution of Eq.(8.375) normalizing physical quantities for $0.1 < \xi < 0.9$ and $\zeta = 0.9$

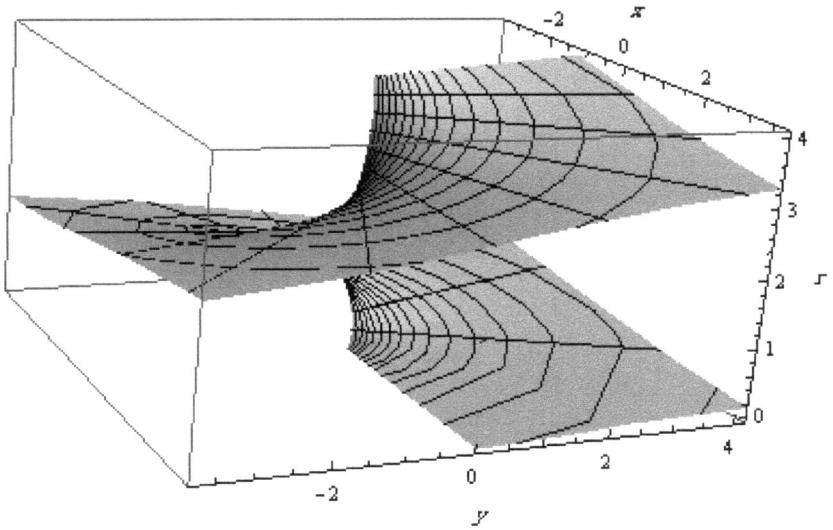

Fig. 8.44: Graph of solution of Eq.(8.375) normalizing physical quantities for $0.1 < \xi < 0.9$ and $\zeta = 0.7$

8.34 The Boltzmann/Vlasov-Boltzmann on fractal time

For the fractal smooth Lagrangian $\mathbf{L} : F \times \mathbf{TMF} \to R$ where $\mathbf{TMF}$ is a the fractal analogous tangent bundle of F, and $\mathbf{L}$ assumed to be a $C^{2\beta}$-function with all its arguments, the fractal action is defined by [El-Nabulsi and Golmankhaneh (2021a)]:

$$\mathbf{S}[\mathbf{q}(t)] = \int_a^t \mathbf{L}(D_t^\beta \mathbf{q}(t), \mathbf{q}(t), t) d_F^\beta t \qquad t \in F. \tag{8.380}$$

where β is the fractal dimension of the fractal time set. For the simplicity, let $D_t^\beta \mathbf{q}(t) = D_t^\beta \mathbf{q}(t)$. Using the fractal calculus of variations for finding extremum of Eq.(8.380) we obtain fractal Euler-Lagrange equation as:

$$\frac{\partial \mathbf{L}}{\partial \mathbf{q}} - D_t^\beta \left(\frac{\partial \mathbf{L}}{\partial \mathbf{q}_t} \right) = \left(\frac{\partial \mathbf{L}}{\partial \mathbf{q}_t} \right), \tag{8.381}$$

where $\mathbf{q}_t = \left(D_t^\beta \mathbf{q} \right)$. Then corresponding Hamilton's equations by using Legendre transformation , and supposing the $\mathbf{L}$ is a convex function one can get:

$$(\mathbf{q}_t, \mathbf{p}_t) = \left(\frac{\partial \mathbf{H}}{\partial \mathbf{p}}, -\frac{\partial \mathbf{H}}{\partial \mathbf{q}} \right). \tag{8.382}$$

where,

$$\mathbf{p} = \frac{\partial \mathbf{L}}{\partial \mathbf{q}_t}, \qquad \mathbf{H} = \mathbf{q}_t \mathbf{p} - \mathbf{L}. \tag{8.383}$$

where $\mathbf{H}$ is might called fractal Hamiltonian. Also, $\mathbf{p}$ and $\mathbf{q}$ is called generalized coordinate and conjugate momentum. If the density of gas molecules in the phase space has a function $\mathbf{f}(\vec{\mathbf{x}}, \vec{\mathbf{p}}, t) : R^3 \times R^3 \times F \to R^+$ where $\vec{\mathbf{x}}$ and $\vec{\mathbf{p}}$ are position and momentum, respectively, then the fractal Liouville's theorem of incompressible phase space flow is as follows:

$$d^\beta \mathbf{f} = D_t^\beta \mathbf{f} + \sum_{i=1}^{3} \frac{\partial \mathbf{f}}{\partial \mathbf{x}^i} x_t^i + \frac{\partial \mathbf{f}}{\partial \mathbf{p}^i} p_t^i \tag{8.384}$$

Using Eq.(8.382) we obtain [El-Nabulsi and Golmankhaneh (2021a)]:

$$d^\beta \mathbf{f} = D_t^\beta \mathbf{f} + \sum_{i=1}^{3} \frac{\partial \mathbf{f}}{\partial \mathbf{x}^i} \frac{\partial \mathbf{H}}{\partial \mathbf{p}^i} - \frac{\partial \mathbf{f}}{\partial \mathbf{p}^i} \frac{\partial \mathbf{H}}{\partial \mathbf{x}^i}$$

$$= D_t^\beta \mathbf{f} + \frac{\partial \mathbf{H}}{\partial \vec{\mathbf{p}}} . \nabla_{\vec{x}} \mathbf{f}(\vec{x}, \vec{\mathbf{p}}, t) - \frac{\partial \mathbf{H}}{\partial \vec{\mathbf{x}}} . \nabla_{\vec{\mathbf{p}}} \mathbf{f}(\vec{x}, \vec{\mathbf{p}}, t). \tag{8.385}$$

In view of equation of $\mathbf{H} = \frac{\vec{\mathbf{p}}^2}{2m} + \mathbf{V}(\vec{\mathbf{x}}, t)$ we have:

$$D_t^\beta \mathbf{f} + \frac{\vec{\mathbf{p}}}{m} . \nabla_{\vec{\mathbf{x}}} \mathbf{f}(\vec{\mathbf{x}}, \vec{\mathbf{p}}, t) - \nabla_{\vec{\mathbf{x}}} \mathbf{V}(\vec{\mathbf{x}}, t) . \nabla_{\vec{\mathbf{p}}} \mathbf{f}(\vec{\mathbf{x}}, \vec{\mathbf{p}}, t) \equiv (D_t^\beta \mathbf{f})_{collisions}, \tag{8.386}$$

which might be called fractal Boltzmann/Vlasov equation. One can rewrite Eq.(8.386) as follows:

$$D_t^\beta \mathbf{f} + \vec{\mathbf{v}}^\beta . \nabla_{\vec{\mathbf{x}}} \mathbf{f}(\vec{\mathbf{x}}, \vec{\mathbf{v}}, t) - \frac{1}{m} \nabla_{\vec{\mathbf{x}}} \mathbf{V}(\vec{\mathbf{x}}, t) . \nabla_{\vec{\mathbf{v}}^\beta} \mathbf{f}(\vec{\mathbf{x}}, \vec{\mathbf{v}}, t) = 0, \tag{8.387}$$

where,

$$\vec{\mathbf{v}}^\beta = D_t^\beta x(t)\hat{i} + D_t^\beta y(t)\hat{j} + D_t^\beta z(t)\hat{k}, \qquad D_t^\beta \vec{\mathbf{p}} = \vec{\mathbf{F}}, \qquad \vec{\mathbf{p}} = m\vec{\mathbf{v}}^\beta. \tag{8.388}$$

and the fractal Lorentz force is defined by:

$$\vec{\mathbf{F}} = -\frac{1}{m} \nabla_{\vec{\mathbf{x}}} \mathbf{V}(\vec{\mathbf{x}}, t) = \frac{q_i}{m_i} (\vec{E} + \vec{\mathbf{v}}^\beta \times \vec{B}). \tag{8.389}$$

It is easy to verify using the previous equations that:

$$D_t^\beta \mathbf{f}_i + \vec{\mathbf{v}}^\beta . \nabla_{\vec{\mathbf{x}}} \mathbf{f}_i(\vec{\mathbf{x}}, \vec{\mathbf{v}}, t) + \frac{q_i}{m_i} (\vec{E} + \vec{\mathbf{v}}^\beta \times \vec{B}) . \nabla_{\vec{\mathbf{v}}^\beta} \mathbf{f}_i(\vec{\mathbf{x}}, \vec{\mathbf{v}}, t) = 0. \tag{8.390}$$

The fractal time Maxwell's equations are suggested as follows [El-Nabulsi and Golmankhaneh (2021a)]:

$$\nabla_{\bar{\mathbf{x}}} \times \vec{E} = -D_t^\beta \vec{B}, \quad \nabla_{\bar{\mathbf{x}}} \times \vec{B} = \mu_0 \vec{j} + \frac{1}{c^2} D_t^\beta \vec{E},$$

$$\nabla_{\bar{\mathbf{x}}} . \vec{E} = \frac{\rho}{\epsilon_0}, \quad \nabla_{\bar{\mathbf{x}}} . \vec{B} = 0. \tag{8.391}$$

8.35   Fluids equation on fractal time

In this section, we will derive the fractal fluid equation [El-Nabulsi and Golmankhaneh (2021a)]. By fractal integrating of Eq.(8.390) we have:

$$\int D_t^\beta \mathbf{f}_i d^3 v + \int \vec{\mathbf{v}}^\beta . \nabla_{\bar{\mathbf{x}}} \mathbf{f}_i(\bar{\mathbf{x}}, \vec{\mathbf{v}}^\beta, t) d^3 v \tag{8.392}$$

$$+ \int \frac{q_i}{m_i} (\vec{E} + \vec{\mathbf{v}}^\beta \times \vec{B}) . \nabla_{\vec{\mathbf{v}}^\beta} \mathbf{f}_i(\bar{\mathbf{x}}, \vec{\mathbf{v}}^\beta, t) d^3 v = 0.$$

Let:

$$\int D_t^\beta \mathbf{f}_i d^3 v = D_t^\beta n_i, \tag{8.393}$$

$$\int \vec{\mathbf{v}}^\beta . \nabla_{\bar{\mathbf{x}}} \mathbf{f}_i(\bar{\mathbf{x}}, \vec{\mathbf{v}}^\beta, t) d^3 v = \nabla_{\bar{\mathbf{x}}} . (n_i \bar{\vec{\mathbf{v}}}_i^\beta) \tag{8.394}$$

$$\int (\vec{E} + \vec{\mathbf{v}}^\beta \times \vec{B}) . \nabla_{\vec{\mathbf{v}}^\beta} \mathbf{f}_i(\bar{\mathbf{x}}, \vec{\mathbf{v}}^\beta, t) d^3 v = 0, \tag{8.395}$$

where,

$$n_i = \int \mathbf{f}_i(\bar{\mathbf{x}}, \vec{\mathbf{v}}^\beta, t) d^3 v, \quad \bar{\vec{\mathbf{v}}}_i^\beta = \frac{\int \vec{\mathbf{v}}^\beta \mathbf{f}_i(\bar{\mathbf{x}}, \vec{\mathbf{v}}^\beta, t) d^3 v}{\int \mathbf{f}_i(\bar{\mathbf{x}}, \vec{\mathbf{v}}^\beta, t) d^3 v}. \tag{8.396}$$

Then we get after simplification:

$$D_t^\beta n_i + \nabla_{\bar{\mathbf{x}}} . (n_i \bar{\vec{\mathbf{v}}}_i^\beta) = 0. \tag{8.397}$$

which might be called the fractal continuity equation . Multiply both sides of Eq.(8.392) by $\vec{\mathbf{v}}^\beta$ and taking fractal integral we get:

$$\int \vec{\mathbf{v}}^\beta D_t^\beta \mathbf{f}_i d^3 v + \int \vec{\mathbf{v}}^\beta (\vec{\mathbf{v}}^\beta . \nabla_{\bar{\mathbf{x}}}) \mathbf{f}_i(\bar{\mathbf{x}}, \vec{\mathbf{v}}, t) d^3 v$$

$$+ \int \vec{\mathbf{v}}^\beta \left(\frac{q_i}{m_i} (\vec{E} + \vec{\mathbf{v}}^\beta \times \vec{B}) \right) . \nabla_{\vec{\mathbf{v}}^\beta} \mathbf{f}_i(\bar{\mathbf{x}}, \vec{\mathbf{v}}, t) d^3 v = 0. \tag{8.398}$$

Using the following results:

$$\int \vec{\mathbf{v}}^{\beta} D_t^{\beta} \mathbf{f}_i d^3 v = D_t^{\beta}(n_i \vec{\mathbf{v}}_i^{\beta}),$$

$$\int \vec{\mathbf{v}}^{\beta}(\vec{\mathbf{v}}^{\beta}.\nabla_{\vec{\mathbf{x}}}) \mathbf{f}_i(\vec{\mathbf{x}}, \vec{\mathbf{v}}^{\beta} \vec{\mathbf{v}}, t) d^3 v = \nabla_{\vec{\mathbf{x}}}.(n_i \overline{\vec{\mathbf{v}}^{\beta} \vec{\mathbf{v}}_i^{\beta}})$$

$$\int \vec{\mathbf{v}}^{\beta} \vec{E}.\nabla_{\vec{\mathbf{v}}^{\beta}} \mathbf{f}_i(\vec{\mathbf{x}}, \vec{\mathbf{v}}, t) = -n_i \vec{E}$$

$$\int \vec{\mathbf{v}}^{\beta}(\vec{\mathbf{v}}^{\beta} \times \vec{B})).\nabla_{\vec{\mathbf{v}}^{\beta}} \mathbf{f}_i(\vec{\mathbf{x}}, \vec{\mathbf{v}}, t) d^3 v = -n_i \vec{\mathbf{v}}_i^{\beta} \times \vec{B}. \tag{8.399}$$

We find:

$$D_t^{\beta}(n_i \vec{\mathbf{v}}_i^{\beta}) + \nabla_{\vec{\mathbf{x}}}.(n_i \overline{\vec{\mathbf{v}}^{\alpha} \vec{\mathbf{v}}_i^{\alpha}}) - \frac{q_i}{m_i} n_i(\vec{E} + \vec{\mathbf{v}}_i^{\beta} \times \vec{B}) = 0. \tag{8.400}$$

If $\vec{\mathbf{v}}^{\beta} = \vec{\mathbf{v}}_i^{\beta} + \vec{\mathbf{w}}^{\beta}$, then we have:

$$\nabla_{\vec{\mathbf{x}}}.(n_i \overline{\vec{\mathbf{v}}^{\beta} \vec{\mathbf{v}}_i^{\beta}}) = n_i \vec{\mathbf{v}}_i^{\beta}.\nabla_{\vec{\mathbf{x}}} \vec{\mathbf{v}}_i^{\beta} + \vec{\mathbf{v}}_i^{\beta} \nabla_{\vec{\mathbf{x}}}.(n_i \vec{\mathbf{v}}_i^{\beta}) + \nabla_{\vec{\mathbf{x}}}.(n_i \overline{\vec{\mathbf{w}}^{\beta} \vec{\mathbf{w}}_i^{\beta}}), \tag{8.401}$$

where $\overline{\vec{\mathbf{w}}^{\beta} \vec{\mathbf{w}}_i^{\beta}} = (T_i/m_i)I$, $T_i = P_i/n_i$. We obtain accordingly :

$$D_t^{\beta}(\vec{\mathbf{v}}_i^{\beta}) + \vec{\mathbf{v}}_i^{\beta}.\nabla_{\vec{\mathbf{x}}} \vec{\mathbf{v}}_i^{\beta} + \frac{1}{n_i m_i} \nabla_{\vec{\mathbf{x}}} P_i - \frac{q_i}{m_i}(\vec{E} + \vec{\mathbf{v}}_i^{\beta} \times \vec{B}) = 0, \tag{8.402}$$

which might be called fractal Navier-Stokes equation and hence may describe the fractal flow of incompressible fluids. Then, we can rewrite Eq.(8.402) in the following form:

$$\varrho_i D_t^{\beta}(\vec{\mathbf{v}}_i^{\beta}) + \vec{\mathbf{v}}_i^{\beta}.\nabla_{\vec{\mathbf{x}}} \vec{\mathbf{v}}_i^{\beta} + \nabla_{\vec{\mathbf{x}}} P_i - q_i n_i(\vec{E} + \vec{\mathbf{v}}_i^{\beta} \times \vec{B}) = 0, \tag{8.403}$$

where $\varrho_i = n_i m_i$.

8.36 Fractal free unmagnetized electrons

Considering again the dynamics of free electrons in the absence of the magnetic field, we can write based on Eq.(8.390) and the free electrons in the absence of the magnetic field, we can write [El-Nabulsi and Golmankhaneh (2021a)]:

$$D_t^{\beta} f_1 + v^{\beta} \frac{\partial f_1}{\partial x} + \frac{e}{m} E \frac{\partial f_0}{\partial v^{\beta}} = 0, \tag{8.404}$$

where $f = f_0 + f_1$ and $v^{\beta} = |\vec{\mathbf{v}}^{\beta}|$. If we set:

$$f_1(x, v^{\beta}, t) = \hat{f}_1(v^{\beta}) \exp\left(j(kx - \int_0^t \omega(t'), d_F^{\beta} t') \right), \tag{8.405}$$

and

$$E(x,t) = \hat{E} \exp\left(j(kx - \int_0^t \omega(t')d_F^\alpha t')\right), \qquad (8.406)$$

then we have:

$$\hat{f}_1(v^\beta) = j\left\{\frac{eE}{m}\frac{1}{(S_F^\beta(\omega)) - kv^\beta}\right\}\frac{\partial f_0}{\partial v^\beta}. \qquad (8.407)$$

where $S_F^\beta(*)$ is the integral staircase function. In the same manner was given in the standard model we can see that the fractal dispersion relation is given by:

$$K^\beta(k,\omega,t) = 1 - \frac{S_F^\beta(\omega_p)^2}{(S_F^\beta(\omega))^2}\left(1 + \frac{2k^2 v_{th}^2}{2(S_F^\beta(\omega))^2}\right) = 0, \qquad (8.408)$$

$$S_F^\beta(\omega))^2 \approx S_F^\beta(\omega_p)^2 + \frac{3}{2}k^2 v_{th}^2. \qquad (8.409)$$

Using $a_1\omega^\beta < S_F^\beta(\omega) < a_2\omega^\beta$, one can write:

$$\omega^{2\beta} \approx \omega_p^{2\beta} + \frac{3}{2}k^2 v_{th}^2$$

$$\omega \approx \left(\omega_p^{2\beta} + \frac{3}{2}k^2 v_{th}^2\right)^{1/2\beta} \qquad (8.410)$$

The group velocity is given in that case by:

$$v_g \approx \frac{3k}{2\beta}\left(\omega_p^{2\beta} + \frac{3}{2}k^2 v_{th}^2\right)^{\frac{1}{2\beta}-1} v_{th}^2, \qquad (8.411)$$

and for $\sqrt{\frac{3}{2}}kv_{th} > \omega_p^\beta$, then:

$$v_g \rightarrow \frac{3k}{2\beta}\left(\frac{3}{2}k^2\right)^{\frac{1}{2\beta}-1} v_{th}^{\frac{1}{\beta}} \propto k^{\frac{1}{\beta}-1} \qquad (8.412)$$

We observe that at large k (small wavelength), information travels faster than the thermal velocity for $1/2 < \beta < 1$ and lower for $0 < \beta < 1/2$. Electromagnetic waves are hardly affected by fractal kinetic theory treatment in unmagnetized plasma .

8.37 Fractal wave in unmagnetized plasma

Moving particle in uniform plasma is modeled by [El-Nabulsi and Gol-mankhaneh (2021a)]:

$$D_t^\beta n_{i1} + n_{i1} \nabla_{\bar{x}} \vec{v}_{i1}^\beta = 0$$

$$D_t^\beta \vec{v}_{i1}^\beta = -\frac{1}{n_{i1} m_i} \nabla_{\bar{x}} P_{i1} + \frac{q_i}{m_i} \vec{E}_1$$

$$\nabla_{\bar{x}} \times \vec{E}_1 = -D_t^\beta \vec{B}_1,$$

$$\nabla_{\bar{x}} \times \vec{B}_1 = \mu_0 \vec{j}_1 + \frac{1}{c^2} D_t^\beta \vec{E}_1,$$

$$\nabla_{\bar{x}} . \vec{E}_1 = \frac{\rho}{\epsilon_0},$$

$$\nabla_{\bar{x}} . \vec{B}_1 = 0. \tag{8.413}$$

This follows that

$$S_F^\beta(\omega)^2 = S_F^\beta(\omega_p)^2 + \frac{k^2 \gamma \bar{T}_{e0}}{m_e}$$

$$\omega \approx \left(\omega_p^{2\beta} + \frac{k^2 \gamma \bar{T}_{e0}}{m_e} \right)^{1/2\beta} \tag{8.414}$$

which is might called the fractal Bohm-Gross formula for electron plasma waves [El-Nabulsi and Golmankhaneh (2021a)]. The fractional Bohm-Gross formula for electron plasma waves is given by

$$\omega = \left(\omega_p^2 + (2 - \alpha) \frac{k^2 \gamma \bar{T}_{e0}}{m_e} \right)^{1/2}. \tag{8.415}$$

where ω the frequency of electrons, ω_p the plasma frequency, m_e mass of electron, and $\bar{T}_{e0}$ is background temperature [El-Nabulsi and Gol-mankhaneh (2021a)]. The group velocity for electron plasma waves using fractal calculus is given by:

$$v_g \approx \frac{\gamma \bar{T}_{e0} k}{m_e \beta} \left(\omega_p^{2\beta} + k^2 \frac{\gamma \bar{T}_{e0}}{m_e} \right)^{\frac{1}{2\beta} - 1}, \tag{8.416}$$

where k is wave number and for $\sqrt{\frac{\gamma \bar{T}_{e0}}{m_e}} > \omega_p^\beta$, then:

$$v_g \to \frac{\gamma \bar{T}_{e0} k}{m_e \beta} \left(k^2 \frac{\gamma \bar{T}_{e0}}{m_e} \right)^{\frac{1}{2\beta} - 1} \propto k^{\frac{1}{\beta} - 1} \tag{8.417}$$

The group velocity for electron plasma waves using fractional calculus is given by [El-Nabulsi and Golmankhaneh (2021a)]:

$$v_g = \frac{k(2 - \alpha)}{\sqrt{\omega_p^2 + (2 - \alpha) \frac{k^2 \gamma \bar{T}_{e0}}{m_e}}} \frac{\gamma \bar{T}_{e0}}{m_e}. \tag{8.418}$$

In Figures 8.45 and 8.46, we show the effect of using two methods fractional space and fractal calculus which is based on fractal space on deriving the frequency of electrons relation.

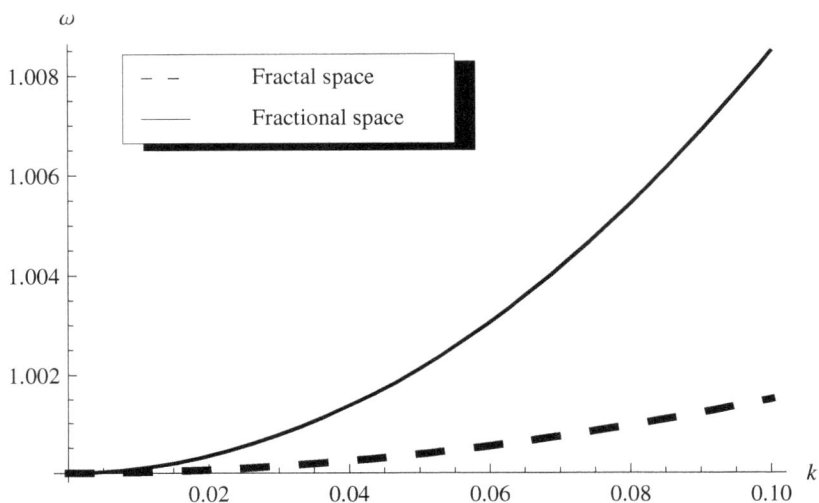

Fig. 8.45: Graph of Eqs.(8.414) and (8.415) for $\alpha = 0.3$ and $\beta = 0.3$

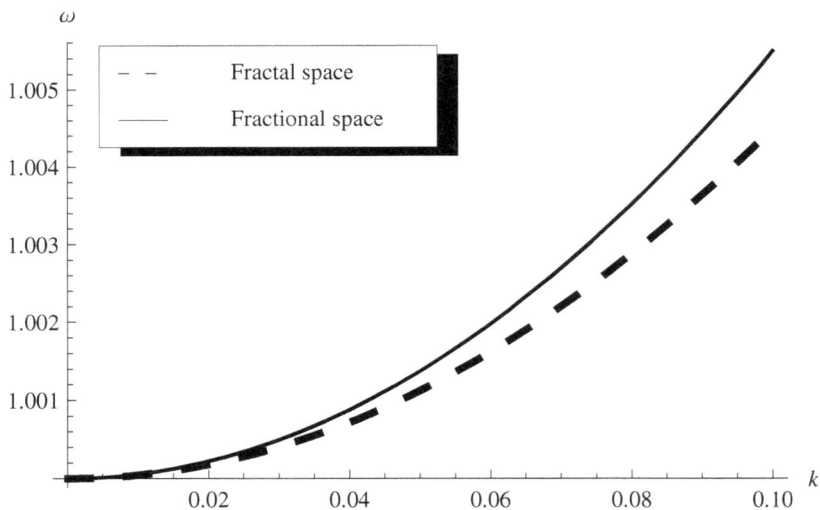

Fig. 8.46: Graph of Eqs.(8.414) and (8.415) for $\alpha = 0.9$ and $\beta = 0.9$

In Figures 8.47 and 8.48, we present the effect of using two methods fractional space and fractal calculus which is based on fractal space on deriving the group velocity for electron plasma waves.

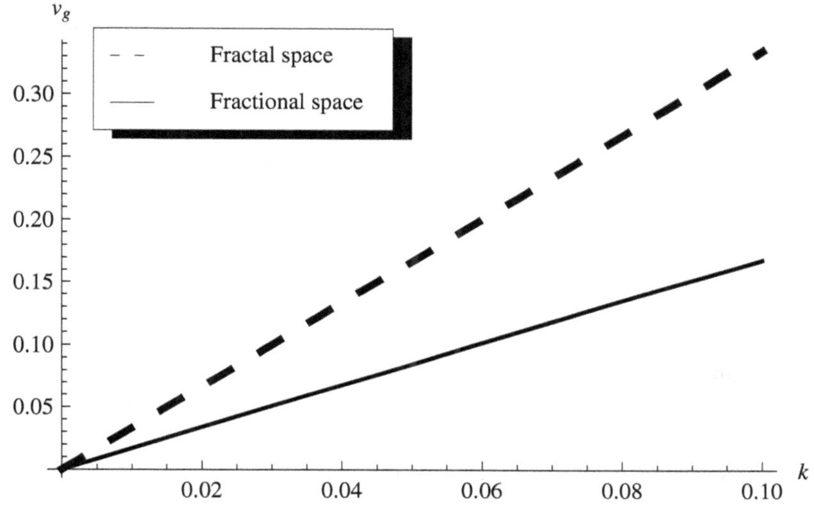

Fig. 8.47: Graph of Eqs.(8.417) and (8.418) for $\alpha = 0.3$ and $\beta = 0.3$

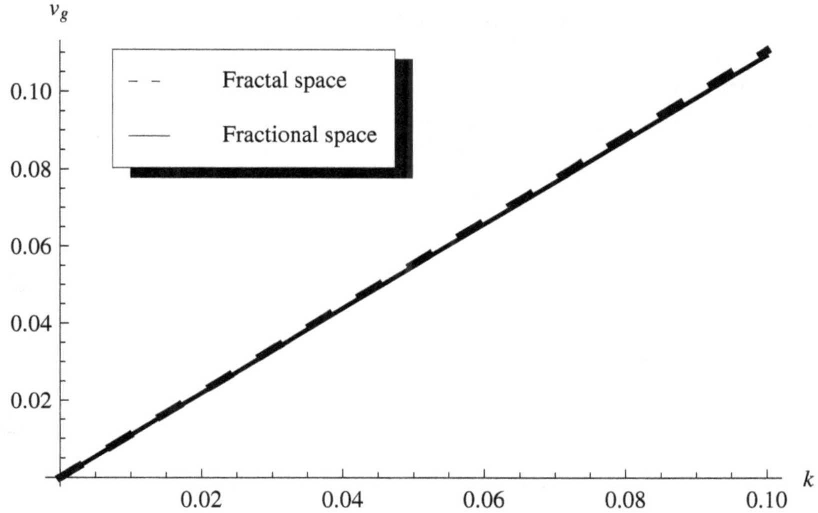

Fig. 8.48: Graph of Eqs.(8.417) and (8.418) for $\alpha = 0.9$ and $\beta = 0.9$

Table 8.3: Comparing of fractional space approach with fractal approach

Approaches	Fractional space	Fractal calculus
Lagrangian	$L(\dot{q}, q, t)$	$\mathbf{L}(D_t^\beta \mathbf{q}(t), \mathbf{q}(t), t)$
Action functional	$S = \frac{1}{\Gamma(\alpha)} \int_{t_1}^{t_2} L(\dot{q}, q, \tau)(t-\tau)^{\alpha-1} d\tau$	$S = \int_{t_1}^{t_2} \mathbf{L}(D_t^\beta \mathbf{q}(t), \mathbf{q}(t), t) dt$
Lagrange's equations	$\frac{\partial L}{\partial q} - \frac{d}{d\tau}\left(\frac{\partial L}{\partial \dot{q}}\right) = \frac{1-\alpha}{t-\tau}\frac{\partial L}{\partial \dot{q}}$	$\frac{\partial \mathbf{L}}{\partial \mathbf{q}} - D_t^\beta\left(\frac{\partial \mathbf{L}}{\partial \dot{\mathbf{q}}}\right) = \left(\frac{\partial \mathbf{L}}{\partial \mathbf{q}_t}\right)$
Hamilton's equations	$(\dot{q}(\tau), \dot{p}(\tau)) = \left(\frac{\partial H}{\partial p}, -\frac{\partial H}{\partial q} - \frac{1-\alpha}{t-\tau}p\right)$	$(\mathbf{q}_t, \mathbf{p}_t) = \left(\frac{\partial \mathbf{H}}{\partial \mathbf{p}}, -\frac{\partial \mathbf{H}}{\partial \mathbf{q}}\right)$
Boltzmann Vlasov equation	$\frac{\partial f(\vec{x}, \vec{v}, \tau)}{\partial \tau}$ $+ \vec{v} \cdot \nabla_{\vec{x}} f(\vec{x}, \vec{v}, \tau)$ $- \frac{1}{m}\nabla_{\vec{x}} V(\vec{x}, \tau) \cdot \nabla_{\vec{v}} f(\vec{x}, \vec{v}, \tau)$ $- \frac{\alpha-1}{\tau-t}\vec{v} \cdot \nabla_{\vec{v}} f(\vec{x}, \vec{v}, \tau) = 0$	$D_t^\beta \mathbf{f} + \vec{v}^\beta \cdot \nabla_{\vec{x}} \mathbf{f}(\vec{x}, \vec{v}, t)$ $- \frac{1}{m}\nabla_{\vec{x}} \mathbf{V}(\vec{x}, t) \cdot \nabla_{\vec{v}^\beta} \mathbf{f}(\vec{x}, \vec{v}, t) = 0$
Navier-Stokes equation	$\frac{\partial \vec{v}_i}{\partial T}$ $+ \frac{1-\alpha}{T}\vec{v}_i + \vec{v}_i \cdot \nabla \vec{v}_i$ $+ \frac{1}{n_i m_i}\nabla P_i - \frac{q_i}{m_i}\left(\vec{E} + \vec{v}_i \times \vec{B}\right) = 0$	$\varrho_i D_t^\beta(\bar{\vec{v}}_i^\beta) + \bar{\vec{v}}_i^\beta \cdot \nabla_{\vec{x}} \bar{\vec{v}}_i^\beta$ $+ \nabla_{\vec{x}} P_i - q_i n_i (\vec{E} + \vec{v}_i^\beta \times \vec{B}) = 0$
Bohm-Gross	$\omega = \left(\omega_p^2 + (2-\alpha)\frac{k^2\gamma T_{e0}}{m_e}\right)^{1/2}$	$\omega \approx \left(\omega_p^{2\beta} + \frac{k^2\gamma T_{e0}}{m_e}\right)^{1/2\beta}$
The group velocity in Unmagnetized Plasma	$v_g = \frac{k(2-\alpha)\frac{\gamma T_{e0}}{m_e}}{\sqrt{\omega_p^2 + (2-\alpha)\frac{k^2\gamma T_{e0}}{m_e}}}$	$v_g \approx \frac{\gamma T_{e0}k}{m_e\beta}\left(\omega_p^{2\beta} + k^2\frac{\gamma T_{e0}}{m_e}\right)^{\frac{1}{2\beta}-1}$

8.38   Casimir effect on fractal spaces

The attraction of two parallel conducting plates in a vacuum is called the Casimir effect/a force from nothing. Many researchers have modeled this effect and references therein) and explained the origin of the force Casimir effect using quantum field theory [Golmankhaneh and Nia (2021)]. In this section, our goal is to relate this force to the fractal space. For this purpose, let us consider electrostatic potential between two metal plates in the fractal space, therefore we can write α-dimension Laplace equation as follows:

$$D_{F,x}^{2\alpha} V(x) = 0, \qquad (8.419)$$

with boundary equation

$$V(0) = 0, \quad V(1) = 0. \qquad (8.420)$$

Then one can find the solution of Eq.(8.419) by using conjugacy of fractal calculus with standard one as follows:

$$V(x) = -S_F^\alpha(x) + \Gamma(\alpha + 1). \qquad (8.421)$$

Since $S_F^\alpha(x) \leq x^\alpha$, then we can write smooth solution as follows:

$$V(x) \approx -x^\alpha + \Gamma(\alpha + 1). \qquad (8.422)$$

Then, we have

$$F(x) = -\frac{dV}{dx} \approx \alpha x^{\alpha-1}. \qquad (8.423)$$

Eq.(8.423) claims that the origin of the Casimir effect is fractal geometry. Also, we see the effect of the dimension of space between plates on their electrostatic potential.

8.39 Fractal models for the viscoelasticity

We generalize the viscoelasticity models to the fractal mediums. The ideal fractal fluid can be modeled and described by Newton's second law of fractal viscosity as follows [Golmankhaneh and Baleanu (2016d)]:

$$\sigma_F = \eta_F D_F^\alpha \epsilon \qquad (8.424)$$

where σ_F is the fractal stress, η_F is the elastic modulus of the fractal material, and ϵ is the strain. But in the nature we have real fractal materials which have properties between the ideal solids and ideal liquids. Then we can model them by using non-local fractal derivatives as follows:

$$\sigma_F \propto_0 \mathcal{D}_t^\beta \epsilon \qquad (8.425)$$

which is called fractal Blair's model. Here, we suggest the fractional non-local fractal order derivative β as an index of memory. Namely, if we choose $\beta = \alpha$ the process is memoryless. But if we choose $0 < \beta < \alpha$, then it shows that the processes are with memory on fractals. In Eq.(8.425) if we choose $\epsilon = \chi_F(t)$ then we obtain

$$\sigma_F \propto \frac{1}{\Gamma(1-\beta)}(S_F^\alpha(t))^\beta$$

$$\propto \frac{1}{\Gamma(1-\beta)}t^{\alpha\beta} \qquad (8.426)$$

where α is the fractal dimension, and β is the index of memory.

Appendix A

Appendix

Now we give algorithms for the mass functions for the fractal sets and curves [Parvate and Gangal (2009); Parvate (2009); Parvate and Gangal (2011)].

A.1 Algorithm for the mass function on fractal sets

The following algorithm try to find out a subdivision P on a scale δ for which the sum is close to the infimum i.e. to $\gamma_\delta^\alpha(F, a, b)$ [Parvate and Gangal (2009); Parvate (2009); Parvate and Gangal (2011)].

Remark A.1. We note that we use finite subdivision in the case of mass function despite the Hausdorff measure.

Figure A.1 shows approximations of $\gamma_{\delta_2}^\alpha(F, a, b)/\gamma_{\delta_1}^\alpha(F, a, b)$ calculated using this algorithm for the two sets described below, $0 < \delta_2 < \delta_1$, for $\alpha \in (0, 1]$. The value of α for which the graph meets 1 is the possible value of the γ-dimension, since it indicates that γ_δ^α is converging to a non-zero finite values as $\delta \to 0$. One can plot Figure A.1 for various pairs of scales (δ_1, δ_2) to confirm it.

A.1.1 The steps of algorithms

Let F be a bounded subsets of R, with $a = \inf$ and $b = \sup F$ [Parvate and Gangal (2009); Parvate (2009); Parvate and Gangal (2011)].

(1) Initial subdivision: Divided the interval $[a, b]$ in equal intervals of size $\tau = (b - a)/N$ for some integer $N > 0$. This forms a subdivision $P = \{x_i = a + ir\}, i = 0, \dots, N$.

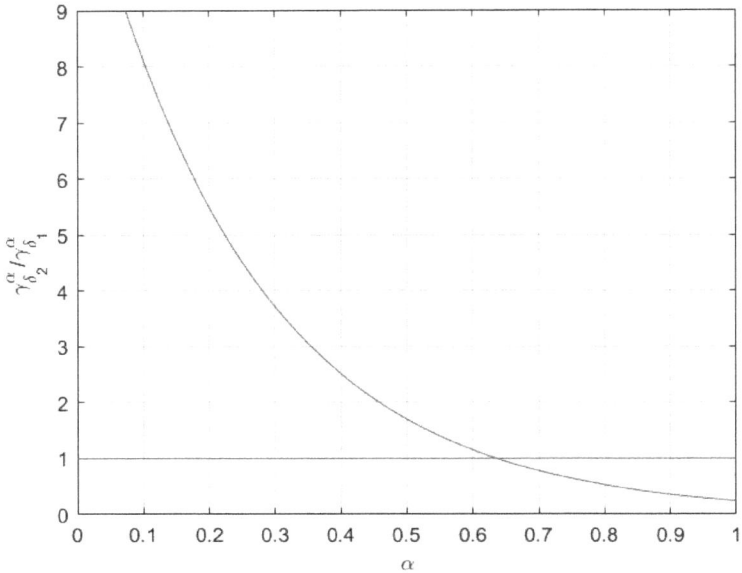

Fig. A.1: Graph for ternary Cantor set

(2) Merging: Form a new subdivision P' consisting of x_0, x_N and only those points x_i, $1 < i < N$ for which one and only one of the intervals of $[x_{i-1}, x_i]$ and $[x_i, x_{i+1}]$ has a nonempty intersection with F. This step is called merging, since the successive components of P containing points of F are merged to form P'. Relabel the points of P' in increasing order by y_i, $0 \leq i \leq M \leq N$. Now it is guaranteed that alternate components of P' have nonempty intersection with F. This step reduces the sum σ^α in view of Jensen's inequality.

(3) Shrinking: For every components $[y_i, y_{i+1}]$ that has nonempty intersection with F, let $z_i = \inf F \cap [y_i, y_{i+1}]$ and $z_{i+1} = \sup F \cap [y_i, y_{i+1}]$. Let Q be the new subdivision formed by the z_i, $0 \leq i \leq M$. In most cases where the dimension is to be calculated computationally, the set F is available as enumeration of points and the inf and sup do not present algorithmic difficulties. Even otherwise, inf and sup can be approximated using methods such as bisection.

(4) Removing isolated points: For components $[z_i, z_{i+1}]$ which contain fewer points of F than $(z_{i+1} - z_i)T/r$ where T is some small threshold value (e.g. 3), we can replace them by components which are arbitrarily small and centered around those points. Thus we can ignore the contribution

from such components while calculating $\sigma^\alpha[F, Q]$. This step takes care of isolated points, since T is the minimum number of points in a bin of size r in order that the particular component contributes.

(5) Analysing convergence for various values α: Obtain two such subdivisions Q_1 and Q_2 for two values of r, viz. r_1 and r_2, $r_2 < r_1$ (i.e. two values N_1, N_2 of N, $N_2 > N_1$). Plot the ratio $\sigma^\alpha[F, Q_2]/\sigma^\alpha[F, Q_1]$ for all $\alpha \in (0, 1]$. The value of α for which this graph meets 1 is the possible value of the γ-dimension, since it indicates that γ_δ^α is converging to a nonzero finite value as $\delta \to 0$. It can be plotted for various scales δ to confirm it. Further, the value of $\sigma^\alpha[F, Q_1]$ or $\sigma^\alpha[F, Q_2]$ is an approximation for the mass.

Remark A.2. We note that with steps (2) to (4), an attempt is being made to form a subdivision Q such that such that $\sigma^\alpha[F, Q]$ is very close to $\gamma_{|Q|}^\alpha(F, a, b)$, while step (1) attempts to control $|Q|$. It is clear that due to step (2) of merging, r is not an upper bound on $|Q|$. But for all practical purpose, $|Q|$ behaves monotonically with r. In particular, it is easily seen that if $r_2 = r_1/m, m \in N$, then $|Q_2| \le |Q_1|$.

A.2 Algorithm for the mass function on fractal curves

In this section, we present the Monte Carlo algorithm for mass function on fractal curves [Parvate *et al.* (2011)]. Let recall the mass function on fractal curve as

$$\gamma^\alpha(F, a, b) = \lim_{\delta \to 0} \inf_{P:|P| \le \delta} \sum_{i=0}^{n-1} \frac{|\mathbf{w}(t_{i+1}) - \mathbf{w}(t_i|^\alpha)}{\Gamma(\alpha + 1)}. \tag{A.1}$$

As pervious algorithm, our aim is to find a subdivision P such that $\sigma^\alpha[F, P]$ is close to the infimum. Let $[a, b]$ be the domain of $\mathbf{w}$, and randomly means with uniform probability. We begin with uniform subdivision P such that $|P| = \delta/4$, and iteratively improve it using the following prescription.

(1) Choose two number $x, y \in [a, b]$ randomly, and relabel them if necessary so that $x \le y$. Then $[x, y] \subset [a, b]$. Let $P' = \{t_i : 0 \le i \le m\}$ denote the set of all points of $P \cap [x, y]$. We now modify P' in one of the following ways with equal probability, and denote the resultant by P'':

(a) With a probability $p_c = \min(1, \delta/(y - x))$, we shift each point t_i (except t_0 and t_m) by a random amount between $[-\delta/2, \delta/2]$, if the resultant subdivision P'' still satisfies $|P''| \le \delta$.

(b) With a probability $p_d = \min(1, \delta/(y-x))$, we remove each point t_i (expect t_0 and t_m) from P', if the resultant subdivision P'' still satisfies $|P''| \leq \delta$.

(c) With a probability $p_i = \min(1, \delta/(y-x))$, we insert a point between each t_i and t_{i+1} which is chosen randomly from $[t_i, t_{i+1}]$. (However, to avoid accumulating too much of rounding error, we insert the point only if the distance between any two resultant successive point is greater than $\delta/10$).

(2) Form a new subdivision $P_1 = (P \cap [a, x)) \cup P'' \cup (P \cap (y, b])$, i.e. the subdivision of which the points belonging to $[x, y]$ are changed by the above procedure. If $\sigma^\alpha[F, P_1] < \sigma^\alpha[F, P]$, then we consider P_1 as the current subdivision which will be possibly improved further using above steps. Otherwise we consider P again for the purpose.

Remark A.3. As the sum $\sigma^\alpha[F, P1]$ approaches the infimum, many of the newly formed subdivision P' are rejected since they sum up higher than P. Thus near the infimum, the sum remains constant for many consecutive iterations and changes only intermittently. Therefore the usual convergence of terminating iteration when the difference between successive iterations or every K iterations (K being a suitable large integer) goes below a certain small number, is not useful in this case. Instead, after examining the sum over a large number of iterations we observe that the sum stops making significant progress between $N' = 1000$ to $N' = 2000$, where $N' = N/2$ is the number of iterations N normalized by the current subdivision size n. Further, we need to go through all these iterations more than once, just to ensure that subdivision is really optimal. Occasionally, it may happen that the sum settles a little above the optimal value, getting trapped in a local minimum. The results of this algorithm as applied on the von Koch curve, parameterized as in Eq.(4.261). It turns out that the mass of the entire von Koch curve is a little less than $0.51/\Gamma(\alpha+1)$, $\alpha = \ln(4)/\ln(3)$.

A.3 Algorithm for α-dimension of fractal curves

The above description assumes that the value of α is the same as the γ-dimension of the set F, say α_0 [Parvate *et al.* (2011)]. One expect δ-independence in the values of $\sigma^\alpha[F, P(\delta)]$ where $P(\delta)$ denotes the resultant subdivision of the algorithm at the scale δ, since the value of γ_δ^α converges to a finite nonzero value. This is what we observe from the values of $\sigma^\alpha[F, P(\delta)]$ obtained for various values of δ. Now consider $\alpha \neq \alpha_0$.

Let $0 < \delta_1 < \delta_2$. If $\alpha < \alpha_0$ then $\gamma^\alpha(F, a, b) = \infty$. Therefore we expect that $R(\alpha) = \sigma^\alpha[F, P(\delta_1)]/\sigma^\alpha[F, P(\delta_2)] > 1$. Similarly since $\alpha > \alpha_0$ implies $\gamma^\alpha(F, a, b) = 0$, we expect that $R(\alpha) < 1$. Namely, we need to find the number α_0 such that $R(\alpha_0) = 1$. We note that $\alpha_0 \in [1, m]$ m being the embedding dimension, since $F \in R^m$ is a curve. Treating this as the initial bracket of values for α_0, one just need to use some algorithm such as bisection to shrink this bracket to sufficient accuracy.

A.4 Comparison between γ-dimension and box dimension

In this section we present comparison between γ-dimension and box dimension [Parvate *et al.* (2011)]. Let $\dim_{F\cap[a,b]} = \alpha$. Then γ^β diverges for any $\beta < \alpha$. Thus for any $k > 0$ there exists $\delta_0 > 0$ such that

$$\delta < \delta_0 \Rightarrow \gamma_\delta^\beta(F, a, b) > k. \tag{A.2}$$

Let P be any subdivision such that $|P| \leq \delta$, and let $N_\delta(F \cap [a, b])$ be the number of nonzero terms in the sum $\sigma^\alpha[F, P]$. Thus for arbitrary but fix $k > 0$ and $\delta < \delta_0$,

$$k < \gamma_\delta^\beta(F, a, b) \leq \frac{N_\delta(F \cap [a, b])\delta^\beta}{\Gamma(\beta + 1)}, \tag{A.3}$$

where $0 < \beta < \alpha \leq 1$. Thus,

$$\ln(k) \leq \ln N_\delta(F \cap [a, b]) + \beta \ln(\delta) - \ln(\Gamma(\beta + 1)), \tag{A.4}$$

which implies

$$-\beta \ln(\delta) \leq \ln(N_\delta(F \cap [a, b])) - \ln(k) - \ln(\Gamma(\beta + 1)). \tag{A.5}$$

Dividing by $-\ln(\delta)$ which is positive for $\delta < 1$, we have

$$\beta \leq \frac{\ln(N_\delta(F \cap [a, b]))}{-\ln(\delta)} - \frac{\ln(k) - \ln(\Gamma(\beta + 1))}{-\ln(\delta)} \tag{A.6}$$

Taking limit as $\delta \to 0$ and noting that the first term of (A.6) is the definition of the box dimension $\dim_B(F \cap [a, b])$ in the limit and the denominator of the second term of (A.6) diverges, we obtain

$$\beta \leq \dim_B(F \cap [a, b]) = \lim_{\delta \to 0} \frac{\ln(N_\delta(F \cap [a, b]))}{-\ln(\delta)} \tag{A.7}$$

Since this is true for any $\beta < \alpha = \dim_\gamma(F \cap [a, b])$, so that

$$\dim_\gamma(F \cap [a, b]) \leq \dim_B(F \cap [a, b]), \tag{A.8}$$

we see that the γ-dimension of a set is at most equal to its box dimension.

Example A.1. The box dimension and γ-dimension of the middle-$\frac{1}{3}$ Cantor set C are equal.

$$\dim_\gamma C = \dim_B C = \frac{\ln 2}{\ln 3}. \tag{A.9}$$

Example A.2. The box dimension of $K = \{0, 1, \frac{1}{2}, \frac{1}{3}, \ldots\}$ differ from γ-dimension. Namely,

$$\dim_\gamma K = 0, \tag{A.10}$$

since K is compact. But box dimension of K is $\dim_B K = 0.5$.

A.5 Comparison between γ-dimension and Hausdorff dimension

We recall that $H^\alpha_\delta(E)$ denotes the coarse grained Hausdorff measure of a subset E of R, and $H^\alpha(E)$ denotes the Hausdorff measure [Parvate *et al.* (2011)]. Let P be a subdivision with $|P| \leq \delta$. Those components $[x_i, x_{i+1}]$ of P for which $\theta(F, [x_i, x_{i+1}])$ is nonzero, form a δ-cover of $F \cap [a, b]$. Thus

$$\sigma^\alpha[F, P] \geq \Gamma(\alpha + 1) H^\alpha_\delta(F \cap [a, b]). \tag{A.11}$$

As this is true for any P such that $|P| \leq \delta$, it follows that

$$\gamma^\alpha_\delta(F, a, b) \geq \Gamma(\alpha + 1) H^\alpha_\delta(F \cap [a, b]) \tag{A.12}$$

for each $\delta > 0$. So taking limit as $\delta \to 0$,

$$\gamma^\alpha(F, a, b) \geq \Gamma(\alpha + 1) H^\alpha_\delta(F \cap [a, b]) \tag{A.13}$$

which also implies

$$\dim_H(F \cap [a, b]) \leq \dim_\gamma(F \cap [a, b]), \tag{A.14}$$

we see that the γ-dimension of a set is at least equal to its Hausdorrf dimension.

Theorem A.1. *For a compact set $F \subset R$*

$$\gamma^\alpha(F, a, b) = \Gamma(\alpha + 1) H^\alpha_\delta(F \cap [a, b]) \tag{A.15}$$

Proof. For $\delta > 0$, let $\{A_i, i = 1, 2, \ldots\}$ be any countable cover of $F \cap [a, b]$ such that $diam A_i \leq \delta/2$ for all i The sets A_i need not be open or closed. Then

$$H^\alpha_{\delta/2}(F \cap [a, b]) \leq \sum_i (diam A_i)^\alpha. \tag{A.16}$$

Consider closed intervals $B_i = [u_i, v_i]$ where $u_i = \inf A_i$ and $v_i = \sup A_i$. Then $A_i \subset B_i$ and $diam B_i = diam A_i$. Thus $\{B_i\}$ forms a cover of $F \cap [a, b]$ and

$$\sum_i (diam B_i)^\alpha = \sum_i (diam A_i)^\alpha \geq H_{\delta/2}^\alpha (F \cap [a, b]). \tag{A.17}$$

Given $\epsilon \in (0, (\delta/2)^\alpha)$, let $\{C_i\}_{i=1}^\infty$ be the open intervals which is defined by

$$C_i = \left(u_i - \frac{1}{2}\left(\frac{\epsilon}{2^i}\right)^{\frac{1}{\alpha}}, v_i + \frac{1}{2}\left(\frac{\epsilon}{2^i}\right)^{\frac{1}{\alpha}} \right). \tag{A.18}$$

The class $\{C_i\}_{i=1}^\infty$ thus forms an open cover of $F \cap [a, b]$ and

$$diam C_i = diam A_i + \left(\frac{\epsilon}{2^i}\right)^{\frac{1}{\alpha}} < \delta \tag{A.19}$$

so that

$$\sum (diam C_i)^\alpha = \sum \left(diam A_i + \left(\frac{\epsilon}{2^i}\right)^{\frac{1}{\alpha}} \right)^\alpha. \tag{A.20}$$

In view of Jensen's inequality for the case of two variables assures that $(s_1 + s_2)^t \leq s_1^t + s_2^t$, for $s_1, s_2 > 0$ and $0 < t < 1$, is that

$$\sum (diam C_i)^\alpha \leq \sum (diam A_i)^\alpha + \epsilon \sum \frac{1}{2^i} = \sum (diam A_i)^\alpha + \epsilon. \tag{A.21}$$

We see that a finite cover consisting of closed intervals can be constructed. As F is compact, so is $F \cap [a, b]$. Therefore a finite subset of $\{C_i\}$ covers $F \cap [a, b]$. We denote this finite subcover by $\{D_i, i = 1, \ldots, n\}$. The D_i are open intervals of the form (a_i, b_i). Without loss of generality one can choose this finite subcover $\{D_i\}$ such that $D_i \not\subset D_j$ whenever $i \neq j$. Further, the sets are labeled such that $a_i < a_{i+1}$. But as $D_i \not\subset D_{i+1}$ and $D_{i+1} \not\subset D_i$, it implies that $a_i < a_{i+1}$ and $b_i < b_{i+1}$. Consider the closures $\bar{D}_i$ of D_i. As $\{D_i\}$ is a finite subcover out of $\{C_i\}$ and $\{\bar{D}_i\}$ have the same diameters as D_i, it follows from Eq.(A.21) that

$$\sum (diam \bar{D}_i)^\alpha = \sum (diam A_i)^\alpha + \epsilon \tag{A.22}$$

Let $I_1 = \bar{D}_1$ and $I_i = \bar{D}_i/D_{i-1}$ for $2 \leq i \leq n$. The collection $\{I_i\}$ forms a finite cover of $F \cap [a, b]$ by closed intervals, and

$$\sum (diam I_i)^\alpha \leq \sum (diam A_i)^\alpha + \epsilon. \tag{A.23}$$

The closed intervals I_i share at most endpoints. The set of all the endpoints of I_i, $1 \leq i \leq n$ forms a subdivision P of $[a, b]$ which can be refined to a subdivision Q such that $|Q| \leq \delta$ and

$$\frac{1}{\Gamma(\alpha + 1)} \sigma^\alpha [F, Q] = \sum (diam I_i)^\alpha \leq \sum (diam I_i)^\alpha \leq \sum (diam A_i)^\alpha + \epsilon. \tag{A.24}$$

Therefore,

$$\frac{1}{\Gamma(\alpha + 1)} \gamma_\delta^\alpha (F, a, b) \leq \sum (diam I_i)^\alpha \leq \sum (diam A_i)^\alpha + \epsilon. \tag{A.25}$$

Since this relation holds for any countable cover $\{A_i\}$ of $F \cap [a, b]$ such that $diam A_i \leq \delta/2$ and for arbitrary $\epsilon > 0$, it follows that

$$\frac{1}{\Gamma(\alpha + 1)} \gamma_\delta^\alpha (F, a, b) \leq H_{\delta/2}^\alpha (F \cap [a, b]) \leq H^\alpha (F \cap [a, b]). \tag{A.26}$$

consequently in the limit as $\delta \to 0$

$$\frac{1}{\Gamma(\alpha + 1)} \gamma^\alpha (F, a, b) \leq H^\alpha (F \cap [a, b]). \tag{A.27}$$

Eq.(A.13) and (A.27) together imply the required equality which completes the proof. □

Corollary A.1. *If $F \subset R$ is compact, then* $\dim_\gamma F = \dim_H F$.

Example A.3. Consider the Cantor set. This set is compact and has a Hausdorff dimension $\alpha = \log(2)/\log(3)$. By Theorem A.1 $H^\alpha(C, [a, b]) = \frac{1}{\Gamma(\alpha+1)} \gamma^\alpha(C, a, b)$ and $\dim_\gamma C = \dim_H C = \alpha = \log(2)/\log(3)$. Using the self-similarity of C and the monotonicity as well as scaling and translation properties of the mass function, we can calculate S_F^α at each point.

Remark A.4. We note that in general γ-dimension and Hausdorff dimension are significantly different. The difference arise because of the use of subdivision are finite, instead of arbitrary countable covers as in Hausdorff measure. In particular, it can be easly seen that $\dim_\gamma \bar{F} = \dim_\gamma F$, where $\bar{F}$ denotes the closure of the set F. This property is comparable to a similar property of box dimension. For example, consider the set $Q' = Q \cap [0, 1]$, the set of all rational numbers in the interval $[0, 1]$. For this set, $\dim_\gamma Q' = \dim_B Q' = 1$, whereas $\dim_H Q' = 0$.

Example A.4. Let C_i be the set of the endpoints of the intervals which are formed at the ith state of construction of the middle-$\frac{1}{3}$ Cantor set C.

Consider $\{[0, \frac{1}{3}], [\frac{2}{3}, 1]\}$ as the first stage, C_i contains 2^{i+2} points. Let $C' = \bigcup_{i=1}^{\infty} C_i$. This set is countable, and dense in C. Therefore,

$$\dim_\gamma C' = \dim_\gamma \bar{C}' = \dim_\gamma C = \frac{\ln(2)}{\ln(3)} \qquad (A.28)$$

whereas

$$\dim_H C' = 0. \qquad (A.29)$$

Remark A.5. The γ-dimension is finer than the box dimension, but not than Hausdorff dimension. For a set $E \subset R$ we have

$$\dim_B E \geq \dim_\gamma E \cap [a, b] \geq \dim_H E. \qquad (A.30)$$

Specifically, the γ-dimension is unaffected by clusters of points unlike the box dimension, as the example of the set $K = \{0, 1, 1/2, 1/3, \ldots\}$ in the previous subsection demonstrates. On the other hand, it is sensitive to countable but dense sets such as rationales.

A.6 F^α-integrating of a function

As an example of F^α-integration, we calculate integral of $f(x) = x\chi_C(x)$ as follows [Parvate *et al.* (2011)]:

$$g(y) = \int_0^y x\chi_C(x)d_C^\alpha x = \int_0^y x d_C^\alpha x, \qquad (A.31)$$

where C is the middle $\frac{1}{3}$ Cantor set, and $\alpha = \ln(2)/\ln(3)$ is its γ-dimension. The function $f(x) = x\chi_C(x)$ is C-continuous on $[0, 1]$, hence it is C^α-integrable. The set $P_n = \{x_i = i/n : 0 \leq i \leq n]\}$ is a subdivision of $[0, 1]$. For any component $[x_i, x_{i+1}]$ of P_n $x_i \leq m[f, F, [x_i, x_{i+1}]]$ and $x_{i+1} \geq M[f, F, [x_i, x_{i+1}]]$ if $F \cap [x_i, x_{i+1}] \neq \emptyset$. Therefore,

$$\underline{g(1)} = \lim_{n\to\infty} \sum_{i=0}^n \left\{ \frac{i}{n} \left[S_F^\alpha \left(\frac{i+1}{n} \right) - S_F^\alpha \left(\frac{i}{n} \right) \right] \right\} \leq L^\alpha[f, F, P_n] \qquad (A.32)$$

and

$$\overline{g(1)} = \lim_{n\to\infty} \left\{ \frac{i+1}{n} \left[S_F^\alpha \left(\frac{i+1}{n} \right) - S_F^\alpha \left(\frac{i}{n} \right) \right] \right\} \leq L^\alpha[f, F, P_n] \geq U^\alpha[f, F, P_n]. \qquad (A.33)$$

Further, it can be seen that

$$\lim_{n\to\infty} [\overline{g(1)} - \underline{g(1)}] = 0. \qquad (A.34)$$

Thus, $g(1)$ can be calculated using the limit

$$g(1) = \lim_{n\to\infty} \left\{ \frac{i}{n} \left[S_F^\alpha \left(\frac{i+1}{n} \right) - S_F^\alpha \left(\frac{i}{n} \right) \right] \right\}. \tag{A.35}$$

Similarly for integers $m > 0$,

$$g\left(\frac{1}{3^m} \right) = \lim_{n\to\infty} \left\{ \frac{i}{3^m n} \left[S_F^\alpha \left(\frac{i+1}{3^m n} \right) - S_F^\alpha \left(\frac{i}{3^m n} \right) \right] \right\} \tag{A.36}$$

Using the self-similarity of C and scaling of S_F^α, we see from Eq.(A.35) and Eq.(A.36) that

$$g\left(\frac{1}{3^m} \right) = \frac{1}{3^{m(1+\alpha)}} g(1) = \frac{1}{6^m} g(1). \tag{A.37}$$

We make use of the ternary representation of numbers which simplifies many calculations involving the Cantor set. Any number $y \in [0,1]$ can be represented by the series

$$y = \sum_{i=1}^{\infty} \frac{t_i(y)}{3^i} \tag{A.38}$$

where $t_i(y) = 0, 1$, or 2 is the ith ternary digit of y after ternary point. The number y belongs to C if and only if y has a representation of the form Eq.(A.38) where $t_i(y) = 0$ or 2 for all i. An approximation of $y \in [0,1]$ by a finite of digits is denoted by

$$T_0 = 0, \quad and \quad T_n(y) = \sum_{i=1}^{n} \frac{t_i(y)}{3^i}. \tag{A.39}$$

The sequence $\{T_n(y)\}_{n=0}^{\infty}$ is a monotonically (but not strictly) increasing sequence whose limit is y. Hence we can write

$$g(y) = \sum_{i=1}^{\infty} \int_{T_{i-1}(y)}^{T_i(y)} x\chi(x) d_C^\alpha x = \sum_{i=1}^{\infty} I_i(y) \tag{A.40}$$

where

$$I_i(y) = \int_{T_{i-1}(y)}^{T_i(y)} x\chi(x) d_C^\alpha x \tag{A.41}$$

The quantities $I_i(y)$ is calculated using the self-similarly of C, the scaling and translation properties of S_F^α and Eq.(A.37). Let $y \in [0,1]$ and let n be any integer such that $i < n \Rightarrow t_i(y) = 0$, or 2. Then $i < n \Rightarrow T_i(y) \in C$.

For calculating $I_n(y)$, there are three cases corresponding to three possible values of $t_n(y)$:

(1) $t_n(y) = 0$, where $T_{n-1}(y) = T_n(y)$ and $I_n(y) = 0$.
(2) $t_n(y) = 1$, in this case we have

$$T_n(y) - T_{n-1}(y) = \frac{1}{3^n} = \sum_{i=n+1}^{\infty} \frac{2}{3^i} \qquad (A.42)$$

so that there is another sequence $\{t_i(T_n(y))\}$ which does not contain the digit 1, hence $T_n(y) \in C$. The set $[T_{n-1}(y), T_n(y)] \cap C$ is written as

$$\{z : i < n \Rightarrow t_i(z) = t_i(y); t_n(z) = 0; i > n \Rightarrow t_i(z) = 0, \text{ or } 2\} \quad (A.43)$$

Thus it is a scaled down version of C by a factor $1/3^n$ and translated by $T_{n-1}(y)$. Hence

$$
\begin{aligned}
I_n(y) &= T_{n-1}(y) \int_{T_{n-1}(y)}^{T_n(y)} \chi_C d_C^{\alpha} x + \int_{T_{n-1}(y)}^{T_n(y)} (x - T_{n-1}(y)) \chi_C d_C^{\alpha} x \\
&= T_{n-1}(y) \int_0^{1/3^n} \chi_C d_C^{\alpha} x + \int_0^{1/3^n} x \chi_C d_C^{\alpha} x \\
&= \Gamma(\alpha+1) \left[\frac{T_{n-1}(y)}{3^{\alpha n}} \right] + \frac{g(1)}{3^{n(1+\alpha)}} \\
&= \Gamma(\alpha+1) \left[\frac{T_{n-1}(y)}{2^n} \right] + \frac{g(1)}{6^n} \qquad (A.44)
\end{aligned}
$$

If $y = T_n(y)$ then $y \in C$. But if $y > T_n(y)$, then as $t_n(y) = 1$ and $t_i \neq 0$ for some $i > n$, therefore $y \notin C$. Therefore the half open interval $(T_n(y), y]$ does not intersect C implying that $I_k(y) = 0$ for all $k > n$.
(3) $t_n(y) = 2$, here, $T_n(y)$ clearly belong to C. If D is the set:

$$D = \{z : i < n \Rightarrow t_i(z) = t_i(y); t_n(z) = 0; i > n \Rightarrow t_i(z) = 0, \text{ or } 2\} \qquad (A.45)$$

then D is a scaled down version of C by a factor $1/3^n$, $D \subset [T_{n-1}(y), T_n(y)]$, and more specifically, $[T_{n-1}(y), T_n(y)] \cap C = D \cup \{T_n(y)\}$. Therefore by arguments similar to the case $t_n(y) = 1$,

$$I_n(y) = \Gamma(\alpha+1) \left[\frac{T_{n-1}(y)}{2^n} \right] + \frac{g(1)}{6^n} \qquad (A.46)$$

But unlike the case $t_n(y) = 1$, there is a possibility that $C \cap (T_n(y), y]$ is nonempty so that $I_k(y)$ need not be zero for all $k > n$.

Remark A.6. We can summarize the results above as follows:

$$I_n(y) = \begin{cases} 0, & \text{if } t_n(y) = 0, \text{ or } t_i(y) = 1 \text{ for some } i < n; \\ \Gamma(\alpha + 1)\left[\frac{T_{n-1}(y)}{2^n}\right] + \frac{g(1)}{6^n}, & \text{otherwise.} \end{cases}$$

(A.47)

This description requires the value of $g(1)$. It can be found out by putting $y = 1$ in Eq.(A.47). If $y = 1$, then $t_i(y) = 2$ for all i. Also,

$$T_n(1) = \sum_{i=1}^{n} \frac{2}{3^i} = 1 - 3^{-n}.$$

(A.48)

Therefore,

$$I_n(1) = \Gamma(\alpha + 1)\left[\frac{(1 - 3^{-(n-1)})}{2^n} + \frac{g(1)}{6^n}\right].$$

(A.49)

Substituting Eq.(A.49) into Eq.(A.47) and solving Eq.(A.40) for $g(1)$, we get

$$g(1) = \frac{\Gamma(\alpha + 1)}{2}$$

(A.50)

Thus,

$$g(y) = \int_0^y x\chi_C(x)d_F^\alpha x = \sum_{n=1}^{\infty} I_n(y)$$

(A.51)

where

$$I_n(y) = \begin{cases} 0, & \text{if } t_n(y) = 0, \text{ or } t_i(y) = 1 \\ & \text{for some } i < n; \\ \Gamma(\alpha + 1)\left[\frac{T_{n-1}(y)}{2^n} + \frac{1}{2 \times 6^n}\right], & \text{otherwise.} \end{cases}$$

(A.52)

A.7   Repeated F^α-derivative

Dynamical systems are modeled by differential equation involving second and higher order derivatives. The successive operation of the D_F^α operator is also possible and gives meaningful results [Parvate *et al.* (2011)].

Example A.5. Consider a function $g(x) = (S_F^\alpha(x))^2$ where $F \subset R$ is an α-perfect set. Then its fractal derivative is as follows:

(1) If $x \notin F$

$$D_F^\alpha g(x) = 0.$$

(A.53)

(2) If $x \in F$, then we have

$$D_F^\alpha g(x) = F - \lim_{y \to x} \frac{S_F^\alpha(y))^2 - S_F^\alpha(x))^2}{S_F^\alpha(y)) - S_F^\alpha(x))}$$
$$= 2S_F^\alpha(x) \qquad (A.54)$$

Eq.(A.53) and (A.54) are combined to give

$$D_F^\alpha g(x) = 2\chi_F(x)S_F^\alpha(x). \qquad (A.55)$$

where χ_F is the characteristic function of F. Now we take the second F^α-derivative of g we have:

$$D_F^\alpha (S_F^\alpha(x))^2 = D_F^\alpha (2S_F^\alpha(x)\chi_F(x)$$
$$= 2D_F^\alpha S_F^\alpha(x)$$
$$= 2\chi_F(x). \qquad (A.56)$$

Remark A.7. Apart from the γ-dimension of F, the order α also has another significance. This will be clear from Example A.5. If C is the Cantor set, since $S_C^\alpha(x)$ is bounded by the power 2α of x:

$$ax^{2\alpha} \le g(x) = (S_C^\alpha(x))^2 \le bx^{2\alpha} \qquad (A.57)$$

so that

$$x \in C \Rightarrow 2ax^\alpha \le D_C^\alpha g(x) \le 2bx^\alpha \qquad (A.58)$$

and

$$x \in C \Rightarrow (D_C^\alpha)^2 g(x) = 2. \qquad (A.59)$$

We note that the F^α-differentiation reduces the power of bounds by α.

Remark A.8. We note that to generalized the result of Example A.5 we can write

$$D_F^\alpha (S_F^\alpha(x))^n = nS_F^\alpha(x))^{n-1}\chi_F(x) \qquad (A.60)$$

for any integer $n > 0$.

A.8 Repeated F^α-integration

The F^α-integration can also performed sequentially in succession. Let F be an α [Parvate *et al.* (2011)]. As we know that F^α-integration of $\chi_F(x)$ is $S_F^\alpha(x)$:

$$\int_a^{x'} \chi_F(x)d_F^\alpha x = S_F^\alpha(x'), \qquad (A.61)$$

where for simplicity we have taken $S_F^\alpha(a) = 0$. Let

$$g_1(x') = \int_a^{x'} S_F^\alpha(x) d_F^\alpha x = \frac{1}{2}(S_F^\alpha(x'))^2, \qquad (A.62)$$

where Eq.(A.55) and the fundamental theorem 4.13 are used. Next, by using Eq.(A.60) it easy to obtain

$$\int_a^{x'} (S_F^\alpha(x))^n d_F^\alpha x = \frac{1}{n+1}(S_F^\alpha(x'))^{n+1}. \qquad (A.63)$$

A.9 A few analogies between F^α-calculus and ordinary calculus

The F^α-calculus is a generalization of ordinary calculus with Riemann approach [Parvate *et al.* (2011)]. Table A.1 summarizes a few analogies between various quantities in respective frameworks.

Table A.1: Comparison of ordinary calculus with fractal calculus

Ordinary calculus	Fractal calculus
$\mathbf{R}$(real-line)	An α-perfect set
limit	Fractal limit(F-lim)
Continuity	Fractal Continuity
$\int_0^y x^n dx = \frac{1}{n+1}y^{n+1}$	$\int_0^y (S_F^\alpha(x))^n d_F^\alpha x = \frac{1}{n+1}S_F^\alpha(y)^{n+1}$
$\frac{d}{dx}x^n = nx^{n-1}$	$D_F^\alpha S_F^\alpha(x)^n = nS_F^\alpha(x)^{n-1}\chi_F(x)$

Bibliography

Ashrafi, S. and Golmankhaneh, A. K. (2018). Dimension of quantum mechanical path, chain rule, and extension of landau's energy straggling method using f^α-calculus, *Turkish Journal of Physics* **42**, 2, pp. 104–115.

Balankin, A. S. (2015). A continuum framework for mechanics of fractal materials i: from fractional space to continuum with fractal metric, *Eur. Phys. J. B* **88**, 4.

Banchuin, R. (2022). Noise analysis of electrical circuits on fractal set, *COMPEL-The international journal for computation and mathematics in electrical and electronic engineering* .

Barlow, M. T. and Perkins, E. A. (1988). Brownian motion on the sierpinski gasket, *Probab. Theory Rel.* **79**, 4, pp. 543–623.

Barnsley, M. F. (2014). *Fractals everywhere* (Academic press).

Ben-Avraham, D. and Havlin, S. (2000). *Diffusion and reactions in fractals and disordered systems* (Cambridge university press).

Bouchaud, J.-P. and Georges, A. (1990). Anomalous diffusion in disordered media: statistical mechanisms, models and physical applications, *Physics reports* **195**, 4-5, pp. 127–293.

Bruckner, A. M., Bruckner, J. B., and Thomson, B. S. (1997). *Real analysis* (Prentice Hall).

Bunde, A. and Havlin, S. (2013). *Fractals in science* (Springer).

Cetinkaya, F. A. and Golmankhaneh, A. K. (2021). General characteristics of a fractal sturm-liouville problem, *Turk. J. Math.* **45**, 4.

Czachor, M. (2019). Waves along fractal coastlines: From fractal arithmetic to wave equations, *Acta Phys. Pol. B* **50**, 4, p. 813.

DiMartino, R. and Urbina, W. (2014a). On cantor-like sets and cantor-lebesgue singular functions, *arXiv preprint arXiv:1403.6554* .

DiMartino, R. and Urbina, W. O. (2014b). Excursions on cantor-like sets, *arXiv preprint arXiv:1411.7110*.

Edgar, G. A. (1998). *Integral, Probability, and Fractal Measures* (Springer New York).

El-Nabulsi, R. A. and Golmankhaneh, A. K. (2021a). Dynamics of particles in cold electrons plasma: fractional actionlike variational approach versus fractal spaces approach, *Waves in Random and Complex Media* **0**, 0, pp. 1–22, doi:10.1080/17455030.2021.1909779, https://doi.org/10.1080/17455030.2021.1909779.

El-Nabulsi, R. A. and Golmankhaneh, A. K. (2021b). On fractional and fractal einstein's field equations, *Mod. Phys. Lett. A* **36**, 05, p. 2150030.

El-Nabulsi, R. A. and Golmankhaneh, A. K. (2022). Nonstandard and fractal electrodynamics in finsler-randers space, *Int. J. Geom. Methods M.* , p. 2250080.

Falconer, K. (2004). *Fractal geometry: mathematical foundations and applications* (John Wiley & Sons).

Fonda, A., Fonda, A., and Heine (2018). *Kurzweil-Henstock Integral for Undergraduates* (Springer).

Freiberg, U. and Zähle, M. (2002). Harmonic calculus on fractals-a measure geometric approach i, *Potential analysis* **16**, 3, pp. 265–277.

Golmankhaneh, A. and Ashrafi, S. (2017). Energy straggling function by f^α-calculus, *ASME J. Comput. Nonlin. Dyn* **12**, 5, p. 051010.

Golmankhaneh, A. K. (2010). *Investigation in dynamics: with focus on fractional dynamics and application to classical and quantum mechanical processes.*, Ph.D. thesis, University of Pune, India.

Golmankhaneh, A. K. (2017). On the calculus of parameterized fractal curves, *Turk. J. Phys.* **41**, 5, pp. 418–425.

Golmankhaneh, A. K. (2018). About kepler's third law on fractal-time spaces, *Ain Shams Eng. J.* **9**, 4, pp. 2499–2502.

Golmankhaneh, A. K. (2019a). On the fractal langevin equation, *Fractal Fract.* **3**, 1, p. 11.

Golmankhaneh, A. K. (2019b). Statistical mechanics involving fractal temperature, *Fractal Fract.* **3**, 2, p. 20.

Golmankhaneh, A. K. (2021). Tsallis entropy on fractal sets, *Journal of Taibah University for Science* **15**, 1, pp. 543–549.

Golmankhaneh, A. K., Ali, K., Yilmazer, R., and Kaabar, M. (2021a). Local fractal fourier transform and applications, *Comput. Methods Differ. Equ.* , pp. –doi:10.22034/cmde.2021.42554.1832.

Golmankhaneh, A. K. and Balankin, A. S. (2018). Sub-and super-diffusion on cantor sets: Beyond the paradox, *Phys. Lett. A.* **382**, 14, pp. 960–967.

Golmankhaneh, A. K. and Baleanu, D. (2016a). Diffraction from fractal grating cantor sets, *J. Mod. Optic.* **63**, 14, pp. 1364–1369.

Golmankhaneh, A. K. and Baleanu, D. (2016b). Fractal calculus involving gauge function, *Commun. Nonlinear Sci.* **37**, pp. 125–130.

Golmankhaneh, A. K. and Baleanu, D. (2016c). New derivatives on the fractal subset of real-line, *Entropy* **18**, 2, p. 1.

Golmankhaneh, A. K. and Baleanu, D. (2016d). Non-local integrals and derivatives on fractal sets with applications, *Open Physics* **14**, 1, pp. 542–548.

Golmankhaneh, A. K. and Cattani, C. (2019). Fractal logistic equation, *Fractal Fract.* **3**, 3, p. 41.

Golmankhaneh, A. K., Fazlollahi, V., and Baleanu, D. (2013a). Newtonian mechanics on fractals subset of real-line, *Rom. Rep. Phys.* **65**, 1, pp. 84–93.

Golmankhaneh, A. K. and Fernandez, A. (2018). Fractal calculus of functions on cantor tartan spaces, *Fractal Fract.* **2**, 4, p. 30.

Golmankhaneh, A. K. and Fernandez, A. (2019). Random variables and stable distributions on fractal cantor sets, *Fractal Fract.* **3**, 2, p. 31.

Golmankhaneh, A. K., Fernandez, A., Golmankhaneh, A. K., and Baleanu, D. (2018). Diffusion on middle-ξ cantor sets, *Entropy* **20**, 7, p. 504.

Golmankhaneh, A. K., Golmankhaneh, A. K., and Baleanu, D. (2013b). About maxwell's equations on fractal subsets of r^3, *Open Physics* **11**, 6, pp. 863–867.

Golmankhaneh, A. K., Golmankhaneh, A. K., and Baleanu, D. (2013c). Lagrangian and hamiltonian mechanics on fractals subset of real-line, *Int. J. Theor. Phys.* **52**, 11, pp. 4210–4217.

Golmankhaneh, A. K., Golmankhaneh, A. K., and Baleanu, D. (2015). About schrödinger equation on fractals curves imbedding in r^3, *Int. J. Theor. Phys.* **54**, 4, pp. 1275–1282.

Golmankhaneh, A. K., K Ali, K., Yilmazer, R., and KA Kaabar, M. (2021b). Economic models involving time fractal, *J. Math. Model. Financ.* **1**, 1, pp. 159–178.

Golmankhaneh, A. K. and Kamal Ali, K. (2021). Fractal kronig-penney model involving fractal comb potential, *J. Math. Model.* **9**, 3, pp. 331–345.

Golmankhaneh, A. K., Kamal Ali, K., Yilmazer, R., and Welch, K. (2021c). Electrical circuits involving fractal time, *Chaos* **31**, 3, p. 033132.

Golmankhaneh, A. K. and Nia, S. M. (2021). Laplace equations on the fractal cubes and casimir effect, *Eur. Phys. J. Special Topics* **230**, 21, pp. 3895–3900.

Golmankhaneh, A. K. and Sibatov, R. T. (2021). Fractal stochastic processes on thin cantor-like sets, *Mathematics* **9**, 6, p. 613.

Golmankhaneh, A. K. and Tunç, C. (2017). On the lipschitz condition in the fractal calculus, *Chaos, Solitons & Fractals* **95**, pp. 140–147.

Golmankhaneh, A. K. and Tunç, C. (2019a). Sumudu transform in fractal calculus, *Appl. Math. Comput.* **350**, pp. 386–401.

Golmankhaneh, A. K. and Tunç, C. (2020). Stochastic differential equations on fractal sets, *Stochastics* **92**, 8, pp. 1244–1260.

Golmankhaneh, A. K., Tunç, C., and Şevli, H. (2021d). Hyers–ulam stability on local fractal calculus and radioactive decay, *Eur. Phys. J. Special Topics* **230**, 21, pp. 3889–3894.

Golmankhaneh, A. K. and Welch, K. (2021). Equilibrium and non-equilibrium statistical mechanics with generalized fractal derivatives: A review, *Mod. Phys. Lett. A* **36**, 14, p. 2140002.

Golmankhaneh, K. A. and Tunç, C. (2019b). Analogues to lie method and noether's theorem in fractal calculus, *Fractal Fract.* **3**, 2, p. 25.

Gowrisankar, A., Golmankhaneh, A. K., and Serpa, C. (2021). Fractal calculus on fractal interpolation functions, *Fractal Fract.* **5**, 4, p. 157.

Hartman, S. and Mikusinski, J. (2014). *The theory of Lebesgue measure and integration* (Elsevier).

Kamal, K. A., Golmankhaneh, A. K., and Yilmazer, R. (2021). Battery discharging model on fractal time sets, *Int. J. Nonlin. Sci. Num.*, p. 000010151520200139.

Kamal Ali, K., Golmankhaneh, A. K., Yilmazer, R., and Ashqi Abdullah, M. (2022). Solving fractal differential equations via fractal laplace transforms, *J. Appl. Anal.* doi:10.1515/jaa-2021-2076, https://doi.org/10.1515/jaa-2021-2076.

Kigami, J. (2001). *Analysis on Fractals* (Cambridge University Press).

Klafter, J., Lim, S., and Metzler, R. (eds.) (2012). *Fractional dynamics: recent advances* (World Scientific).

Kolwankar, K. M. and Gangal, A. D. (1998). Local fractional fokker-planck equation, *Phys. Rev. Lett.* **80**, 2, p. 214.

Kolwankar, K. M. and Gangal, A. D. (1999). Local fractional calculus: a calculus for fractal space-time, in M. Dekking, J. L. Véhel, E. Lutton, and C. Tricot (eds.), *Fractals* (Springer London, London), pp. 171–181.

Kufner, A., John, O., and Fucik, S. (1977). *Function spaces*, Vol. 3 (Springer Science & Business Media).

Kunze, H., La Torre, D., Mendivil, F., and Vrscay, E. R. (2019). Differential equations using generalized derivatives on fractals, in *International Conference on Applied Mathematics, Modeling and Computational Science* (Springer), pp. 81–91.

Lapidus, M. L., Radunović, G., and Žubrinić, D. (2017). *Fractal Zeta Functions and Fractal Drums* (Springer International Publishing).

Mandelbrot, B. B. (1982). *The fractal geometry of nature* (WH freeman New York).

Metzler, R., Glöckle, W. G., and Nonnenmacher, T. F. (1994). Fractional model equation for anomalous diffusion, *Physica A* **211**, 1, pp. 13–24.

Nottale, L. and Schneider, J. (1984). Fractals and nonstandard analysis, *J. Math. Phys.* **25**, 5, pp. 1296–1300.

Parvate, A. (2009). *Calculus on Fractal Subsets of Real Line: Formulation, Techniques and Applications in Physics*, Ph.D. thesis, University of Pune, Pune, India.

Parvate, A. and Gangal, A. (2011). Calculus on fractal subsets of real line ii: Conjugacy with ordinary calculus, *Fractals* **19**, 03, pp. 271–290.

Parvate, A. and Gangal, A. D. (2009). Calculus on fractal subsets of real line—i: Formulation, *Fractals* **17**, 01, pp. 53–81.

Parvate, A., Satin, S., and Gangal, A. (2011). Calculus on fractal curves in r^n, *Fractals* **19**, 01, pp. 15–27.

Podlubny, I. (1998). *Fractional differential equations: an introduction to fractional derivatives, fractional differential equations, to methods of their solution and some of their applications* (Elsevier).

Ri, S.-I., Drakopoulos, V., and Nam, S.-M. (2021). Fractal interpolation using harmonic functions on the koch curve, *Fractal Fract.* **5**, 2, p. 28.

Royden, H. L. and Fitzpatrick, P. (1988). *Real analysis*, Vol. 32 (Macmillan New York).

Sandev, T. (2020). *Fractional Equations and Models: Theory and Applications* (Springer).

Satin, S. and Gangal, A. (2016). Langevin equation on fractal curves, *Fractals* **24**, 03, p. 1650028.

Satin, S. and Gangal, A. (2019). Random walk and broad distributions on fractal curves, *Chaos, Solitons & Fractals* **127**, pp. 17–23.

Satin, S. E., Parvate, A., and Gangal, A. (2013). Fokker–planck equation on fractal curves, *Chaos, Solitons & Fractals* **52**, pp. 30–35.

Shlesinger, M. F. (1988). Fractal time in condensed matter, *Annu. Rev. Phys. Chem.* **39**, 1, pp. 269–290.

Spiegel, M. R. (1969). *Schaum's outline of theory and problems of real variables: Lebesgue measure and integration with applications to Fourier series* (Schaum's Outline Series).

Stillinger, F. H. (1977). Axiomatic basis for spaces with noninteger dimension, *J. Math. Phys.* **18**, 6, pp. 1224–1234.

Swartz, C. W. (2001). *Introduction to gauge integrals* (World Scientific).

Tarasov, V. E. (2018). No nonlocality. no fractional derivative, *Commun. Nonlinear. Sci.* **62**, pp. 157–163.

Tunç, C., Golmankhaneh, A. K., and Branch, U. (2020). On stability of a class of second alpha-order fractal differential equations, *AIMS Mathematics* **5**, 3, pp. 2126–2142.

Uchaikin, V. V. (2013). *Fractional derivatives for physicists and engineers*, Vol. 2 (Springer).

Valério, D., Ortigueira, M. D., and Lopes, A. M. (2022). How many fractional derivatives are there? *Mathematics* **10**, 5, p. 737.

Vrobel, S. (2011). *Fractal Time* (World Scientific).

Welch, K. (2020). *A Fractal Topology of Time: Deepening into Timelessness* (Fox Finding Press).

Wibowo, S., Indrati, C. R., *et al.* (2021). The relationship between a fractal fα–absolutely continuous function and a fractal bounded p–variation function, in *International Conference on Science and Engineering (ICSE-UIN-SUKA 2021)* (Atlantis Press), pp. 35–38.

Yanovsky, V., Chechkin, A., Schertzer, D., and Tur, A. (2000). Lévy anomalous diffusion and fractional fokker–planck equation, *Physica A* **282**, 1-2, pp. 13–34.

Zaslavsky, G. M. (1994). Fractional kinetic equation for hamiltonian chaos, *Phys. D Nonlinear Phenom.* **76**, 1-3, pp. 110–122.

Index

Variance limited, 253
Vector potential, 236
Vertices, 5

Wedge product, 234
Weierstrass function, 72, 99

Weierstrass-Bolzano, 13
Weight function, 44
White noise, 238
Wiener process, 161

Young's inequality, 51